T0139845

The Latin American Studies Book Series

The Latin American Studies Book Series promotes quality scientific research focusing on Latin American countries. The series accepts disciplinary and interdisciplinary titles related to geographical, environmental, cultural, economic, political and urban research dedicated to Latin America. The series publishes comprehensive monographs, edited volumes and textbooks refereed by a region or country expert specialized in Latin American studies.

The series aims to raise the profile of Latin American studies, showcasing important works developed focusing on the region. It is aimed at researchers, students, and everyone interested in Latin American topics.

Submit a proposal: Proposals for the series will be considered by the Series Advisory Board. A book proposal form can be obtained from the Publisher, Juliana Pitanguy (juliana.pitanguy@springer.com).

More information about this series at http://www.springer.com/series/15104

Mario Pino · Giselle A. Astorga
Editors

Pilauco: A Late Pleistocene Archaeo-paleontological Site

Osorno, Northwestern Patagonia and Chile

Editors
Mario Pino
Facultad de Ciencias, Instituto de Ciencias de la Tierra
Universidad Austral de Chile
Valdivia, Chile

Giselle A. Astorga
Facultad de Ciencias, Instituto de Ciencias de la Tierra
Universidad Austral de Chile
Valdivia, Chile

ISSN 2366-3421 ISSN 2366-343X (electronic)
The Latin American Studies Book Series
ISBN 978-3-030-23920-6 ISBN 978-3-030-23918-3 (eBook)
https://doi.org/10.1007/978-3-030-23918-3

This Springer imprint is published by the registered company Springer Nature Switzerland AG
The registered company address is: Gewerbestrasse 11, 6330 Cham, Switzerland

Preface

Before the sites Monte Verde I and II (Dillehay and Collins 1988, Dillehay et al. 2015), the only validated archaeological evidence were related to early Paleo-Indian sites (Roberts 1940), with ages no older than 13,000 years. The discovery and subsequent research of the Monte Verde sites overturned the Clovis-first Hypothesis opening a new window to study the first human settlers in the Americas. Since then, we have learned a lot about Monte Verde and the ancient living habits of these first Americans.

Pilauco is a promising archaeological and paleontological site contemporaneous to Monte Verde II that is providing an opportunity to refine our understanding of early humans in South America during a period of time marked by strong environmental disruptions, the extinction of megafauna and changes in the configuration of the vegetation.

Interestingly, no ancient sites have yet been discovered north of Osorno ($\sim$ 40° S), and the youngest sites that exist, do not share the degree of preservation of the paleontological remains that exist in Monte Verde II and Pilauco. The most likely explanation of the enhanced preservation potential at these sites is the presence of peat environments that have provided anoxic conditions for the preservation of organic remains. In contrast, some of the very old, much discussed Brazilian sites only contain stone tools (Boeda et al. 2014).

Archeological sites are commonly discovered because of the presence of fossil megafauna, often loosing relevance in front of investigations related to ancient human cultures. At the beginning of the research process at Pilauco, we confronted a similar situation; however, the high diversity of extinct and extant animals, vertebrates and invertebrates, prevented paleontological research at Pilauco from dropping its initial tempo.

This book presents the results of several years of investigations combining different disciplines such as geology, paleobotany, paleontology, archeology and astrophysics at the Pilauco site. Thus, geology tries to explain the local abiotic conditions in which humans and animals coexisted, paleontology and paleobotany provide a view of the local biosphere and their interactions, archeology produces important information to understand to the unifacial edge-trimmed tradition of early settlers in Chilean Northwestern Patagonia, whereas astrophysics research at the site yields strong evidence supporting the cosmic-impact hypothesis as one of the main factors driving the demise of megafauna. However, several important questions remain open. For example, why patterns of extinction of the microfauna are so different from megafauna? How climate seasonality evolved during the Late Glacial Period? Further research at the Pilauco-Los Notros Complex is needed to answer these and other questions that remain open.

We thank many friends and colleagues for the discussions of several aspects included in this book in all the disciplines, but especially to Prof. Tom D. Dillehay for his site visits and fruitful on-site discussions, Dr. Viktoria Georgieva for translating and editing most of the chapters originally written in Spanish. We also thank Devon Barone for improving the language of some chapters originally written in English, Camila Paredes and Martina Klouvobá for proofreading and editing.

We gratefully acknowledge funding from Fondo Nacional de Desarrollo Científico y Tecnológico (FONDECYT), Chile through grants #1100555 and #1150738 to Mario Pino, grant #11140677 to Ana M. Abarzúa, grant #3170958 to Giselle A. Astorga, grant 11170919 to Rafael Labarca and grant TAQUACH/VIDCA, Universidad Austral de Chile to all the authors. We are very grateful for the grant from Ilustre Municipalidad de Osorno "Estudios Paleontológicos y Arqueológicos sitios Pilauco y Los Notros, Patagonia Noroccidental, Osorno, Chile, 2018–2022"; Silvia Constabel thanks grant CORFO INNOVA #13BPC3-19098. The Comet Research Group also greatly appreciates generous research donations from James Marvin and more than 1000 Indiegogo crowdfunding donors. Alejandra Martel-Cea and Francisco Tello also thank the support of CONICYT Doctoral scholarship #21140447 and #21171980, respectively. Omar P. Recabarren, Patricia Canales-Brellenthin and Valentina Alvarez also thank the support of CONICYT Doctoral Becas Chile scholarship #72150064, #72150065 and #72160354, respectively.

Osorno, Chile Mario Pino
Valdivia, Chile Giselle A. Astorga

References

Boëda E, Clemente-Conte I, Fontugne M, Lahaye C, Pino M, Felice GD, Guidon N, Hoeltz S, Lourdeau A, Pagli M, Pessis AM, Viana S, Da Costa A, Douville E (2014) A new late pleistocene archaeological sequence in South America: the vale da pedra furada (Piauí, Brazil). Antiquity 88(341):927–941

Dillehay TD and Collins MB (1988) Early cultural evidence from monte verde in Chile. Nature 332(6160):150

Dillehay TD, Ocampo C, Saavedra J, Sawakuchi AO, Vega RM, Pino M, Collins MB, Scott Cummings L, Arregui I, Villagran XS, Hartmann GA, Mella M, González A, Dix G (2015) New archaeological evidence for an early human presence at monte verde, Chile. PLoS One 10(11):e0141923

Roberts FHH Jr. (1940) Development on the problem of the North American Paleo-Indian, Smithsonian Miscellaneous Collections 30:51–116

Contents

Chapter 1
Pilauco and Los Notros Sites Research: A Narration of Human and Scientific Events

Mario Pino

Abstract The Pilauco Site, located at 159 Río Cachapoal Street, Villa Los Notros, Osorno, was discovered by chance in 1986. We began the research of Pilauco in November 2007. This chapter describes the main human and scientific events that have allowed the development of the geological, paleontological, archeological, and astrophysical research.

Keywords Research history · People · Grants

1.1 History, People, and Projects Involved in the Research

The organization of an archeo-paleontological excavation involves numerous scientific, technical, administrative, financial, logistic and human challenges. Crucial steps span everything from the technical preparations of the excavators to the organization of the kitchen area; some say the success of a task of this magnitude is related to the quantity and quality of the food. The most important thing is that each person working on a paleontological or archeological excavation should understand that digging, in fact, destroys the evidence.

The fossils in Pilauco were discovered by chance in 1986 by workers excavating to install the foundations of a house in Villa Los Notros, Osorno (Figs. 1.1a, b). Our aim of the research in Pilauco was the product of two undergraduate theses carried out by Recabarren (2007) and Montero (2007), who analyzed megafauna fossil remains from the collection of the Historical Museum of the Municipality of Osorno. Before proposing the size and location of the excavation at Villa Los Notros, Pilauco, photographs and a video made by a local TV network in 1986 were analyzed.

M. Pino (✉)
Instituto de Ciencias de la Tierra, Universidad Austral de Chile, Valdivia, Chile
e-mail: mariopinoquivira@gmail.com

Transdisciplinary Center for Quaternary Research in the South of Chile, Universidad Austral de Chile, Valdivia, Chile

M. Pino and G. A. Astorga (eds.), *Pilauco: A Late Pleistocene Archaeo-paleontological Site*, The Latin American Studies Book Series,
https://doi.org/10.1007/978-3-030-23918-3_1

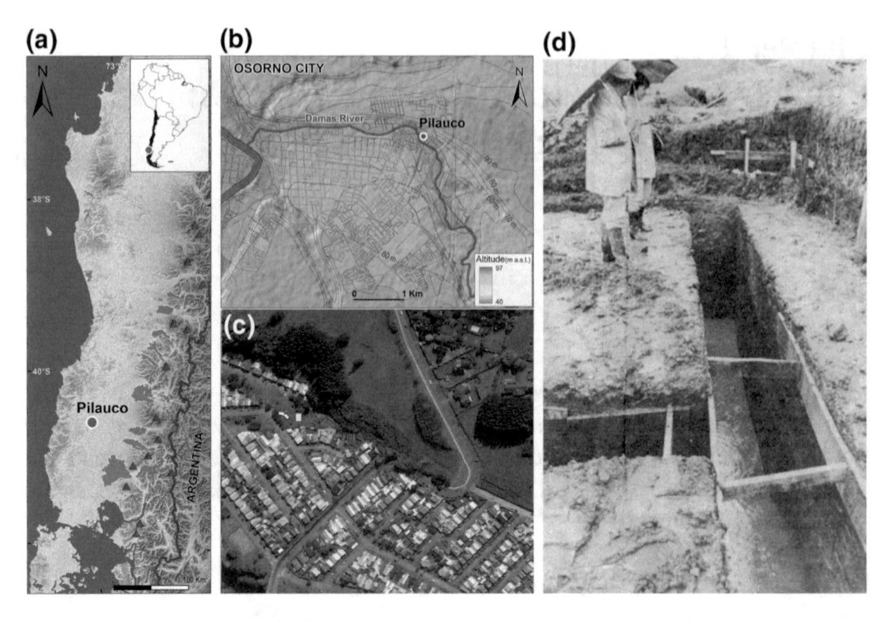

Fig. 1.1 **a–c** Location of the Pilauco Site (reproduced from Google Earth). The red triangles represent active volcanoes. The red and yellow squares the Pilauco and Los Notros sites, respectively. **d** Geologist Patricia Salinas and anthropologist Jorge Inostroza, visiting Pilauco a few days after the discovery of the fossils

In both records, we observed an excavation design of a fork shape, around 80 cm wide and with its three teeth open towards the south.

After winning the public grant "The Palaentological Research of the Pilauco Site", Code BIP 30066849-0 of the National Investment System FNDR, the first formal act was the signing of a Memorandum of Understanding in the Municipal Historical Museum of Osorno on October 19, 2007. Present, among others, were Mr. Gerardo Rosas, Deputy Mayor of the Municipality of Osorno, Mr. Gabriel Peralta, Director of the Historical Museum, and Mr. Claudio Donoso, Director of the Communal Planning Secretariat (SECPLAC), who was also the technical inspector of the project. The main objective of this project was to determine if the findings of Pilauco were random or if they corresponded to a paleontological site.

Simultaneously, the team's archeologist, Dr. Ximena Navarro-Harris, was managing the permission of the Council of National Monuments, which was granted on November 8, 2007 by attestation No. 5833. On the other hand, in order to inform the residents of Villa Los Notros of our future activities and explain to them the possible small inconveniences we would cause, we called them to a meeting in the street in front of the land we would excavate.

On November 1, 2007, we rented a little house located immediately to the west of the site, at Río Cachapoal 159, Villa Los Notros, Pilauco, and began preparing it to host the research team. We hired four workers who had been preselected by the employment office of the Municipality of Osorno: José Fontanilla, Juan Carlos

Fig. 1.2 From left to right José Fontanilla, Juan Carlos Huenchul, Daniel Fritte and René Vargas

Huenchul, Daniel Fritte and René Vargas (Fig. 1.2).[1] Mr. Vargas worked with us until 2017 and Mr. Fritte evolved to a technician and remains among our best collaborators.

Paleontologists Dr. Karen Moreno (who came from Australia) and Dr. Ana María Abarzúa assembled the first team of researchers. The excavation unit was headed by two close collaborators, Izzat Montero and Omar Recabarren, who were responsible for the scientific record and for the personnel hiring, respectively. Undergraduate students of Biological Sciences at Universidad Austral de Chile (UACh), Universidad Católica de Valparaíso and Universidad Católica de Temuco (see details in Pino 2008) excavated at the site. Among other important associates, there were included the journalists Daniela Martin and Francisco Morey, who were in charge of public relations and academic outreach. This hive of activity was achieved despite the fact that the house we rented was never designed for 12 or 14 people to live and work together for 5 months, frequently from Monday to Saturday, 12 h a day.

On the afternoon of November 9, 2007, having already planned the design of the first trench, we held a small ceremony to ask the earth for permission to excavate and reveal its mysteries (Fig. 1.3). I was accompanied by Claudio Donoso, Gabriel Peralta, Carlos Bustos, Alejandra Martel, Leonora Salvadores, Patricio Espinoza, Izzat Montero and Omar Recabarren. On the following day, we removed sediments in a 7-m-long and 1-m-wide trench in the north–south direction (H2–H9 of the grid,

[1] Mr. Rene Vargas died on 21 May 2019.

Fig. 1.3 The site immediately before the first excavations in November 2007 (the ditches from 1986 were not visible)

Fig. 1.4), using a backhoe. This trench allowed us to define the position of an old hill and the existence of a major debris fill deposited after 1986. In the following 2 days, we expanded the excavation to the rows 9 and 10 and the columns D and J of the grid, recognizing the evidence of the 1986 excavations. We recovered the first gomphothere fossil at the coordinate H10 on November 15.

At that time, we were preparing the excavation of the first squares of the grid, looking for the ditches from 1986 (Fig. 1.5). We made a setting of 1 m^2 excavations squares at approximately 50 cm above the ground level. An arbitrary vertical, 10-m-high level was located at the NW edge of the P0 grid (the Z coordinate) and each excavator had to refer his findings to the northern and eastern edge of his square (the coordinates *X* and *Y*, Fig. 1.4).

On November 10, we launched a new website and began live webcam broadcasts of the excavation under the auspices of TELSUR. The portal became a point of interaction with the local, national and global communities and the repository of all our scientific information. The camera could be remotely controlled, which made it easier for me to supervise the excavation when I was not in Pilauco.

The project had initially set out to excavate for a total of 6 weeks in order to register and recover most of the gomphothere skeleton. The excavation became increasingly complex as it progressed, unearthing an enormous diversity of mammal fossils. Claudio Donoso proposed a reformulation of the financing by increasing the number of days we would excavate. The project budget was expanded from 25 to 41 million Chilean Pesos (CLP), excavating from November 10, 2007 until March 30, 2008.

The excavations were conducted across 41 m^2 (Fig. 1.6) since the grids H2–H5 contained only sterile layers from the point of view of the fossil recovery. We excavated another 6 m^2 to the south and west of the main excavation, in order to describe the geological characteristics of the site. We extracted a total of approximately 140 m^3, half of which was removed with the backhoe (the sterile material). The rest was carefully excavated by hand with wooden and plastic spatulas. The final check of the sediments was performed using a sifting screen. 2% of the recorded

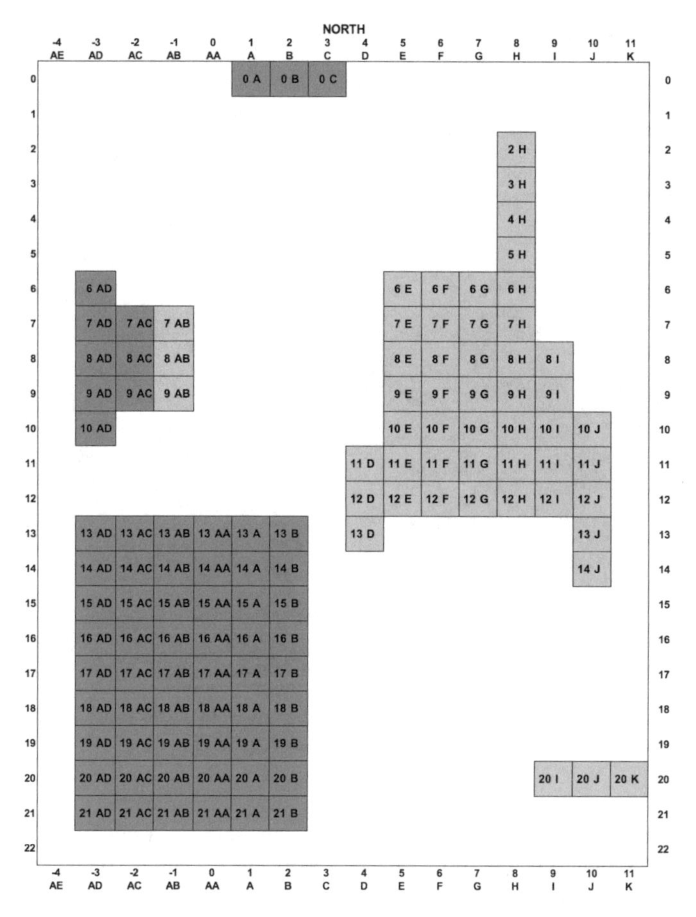

Fig. 1.4 Grid map of excavations in Pilauco. Green represents excavations from 2007 to 2008, blue those ones between 2010 and 2017. Each square is equivalent to 1 m^2

materials were detected on the screen, despite the difficulties in excavating due to the presence of groundwater layers.

At the end of the first campaign, we recovered 684 whole and fractured bones, 37 teeth, 11 coprolites, 15 beetle parts, 348 pieces of wood, 126 seeds, 28 fragments of skin and hair, 909 river-eroded clasts with some type of modifications and 71 sediment samples, along with dozens of other samples of exotic or uncommon character. On March 30, 2008, we covered with a plastic mesh the base and walls of the excavation and sealed it with a help of the backhoe loader, using the opportunity to level the site for the subsequent excavation session, date of which was uncertain at that moment.

From April 2008 to June 2009, we consolidated all of the knowledge we had gathered from the site into an application for a FONDECYT grant, together with Ximena Navarro (archeologist), Ana María Abarzúa (paleobotanist) and Martín Chávez (paleontologist). The project called "The Late Pleistocene—Early Holocene

Fig. 1.5 The first excavation made to find the old ditches from 1986. From left to right Omar Recabarren, Leonora Salvadores, Daniel Fritte and José Fontanilla

Site of Pilauco (south-central Chile): paleoenvironment and taphonomy", FONDECYT 1100555, was selected to be financed between March 2010 and March 2014, receiving a budget of 136,000,000 CLP for 4 years of work. The Municipality of Osorno subsequently purchased the house of Río Cachapoal 159 and granted it as a counterpart to the Pilauco project.

In 2012, Ana María Abarzúa and Martín Chávez obtained postdoctoral and doctoral scholarships, respectively, and had to leave Pilauco, being replaced by paleontologist Karen Moreno and archeo-zoologist Rafael Labarca. At that stage, we re-excavated the site, enlarging the area of investigation (Fig. 1.4).

The specific objectives of the project were to: (A) determine the conditions and variability of the type and levels of sedimentation in different areas of the Pilauco Site, (B) establish and reconstruct the chronology of the climate and vegetation history in the studied area, based on analyses of fossilized pollen and plant remains, (C) determine the faunal assemblages of the Pilauco Site, (D) understand the taphonomic processes involved in the formation of the site, E) propose a systematized methodology for the analysis of lithic artifacts and (F) propose a geomorphological and taphonomic model of formation of Pilauco-type sites that could be used as a prospecting tool for new sites.

As a result of our first research, nine scientific articles were published in mainstream journals: González et al. (2010, 2011, 2014), Labarca et al. (2013, 2014),

Fig. 1.6 The excavation in the summer of 2008, view westwards

Pino et al. (2013), Recabarren et al. (2011, 2014) and Pavez-Fox et al. (2015), and 16° seminar papers were held: Aguilera (2010), Álvarez (2012), Canales (2010), González (2009), Guajardo (2013), Lobos (2013), Macías (2012), Münzenmayer (2018), Pavez-Fox (2011), Pérez (2014), Pinto (2018), Ramirez (2012), Silva (2014), Tello (2013), Soto-Bollmann (2014) and Valenzuela (2017). We described all extinct and extant fauna, a new area was discovered at the site, dozens of radiocarbon dating were performed, the depositional and vegetational environment of the site was reinterpreted and the first lithic artifacts built on exotic rocks were recorded at the site.

In June 2014, a proposal entitled "Exploring the record of late Pleistocene human activity in the context of temporal-spatial variability in microenvironments at the Pilauco site, Osorno, Chile" was submitted at the FONDECYT competition, but was not selected for funding. Between March 2014 and March 2015, we worked with financial support from the Municipality of Osorno and the PEF DID UACH 2014-2 project. During that year, a new and successful FONDECYT project was prepared, which was selected for financing between March 2015 and March 2018 (FONDECYT 1150738, "Geoarchaeological Interpretations in the context of the micro-environmental variability of the Pilauco site, late Pleistocene, Osorno, Chile"; M. Pino, X. Navarro and R. Labarca). Given its inter- and transdisciplinary context, we presented the project to the Archaeology Study Group for the first time, and not to the Earth Science Study Group.

The specific objectives of that project were to: (A) establish the characteristics and variability of local deposits to guide paleontological and archeological prospecting in the Pilauco area, including the search for the source of dacitic volcanic glass in the Cordillera de la Costa Mountains, (B) establish a typology of the lithic record in order to advance techno-morphological characterizations and comparisons between these early non-formalized lithic assemblages, (C) discuss current interpretations of the human use of the space, the type of tasks carried out, especially the centralized exploitation of butchery resources in Pilauco, and (D) develop new taxonomic, taphonomic and zooarcheological studies with the fossil materials of mammals of Pilauco. In this project, we will excavate new areas in Pilauco and two caves that we recently discovered in the vicinity of the site.

In March 2014, the project "Paleotourism Management Model in Southern Chile: Case of the Pilauco Destination, Osorno CORFO INNOVA code 13BPC3-19098" was approved for execution. The PI was Dr. Silvia Constabel from the Tourism Institute, Faculty of Economic and Administrative Sciences of UACh. This was an applied research project to originate a paleotourism model that can be replicated in other areas of the country and it is also the project that finances the production of this book.

Finally, in April 2015, we proposed in an internal competition the creation of a Research Center of Excellence together with several colleagues representing four faculties of UACh, University of Ghent and University of Colorado. We applied for funding of the project "Quaternary Studies Nuclei of the South of Chile" and it was selected for the period July 2015–2018. This project will give a transdisciplinary impetus to Quaternary research in the south of Chile, including the Pilauco Site.

In 2016, we started to excavate a new site in the neighborhood of Villa Los Notros (Fig. 1.7) as a result of geological surveys carried out in January 2008, which detected the presence of a gomphothere tusk. Chapter 14 of this book describes these new discoveries and challenges.

Recently, in this same context of outreach, the Municipality of Osorno inaugurated in 2017 the "Pleistocene Museum Hall" in Chuyaca Park (only 3 blocks from the sites, Fig. 1.8), to house and show to the public the discoveries of the Pilauco and Los Notros Sites. It is a human scale Museum, with two exhibition rooms, a fossil storage, conservation room and a souvenir shop. You are welcome to visit it!

We have demonstrated that in conjunction with the Pleistocene Chuyaca Park, Pilauco and Los Notros Sites can become a tourist attraction. Pilauco has also become an off-campus LabNat of the Faculty of Science of the UACh. Currently, we are funding the research with a grant from Municipalidad de Osorno (yearly US$ 70,000) for the next 5 years.

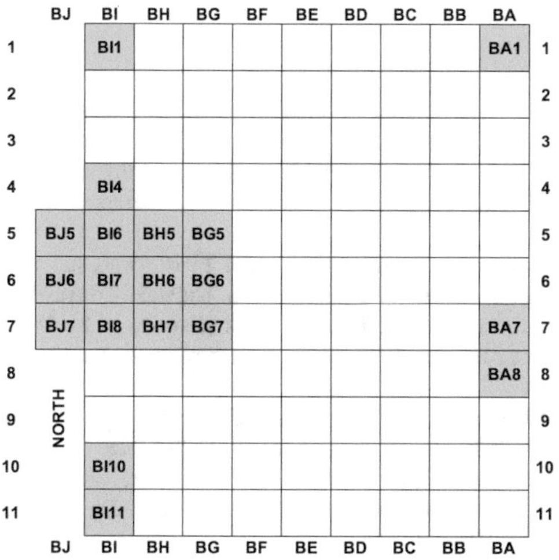

	BJ	BI	BH	BG	BF	BE	BD	BC	BB	BA	
1		BI1								BA1	1
2											2
3											3
4		BI4									4
5	BJ5	BI6	BH5	BG5							5
6	BJ6	BI7	BH6	BG6							6
7	BJ7	BI8	BH7	BG7						BA7	7
8										BA8	8
9	NORTH										9
10		BI10									10
11		BI11									11
	BJ	BI	BH	BG	BF	BE	BD	BC	BB	BA	

Fig. 1.7 Excavation grid map of the Los Notros site

Fig. 1.8 A full-scale reconstructed Xenarthra (Pilosa) on a rainy day at Parque Pleistocénico Chuyaca (The Pleistocene Chuyaca Park)

References

Aguilera F (2010) Análisis de un molar de omnívoro de la megafauna del Pleistoceno tardío, sitio Pilauco, Osorno, Chile. Undergraduate Dissertation. Universidad Austral de Chile

Álvarez V (2012) Análisis de fitolitos recuperados desde coprolitos del sitio Pilauco Osorno-Chile: aproximaciones sobre paleoecologia y paleodieta. Undergraduate Dissertation. Universidad Austral de Chile

Canales P (2010) Identificación de roedor fósil del Pleistoceno superior, sitio paleontológico Pilauco, Osorno, Chile. Undergraduate Dissertation. Universidad Austral de Chile

González E (2009) Sobre la presencia de Conepatus (Carnivora: Mephitidae) en el Pleistoceno superior de Chile: primer registro confirmado. Undergraduate Dissertation. Universidad Austral de Chile

González E, Prevosti FJ, Pino M (2010) Primer registro de Mephitidae (Carnivora: Mammalia) para el Pleistoceno de Chile. Magallania (Chile) 38(2):239–248

González E, Labarca R, Chávez M, Pino M (2014) First fossil record of the smallest deer cf. Pudu Molina, 1792 (ARTIODACTYLA, CERVIDAE) in the Late Pleistocene of South America. J Vertebr Paleontol 34(2):483–488

González E, Pino M, Recabarren O, Busquets T, Canales P, Salvadores L, Chávez M, Vásquez F, Bustos C, Ramos P, Navarro RX (2011) Paleontología social: una experiencia educativa sobre ciencia, patrimonio e identidad. Calid Educ 34:231–245

Guajardo A (2013) Análisis experimental de la presencia de marcas de pisoteo, en fósiles animales del sitio Pilauco, Osorno, Centro-Sur de Chile. Undergraduate Dissertation. Universidad Austral de Chile

Labarca R, Pino M, Recabarren O (2013) Los Lamini (Cetartiodactyla: Camelidae) extintos del yacimiento de Pilauco (Norpatagonia chilena): aspectos taxonómicos y tafonómicos preliminares. Estud Geol (Madr) 69(2):255–269

Labarca R, Recabarren O, Canales-Brellenthin P, Pino M (2014) The gomphotheres (proboscidea: Gomphotheriidae) from Pilauco site: scavenging evidence in the late Pleistocene of the Chilean Patagonia. Quat Int 352:75–84

Lobos V (2013) Análisis paleovegetacional y florístico basado en semillas y frutos fósiles del sitio paleontológico Pilauco, Pleistoceno tardío, Osorno, Centro-Sur de Chile. Undergraduate Dissertation. Universidad Austral de Chile

Macías C (2012) Análisis de posible icnita en Sitio Pilauco, Pleistoceno tardío, Osorno, Centro-Sur de Chile. Undergraduate Dissertation. Universidad Austral de Chile

Montero I (2007) Tafocenosis de Tejidos Blandos, Pleistoceno Superior, Sitio Pilauco Osorno, Chile. Undergraduate Dissertation. Universidad Austral de Chile

Münzenmayer M (2018) Determinación del agente principal de la deposición terrígena en el sitio geoarqueológico Pilauco, Osorno, Patagonia Noroccidental de Chile. Undergraduate Dissertation, Universidad Austral de Chile

Pavez-Fox M (2011) Inferencias sobre el uso del hábitat Pleistocénico por el Pudú (Pudu puda MOLINA, 1782) en el centro-sur de Chile. Undergraduate Dissertation. Universidad Austral de Chile

Pavez-Fox M, Pino M, Corti P (2015) Muzzle morphology and food consumption by pudu (Pudu puda Molina 1782) in south-central Chile. Stud Neotrop Fauna Environ 50(2):107–112

Pérez J (2014) Identificación de pelos registrados en el sitio Pilauco, Pleistoceno Tardío, Osorno: reconocimiento y comparación morfológica de la cutícula. Undergraduate Dissertation. Universidad Austral de Chile

Pino M (2008) La excavación de Pilauco. In: Pino M (ed) Pilauco, un sitio complejo del Pleistoceno tardío. Osorno, Norpatagonia chilena, 1st edn, Universidad Austral de Chile, Imprenta América, Valdivia, Chile, pp 31–35

Pinto B (2018) La presencia de la Cronozona Younger Dryas en el registro del Sitio Pilauco, Osorno, Chile: nuevas interpretaciones basadas en Sporormiella sp. Undergraduate Dissertation. Universidad Austral de Chile

Pino M, Chavez M, Navarro X, Labarca R (2013) The late Pleistocene Pilauco site, south central Chile. Quat Int 299:3–12

Ramirez F (2012 Análisis parasitológico de coprolitos del sitio paleontológico Pilauco, Pleistoceno tardío. Undergraduate Dissertation, Universidad Austral de Chile

Recabarren O (2007) Análisis de restos óseos de gonfoterios del área comprendida entre los 39° 39′ y 42° 49′ S, centro—sur de Chile. Undergraduate Dissertation. Universidad Austral de Chile

Recabarren O, Pino M, Cid I (2011) A new record of Equus (Mammalia: Equidae) from the Late Pleistocene of central-south Chile. Rev Chil Hist Nat 84:535–542

Recabarren O, Pino M, Alberdi MT (2014) La Familia Gomphotheriidae en América del Sur: evidencia de molares al norte de la Patagonia chilena. Estud Geol (Madr) 70(1):1–12

Silva N (2014) Uso de esporas de Sporormiella sp. como proxy para determinar extinción de megafauna en el Sitio Pilauco, Osorno. Undergraduate Dissertation. Universidad Austral de Chile

Tello F (2013) Coleópteros fósiles y reconstrucción paleoambiental de la capa PB-7 del sitio Pilauco, Osorno, Norpatagonia de Chile. Undergraduate Dissertation, Universidad Austral de Chile, 31 p

Soto-Bollmann K (2014). Origen de la fracción arenosa de la capa portadora de megafauna del Sitio Pilauco, Osorno, Norpatagonia de Chile. Undergraduate Dissertation. Universidad Austral de Chile

Valenzuela M (2017) Marcas de arrastre fluvial en huesos: evidencias experimentales. Undergraduate Dissertation, Universidad Austral de Chile

Chapter 2
From Gondwana to the Great American Biotic Interchange: The Birth of South American Fauna

Martín F. Chávez Hoffmeister

Abstract The Great American Biotic Interchange (GABI) is one of the biggest biogeographical events, which has shaped the modern South American fauna. It refers to several migratory pulses of continental fauna between North and South America, beginning during the Miocene and continuing to the present day. It intensified during the Pliocene with the closure of the Isthmus of Panama and has been fundamental in the settling of the Neotropical and Nearctic realms as we know them today. However, this is far from being the only time that South America has been colonized or being part of a biotic exchange with other continental masses. During approximately 400 million years South America was part of the ancient supercontinent Gondwana, being connected with the other large land masses of the Southern Hemisphere. The later break-up of this supercontinent paid an enormous impact in the composition of faunas across the whole hemisphere, and the resulting extended period of South American isolation gives birth to a unique assemblage of animals and plants. This chapter aims to offer an introduction to the biogeographical events and processes that shaped the South American fauna during the Cenozoic, with an emphasis on those that took place before the GABI. It provides a broad context to understand the processes at work during the Late Pleistocene when the Pilauco site was formed and the origin of the animals that inhabited it.

Keywords South America · Biogeography · GABI

2.1 Background

The Great American Biotic Interchange (GABI) (Marshall et al. 1984; Stehli and Webb 1985) is one of the biggest biogeographical events that have shaped the composition of the modern fauna in the Americas. This term refers to the migration process of continental faunas between North and South America, particularly during

M. F. Chávez Hoffmeister (✉)
Instituto de Ciencias de la Tierra, Universidad Austral de Chile, Valdivia, Chile
e-mail: paleoaeolos@gmail.com

M. Pino and G. A. Astorga (eds.), *Pilauco: A Late Pleistocene Archaeo-paleontological Site*, The Latin American Studies Book Series,
https://doi.org/10.1007/978-3-030-23918-3_2

the Late Neogene and the Quaternary. Originally regarded as a relatively short-lived event at the beginning of the Pleistocene, it is currently considered as a complex process involving multiple waves of migration in both directions that initiated during the Miocene (Cione et al. 2007). This process is accelerated significantly by the end of the Pliocene after the formation of the Isthmus of Panama that created a permanent connection between the two continents and ceased the isolation of South America.

The connection between the two continents caused a profound change in the faunal composition of both regions, playing a key role in the formation of the Neotropical eco-zones (South America plus tropical regions of North America) and Neoarctic ecozones (rest of North America). As a result of this exchange, almost half of the species of mammals currently present in South America are descendants of immigrants from the northern hemisphere. However, the GABI is far from being the only significant event that modelled the modern fauna of the continent.

South America was connected to other continental masses for millions of years, allowing the exchange of flora and fauna with currently distant continents such as Africa and Australia. Their subsequent separation and isolation were a cause of great changes in the continental fauna and led to the emergence of globally unique ecosystems.

This chapter aims to provide an updated introduction to the biogeographical processes that shaped the fauna of South America, putting an emphasis on the origin of the fauna that inhabited the continent prior to the GABI. This perspective provides us with a broader context for the origins of the wildlife at the Pilauco site. The aim is to offer an overview and to supply only a selection of crucial and preferably recent references and should not be considered as a comprehensive review of the topic.

2.2 Gondwana (510–110 Ma)

To know the geological history of South America is important for understanding its biogeographic connections with distant continents. The flora and fauna that currently inhabit the continent comprise the Neotropical biogeographic realm, whose northward extension reaches the tropical regions of Central and North America, as well as the islands of the Caribbean and southern Florida. This region shares many common elements with other continents of the southern hemisphere, providing evidence of ancient connections between these continental blocks and the past existence of a single southern continental mass called Gondwana.

The supercontinent Gondwana united South America, Antarctica, Africa, Madagascar, Australia, New Zealand, the Arabian Peninsula and India. The final phases of its original formation took place close to the Ediacaran–Cambrian transition (approx. 570–510 Ma), predating the formation of the supercontinent Pangea (Cawood and Buchan 2007). South America was located along the western margin of Gondwana and conterminous to Africa in the east and to Antarctica in the south. Subsequently, the formation of the supercontinent Pangea was completed at the end of the Carboniferous and throughout the Permian period (approx. 320–250 Ma).

During this time South America was still united with North America, until it began to separate in the Early Jurassic (185 Ma) with the opening of the central Atlantic, marking the beginning of the breakup of Gondwana (Veevers 2004). From then on Gondwana started to progressively break apart and towards the end of the Early Cretaceous (approx. 130–110 Ma) South America began to separate from Africa coeval with the formation of the South Atlantic. The connection between South America and Australia persisted through an Antarctic corridor at least until the Late Cretaceous (circa 80 Ma). South America completed its final transformation into an island continent during the Oligocene (approx. 30 Ma) after the definitive separation of the Antarctic continent and the opening of the Drake Passage.

The altering periods of connection and fragmentation between continental masses imply a wide range of biotic exchanges through different processes of dispersion and transport. Until the end of the Early Cretaceous, the South American flora and fauna were linked with the rest of Gondwana through a system of corridors and still constituted part of the continuous Gondwanan biota. Indeed, this biotic continuity throughout the Permian is one of the most cited evidence which support the existence of the supercontinent Gondwana, represented, for example, by the distribution of plants such as *Glossopteris* and synapsids such as *Cynognathus* and *Lystrosaurus*. The process of continental fragmentation produced successive events of vicariant speciation, interrupted by occasional dispersions over long distances or across bridges. The flora of southern South America, Australia, New Zealand and South Africa share many plants that originate in the former Gondwana and after its fragmentation evolved into endemic species, representing a clear example of vicariant speciation (Fig. 2.1).

The Winteraceae (e.g. Winter's bark) and Proteaceae (e.g. Chilean firetree or notro) are some of the current plants shared by all these regions, while others, such as the genera *Nothofagus* (southern beeches) and *Dicksonia* (tree ferns) are present in all of them except Africa, possibly reflecting a dispersal after the fragmentation of the continent. On the other hand, the fossil record confirms that other plants, such as the Casuarinaceae, possessed a similar distribution in the past (McLoughlin 2001). It also provides evidence for the past existence of Antarctic forests, similar in composition to those found today in New Zealand, Australia and, in particular, southern South America (Poole et al. 2001). The Gondwana supercontinent also acted as a corridor for the dispersion of plants between South America and Oceania, including the *Eucalyptus* genus currently only found in Australasia but whose oldest fossils have been found in Argentinean Patagonia (Gandolfo et al. 2011).

In the case of vertebrates, the marsupials represent a well-documented example for the consequences of the fragmentation of Gondwana (Fig. 2.1). Although the greatest diversity of marsupials is currently found in South America and Australia, the fossil record suggests that the metatherians had their origin in the northern hemisphere. The Early Cretaceous *Sinodelphys* (approx. 125 Ma) from China is the oldest known metatherian (Luo et al. 2003), while the Early Paleocene *Paradectes* (circa 65 Ma) of North America is the oldest known marsupial (O'Leary et al. 2013) and a possible member of the dominant marsupials in South America, the ameridelphians.

In fact, metatherians closely related to marsupials, such as *Kokopellia*, are present in the North American fossil record since the middle of the Cretaceous (approx.

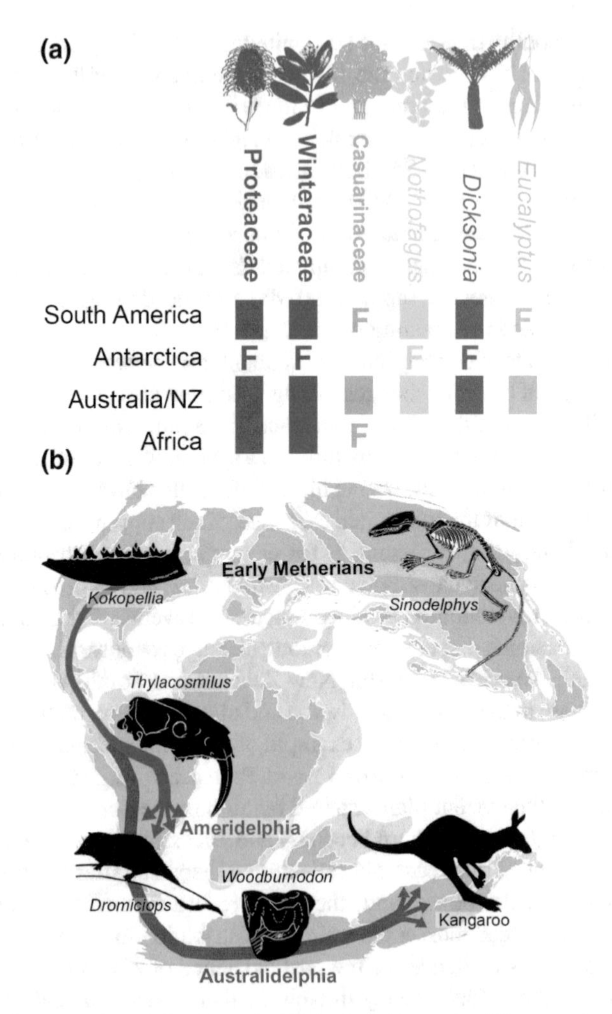

Fig. 2.1 The fragmentation of Gondwana. **a** Plants shared between different regions that were part of Gondwana. Regions where a taxon is only know in the fossil record are indicated with F. **b** Late Cretaceous (94 Ma) map showing the inlet corridors and dispersion pattern inferred of marsupials in Gondwana

100 Ma). On the other hand, the oldest known marsupials in the southern hemisphere are found in the mid-Paleocene record of South America (approx. 60 Ma), while their first appearance in Australia (*Djarthia*) dates from the Early Eocene (approx. 54 Ma) (Beck et al. 2008). This suggests that the arrival of the marsupials in South America could have occurred between the Late Cretaceous and Early Paleocene, from where they might have reached Australia via Antarctica towards the end of the Paleocene. Most of these migrations likely occurred after the onset of continental fragmentation, suggesting a dispersion across temporal corridors such as continental bridges or island arcs. Additionally, this would explain the absence of marsupials in Africa and

New Zealand, which would have separated from the continent prior to their arrival. A similar mechanism could explain the absence of placental mammals in the fossil record of Australia, while they are present in South America and Antarctica.

This hypothesis is supported by molecular analyses, which demonstrate that the 'monito del monte' *Dromiciops gliroides* (Microbiotheria) is currently the most primitive Australidelphian, and only living representative of this group of marsupials outside Australia (Nilsson et al. 2010). Furthermore, the Paleogene fossil record of Antarctica includes Microbiotheriae (e.g. *Woodburnodon*) and Ameridelphiae (Reguero et al. 2002). In fact, the presence of diverse Microbiotheriae during the Paleogene in South America suggests a South American origin of Australidelphiae and their subsequent dispersion and colonization of Australia. Regardless of the centre of origin of the Australidelphiae, there is compelling evidence that the past proximity of continental masses from Gondwana and their subsequent separation played a key role on the dispersion and radiation of marsupials.

2.3 The Isolation of South America and the Ancient Fauna (~55–3 Ma)

The separation of West Antarctica from South America marks the culmination of what George Gaylord Simpson (1980) called a 'splendid isolation'. Separated from other land masses, the South American fauna evolved completely independently, giving rise to a unique assemblage of continental vertebrates. This means that since the mid-Paleocene the South American fauna began to evolve into a plethora of native forms that would dominate the continent for much of the Cenozoic.

The South American fauna during this period of isolation includes opossums (Ameridelphia, Didelphimorphia), large marsupial predators (Sparassodonta), sloths and anteaters (Xenarthra, Pilosa), glyptodonts and armadillos (Xenarthra, Cingulata), a wide range of hoofed herbivores (Notoungulata, Litopterna, Astrapotheria and Pyrotheria), terror birds (Phorusrhacidae), terrestrial crocodiles (Sebecidae) and giant horned tortoises (Meiolaniidae), among others. This fauna represents what Simpson (1980) described as Stratum 1 of South American mammals (Fig. 2.2) that is dominated by the descendants of early immigrants and forms the assemblage of 'original' inhabitants of the continent.

While the process of geographical isolation was not completed until the Oligocene with the onset of deep-water circulation in the Drake Passage, it is possible that faunal isolation had started 25 Ma earlier during the Paleocene–Eocene transition (Reguero et al. 2014). During this initial phase of isolation, both continents were likely separated by a wide shallow sea that inhibited the dispersal of most of the terrestrial animals. This hypothesis is supported by both geological evidence

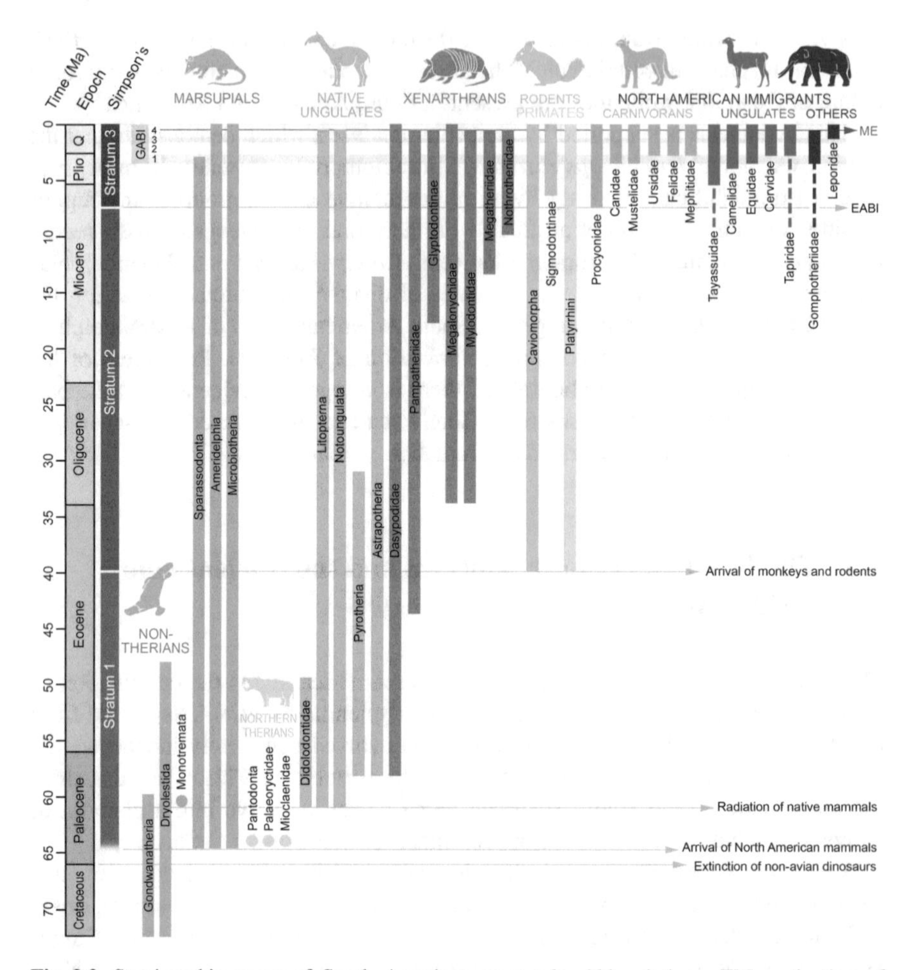

Fig. 2.2 Stratigraphic range of South American mammals. Abbreviations: EM, extinction of megafauna; GABI, Great American Biotic Interchange; EABI, Early American Biotic Interchange

and the high level of endemism amongst Eocene mammals of Seymour Island, Antarctica, which suggests the existence of barriers that would allow vicariant speciation (Reguero et al. 2002, 2014).

The oldest Cenozoic mammalian faunas in South America date back to the Early and Middle Palaeocene and correspond to the Tiupampan and Peligran mammal ages, respectively (Gelfo et al. 2009). These are distinguished by including the first appearance of marsupials and placentals (altogether known as Theria), the last records of non-therian mammals such as the dryolestoids, and the only record of platypus (monotremes). All the therians from the Tiupampian age are closely related to mammals of North America and Eurasia, so they can be considered as early immigrants from the Northern Hemisphere. These include groups of primitive ungulates typical from Paleocene–Eocene North American assemblages (i.e. condylarths) (Gelfo et al. 2009) and ameridelphian marsupials, such as *Roberthoffstetteria*, that are closely related to the North American forms (Goin et al. 2003).

In contrast, the mid-Palaeocene (Peligran) fauna is characterized by the appearance of endemic ungulates like the litopterns and the didolodonts, together with more derived marsupials including the main group of South American mammal predators, the sparassodonts (e.g. *Thylacosmilus*). This suggests that at that time the connection with North America had already been lost allowing a radiation of native species which dominated South America and Antarctica from then on. In fact, the mammal and plants assemblages of Antarctica during the Eocene show great similarities with those from the Paleocene-Eocene of South America, including the presence of litopterns, astrapotheres and non-therian mammals such as the gondwanatherians. By the Late Paleocene Itaboraian mammal age, most of the remaining groups of South American ungulates, such as notoungulates and pyrotheres, had already appeared along with the first armadillos.

As we have seen, there is clear evidence of the northern origin of the ancestors of marsupials and ungulates that comprised the 'original' inhabitants of the continent. However, the superorder Xenarthra seems to be truly endemic to South America. This group of mammals is currently represented by armadillos (Dasypodidae), sloths (Megalonychidae and Bradypodidae) and anteaters (Myrmecophagidae and Cyclopedidae), as well as at least five extinct groups including large armadillos (Glyptodontinae and Pampatheriidae) and giant ground sloths (Megatheriidae, the Mylodontidae and Nothrotheriidae).

Recent analysis based on morphological and molecular data suggests that Xenarthra are the most basal group of living placentals (O'Leary et al. 2013). The oldest known fossil xenarthran comes from the upper Paleocene of Itaboraí, Brazil, and corresponds to the armadillo *Riostegotherium*. This finding, along with its exclusively neotropical occurrence and the abundant fossil records of its radiation in South America, leaves little doubt about the autochthony of this clade.

Accordingly, the fossil record of xenarthrans outside South America is restricted to the Eocene of Antarctica (Carlini et al. 1990) and the Late Miocene of Central and North America (Woodburne 2010), which are interpreted as dispersal events originating from South America. However, alleged fossils of xenarthrans have also been reported in Eurasia during the Paleogene, suggesting possible early dispersion or even an Eurasian origin of at least some xenarthrans. These are *Ernanodon* from the Paleocene of China (Ding 1979) and *Eurotamandua* from the Eocene of Germany, the latter being originally described as an arboreal anteater similar to the modern tamanduas (Carroll 1988). Nevertheless, subsequent studies have shown that both animals lack many diagnostic characters of xenarthrans, including the distinctive vertebral joints that give the name to the group (Gaudin 1999). On the other hand, the phylogenetic analyses demonstrate that *Ernanodon* belongs to the extinct order Palaeanodonta, which is closely related to the order of modern pangolins (Pholidota) which also incorporates *Eurotamandua* (Kondrashov and Agadjanian 2012).

Although continental mammals tend to be the most cited example when referring to the history of South American fauna, similar cases of dispersal and radiation can be observed in other vertebrates. In the case of reptiles, two of the lineages considered endemic in South America are the extinct sebecids (Sebecidae) and alligators (Camaninae). The sebecids were a group of terrestrial crocodylomorphs known from the Palaeocene to Mid-Miocene. These predators were closely related with the baurusuchids of the Late Cretaceous and forms of the European Paleogene such as *Iberosuchus* (Pol and Powell 2011). They were one of the main groups of continental predators in South America and might have invaded the Greater Antilles during the Early Miocene (Brochu and Jiménez 2014). On the other hand, caimans represent the most diverse and abundant crocodylomorphs currently living in South and Central America being amphibious predators, usually intolerant to brackish waters. Although some scattering events between North and South America might have occurred during the early stages of their evolution (Brochu 2011), the oldest and most primitive caimans (*Eocaiman* and *Necrosuchus*) appear on the Early Palaeocene of Argentina. Since then the group has survived on the continent, reaching great morphological diversity during the Miocene with the emergence of highly specialized forms such as the 'duck-like' caiman *Mourasuchus* or the giant *Purussaurus*.

Some groups of birds also evolved under these conditions including rheas and the extinct terror birds. The current distribution of the flightless paleognaths or ratites in the Southern Hemisphere has been traditionally explained as an example of vicariant speciation resulting from the fragmentation of Gondwana. However, this hypothesis has been consistently disproved by molecular studies, which show not only that relationships between ratites do not coincide with the timing of fragmentation of the supercontinent, but neither are the descendants of a single flightless ancestor (Hackett et al. 2008; Mitchell et al. 2014). This scenario suggest that rheas (Rheidae) are one of the most primitive groups of ratites descending from a flying ancestor that separated from the remaining paleognaths at the end of the Cretaceous. The oldest record of rheas corresponds to *Diogenornis* from the Late Palaeocene of Brazil, showing that they had already lost their ability to fly by then (Mayr 2009). The Antarctic fossil record suggests that as in the case of mammals, this group of birds would have invaded the southern continent at the end of the Eocene (Cenizo 2012).

On the other hand, the oldest records of terror birds or phorusrhacids (Phorusrhacidae) also come from the Late Palaeocene of Brazil (Alvarenga 1985). This group of terrestrial birds is considered one of the most characteristic elements of the South American fauna and being one of the main predators of the continent. They belong to the order Cariamiformes, which is currently represented by only two South American species of seriemas (Cariamidae). The group was much more diverse during the Paleogene when it included families in Europe (Idiornithidae) and North America (Bathornithidae). While European fossils possibly related to these families have been wrongly attributed to terror birds in the past, recent findings suggest that the true terror birds were likely present in North Africa and Europe during the Middle Eocene (Mourer-Chauvire et al. 2011; Angst et al. 2013). Although a dispersal from South America into the Old World seems more likely given the presence of older fossil in the South America, these findings raise the possibility that terror birds might represent early invaders from Africa. Whatever the origin of the group, it is clear that the terror birds are part of the original fauna of the South American continent, where they seem to have thrived until the end of the Pleistocene (Alvarenga et al. 2010).

2.4 The Island Hopping Period (40–3 Ma)

For more than 50 Ma, South America was disconnected from the rest of the continental masses, allowing the radiation of native forms. Nevertheless, the isolation did not prevent the arrival of some invaders, who would become an integral part of South American fauna. These long-distance dispersal events characterize the Stratum 2 of Simpson (1980) or the period of the old 'island-hoppers' (Fig. 2.2), during which rodents and primates arrived in South America from Africa.

While such transatlantic dispersions might seem improbable, both the paleontological and molecular evidence suggest that the closest relatives of the South American rodents and primates are from the African continent. This process could have been facilitated by a number of geological factors, such as a greater proximity between both continents and the existence of island arcs in the South Atlantic at the beginning of the Cenozoic (Oliveira et al. 2009). The narrowest point of the Atlantic is currently located between the northeast of Brazil and Senegal (2,575 km). However, during the Eocene, the submarine chains of Rio Grande and Walvis emerged from the sea and formed islands, many of them more than 200 km long, and the distance between the continents was likely reduced to less than half of its current width (approx. 1,000 km). This, combined with the favourable ocean currents between the two continents (Fratantoni et al. 2000), might have allowed the passage of animals on a drift or on natural rafts from one island to another, eventually reaching the new world.

Paleogeographic reconstructions (Oliveira et al. 2009) show that the most favourable conditions for the transatlantic crossing occurred between the Late-Early Eocene and the Early-Late Eocene (50–40 Ma), which seems to coincide with the arrival of rodents and primates to South America. Until recently, there was no fossil evidence suggesting an arrival of both groups before the Oligocene. However, new fossil findings from the Yahuarango formation in the Peruvian Amazon have challenged this scenario. The oldest rodents in South America date back to the Middle or Late Eocene and come from Contamana, where at least five species have been identified (Antoine et al. 2011). They correspond to primitive members of the parvorder Caviomorpha, the largest group of native South American rodents which includes guinea pigs, capybaras and chinchillas among others. Molecular studies incorporating these findings (Voloch et al. 2013) suggest that the separation between the caviomorphs and their Old World relatives likely occurred some 43 Ma ago, shortly after the beginning of a great evolutionary radiation in South America that gave rise to the main living groups at the end of the Eocene. This early diversification is consistent with the Oligocene fossil record of Chile, which includes the genera *Andemys* and *Eoviscaccia* related to agoutis and chinchillas, respectively (Bertrand et al. 2012), and their rapid colonization of the Lesser Antilles around the same time (Velez-Juarbe et al. 2014). At the beginning of the Miocene, the modern caviomorph families were already established, including the Dinomyidae. Currently, this family encompasses only one living species, the pacarana *Dinomys branickii*. However, in the past, it included the largest rodents that ever existed: *Josephoartigasia* and *Phoberomys*, which reached 3 meters in length and weighed more than 700 kg.

The oldest primate in South America correspond to *Perupithecus*. The fossil comes from an outcrop of the Yahuarango formation in the Santa Rosa locality, Peru, and have a similar, or perhaps slightly older age than the rodents of Contamana (Bond et al. 2015). *Perupithecus* is closely related to *Talahpithecus*, a primate from the Middle or Late Eocene of Libya, both representing the closest relatives of the parvorder Platyrrhini that includes all modern New World monkeys. This raises the possibility that the platyrrhines originated and perhaps even radiated in Africa, subsequently becoming extinct and surviving only in South America. *Perupithecus* could also represent a completely independent invasion to the one that eventually led to the development of the true platyrrhines. The next oldest record of primates in South America is *Branisella* from the Late Oligocene of Bolivia (Takai et al. 2000), which is widely recognized as a primitive platyrrhine. Molecular studies suggest that the last common ancestor of all living New World monkeys appeared between the Late Oligocene and Early Miocene (Perez et al. 2013; Kiesling et al. 2015), which is consistent with their diverse fossil record in Argentina during the Early Miocene. This time period also coincides with the colonization of monkeys in the Greater Antilles (MacPhee et al. 2003).

These successful early colonisations left a clear impact on the present: the number of caviomorph and primate species (approx. 370) is more than double than of marsupials and xenarthrans combined (approx. 130). While these are the two best known invasions, other groups of vertebrates also seem to have arrived from Africa during this time period. Molecular studies suggest that turtles, amphisbaenians (Amphisbaenia), and blind snakes (Scolecophidia) crossed the Atlantic between the Palaeocene and Oligocene (Le et al. 2006; Vidal et al. 2007, 2010), while other groups of reptiles arrived during the Neogene.

On the other hand, the lack of permanent connections with other continental land masses was a less effective barrier for flying animals.

The evidence indicates that at least six families of bats colonized South America during the Eocene and three others arrived in several waves from Africa and North America (Lim 2009). Hoatzin birds, hummingbirds and toucans seem to have arrived on the continent during this time period as well. The hoatzin *Opisthocomus hoazin* (Opisthocomiformes) is one of the most characteristic birds of the Neotropical realm, inhabiting the Amazon region and it is famous for having chicks which possess claws on two of their wing digits. However, the oldest records of hoatzins come from the Late Eocene of France and there is evidence of their presence in Africa during the Late Oligocene, suggesting that the order originated in the Old World (Mayr and De Pietri 2014). *Hoazinavis* from the Late Oligocene of Brazil is the oldest record in these birds in South America, representing the minimum time of its arrival to the

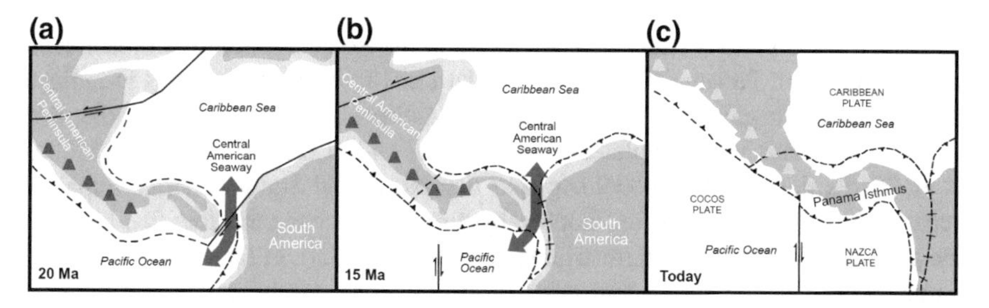

Fig. 2.3 Classic model for the formation of the Isthmus of Panama. **a** Early Miocene initial closure with deep water circulation over the Atrato Channel. **b** Mid-Miocene collision with the South American Plate. **c** Present-day condition established during the Pliocene

continent. Similarly, the initial stages in the evolution of hummingbirds (Trochilidae) took place in Eurasia during the Early Oligocene (Bochenski and Bochenski 2008).

However, molecular studies suggest the last common ancestor of the extant hummingbirds inhabited the lowlands of South America during the Early Miocene (McGuire et al. 2014). These studies reveal that the first invasion of modern hummingbirds to North America occurred during the Middle Miocene, giving origin, amongst others, to the genus *Lampornis* that currently inhabits exclusively North and Central America. Currently, the hummingbirds are completely restricted to the Neotropical realm, as well as toucans (Ramphastidae) and their closest relatives, the American barbets (Capitonidae). Little is known about the early evolution of toucans due to the lack of pre-Pleistocene fossils. It has been suggested that *Rupelramphastoides* from the Late Oligocene of Germany could be related to toucans and barbets, even though their short beaks would place them as a very primitive form (Mayr 2006). Nevertheless, molecular studies support a South American origin for both families during the Middle Miocene (Moore and Migliab 2009; Patel et al. 2011), suggesting that their ancestors arrived at the continent during its period of isolation.

2.5 The Early American Biotic Interchange (EABI, 9–3 Ma)

As we have seen, most of the arrivals during the second Simpson's Stratum came from the Old World. Even if the entry routes of some groups remain a mystery, it is clear that most of them reached the continent through a transatlantic crossing during the Paleogene. After the disappearance of the island arcs and the widening of the South Atlantic, the exchange with Africa began to decline, involving only flying animals like bats or small and resistant animals such as reptiles, which could be easily transported on natural rafts. However, soon a new wave of invaders would arrive to South America, this time from North America (Fig. 2.3).

Even though there is still much debate about the exact chronology of the closure of the Isthmus of Panama (Coates and Stallard 2013), everything seems to indicate that the process began at the beginning of the Miocene. Although recent studies proposed the isthmus was completely closed by the end of the mid-Miocene (Montes et al. 2015), the standard model suggests that North and South America remained separate at least until the Late Pliocene by numerous ocean corridors that hindered the passage of many of the terrestrial vertebrates (Fig. 2.3). In fact, the overall paleontological evidence shows that prior to the Pliocene, faunal exchanges between the two continents were sporadic (Woodburne et al. 2006; Woodburne 2010). This inter-American dispersion predating the GABI is here referred as the Early American Biotic Exchange (EABI) and corresponds to the beginning of the Stratum 3 of Simpson (1980) (Fig. 2.2). The beginning of this stage seems to date back to the Miocene in the case of mammals, but studies on molecular divergence indicate that it could have started even earlier for plants and some other animals (Cody et al. 2010). Eight of the 25 migratory waves identified in plants seem to have occurred more than 20 Ma ago, with the first taking place during the Early Eocene (50 Ma) and only six of them postdate the closure of the isthmus.

The oldest records of mammals with unquestionably South American origin in North America date back to the Late Miocene (8.5–9 Ma) and correspond to the ground sloths *Thinobadistes* and *Pliometanastes*, with the latter possibly giving birth to the later genus *Megalonyx* which is endemic to North America. Shortly after (7.3 Ma), *Cyonasua* would become the first mammal of North American origin settling in South America. It marks the first appearance of raccoons (Procyonidae) on the continent, and gave rise to the extinct *Chapadmalania* (5 Ma), similar to a bear and with a maximum length of 1.5 m. A new family of rodents, Cricetidae, subsequently arrived to South America (6 Ma), where an explosive radiation led to the origin of the vast majority of native New World rats and mice (Sigmodontinae). Five million years ago the second wave of arrivals to North America brought two genera of South American xenarthrans, the pampatheriine *Plaina* and the ground sloth *Glossotherium*, and the emergence of *Titanis*, the only endemic terror bird in North America. The period of one million years predating the GABI lead to the appearance of new xenarthrans and the arrival of capybaras (Hydrochoerinae) in North America, while the peccaries (Tayassuidae) and llamas (Camelidae) make their first confirmed appearance in South America.

If we consider only the earliest records, there seems to be a size bias between North and South American immigrants: while the South American forms arriving in North America were rather large (e.g. ground sloths and terror birds), their North American counterparts were considerably smaller (e.g. raccoons and mice). This could provide clues about the nature of the barriers that prevented the passage from one continent to another. However, this pattern changes significantly if we consider the possible records of large North American mammals in the Peruvian Amazon during the Late Miocene (Campbell et al. 2010). These possible records come from the Madre de Dios formation with an age over 9 Ma, predating even the records of procyonids in South America. This fauna seems to be dominated by families of northern origin, including the gomphothere *Amahuacatherium* (Gomphotheriidae), the peccaries *Sylvochoerus* and *Waldochoerus*, undescribed tapirs, and even *Surameryx*, the only South American record of the family Palaeomerycidae, a group of ruminants that went extinct in North America during the Pliocene. Assuming that the age is correct, this assemblage would suggest that large mammals such as gomphotheres and tapirs have arrived to the continent at the same time that large ground sloths made their way into North America. Unfortunately, both the age and the exact origin of these records have been strongly questioned, since the specimens seem to have been reworked by fluvial channels. Even the validity of the genus *Amahuacatherium* has been disputed and considered indistinguishable from the Pleistocene South American gomphotheres such as *Notiomastodon platensis* (Mothe and Avilla 2015). Consequently, the evidence of the arrival of large North American mammals to South America prior to the Pliocene should be still considered questionable at the moment.

2.6 The Great American Biotic Interchange and the American Schism (GABI, 3 Ma to Present)

During the EABI, some groups of mammals got across the barriers between North and South America to colonize their new habitats with varying degrees of success. This process culminated in the definitive closure of the Isthmus of Panama about 3 Ma ago (Fig. 2.3), which eliminated natural barriers, turning Central America into a permanent bridge for the passage of animals from one continent to another. The Great American Biotic Interchange (GABI) had finally begun and influenced from this moment on the composition of the fauna in the Americas (Fig. 2.4).

While initially it was believed that this cross-dispersion process was quick and simultaneously occurring in both directions after the closure of the isthmus, it is currently considered to have been rather gradual and occurring in multiple north- and southward-directed migration waves. At least four major pulses of reciprocal faunal exchange are recognized (Woodburne 2010). The first one referred as GABI 1 (2.6–2.4 Ma) is associated with the arrival of horses (Equidae), ferrets (Mustelidae), foxes (Canidae) and probably gomphotheres to South America, while armadillos (Dasypodidae) and American porcupines (Erethizontidae) arrived in North America together with a new wave of large xenarthrans. The second pulse (GABI 2) took place about 1.8 Ma ago and marks one of the largest migration waves towards South America including the arrival of the bears (Ursidae), large and small cats (Felidae) including the saber-toothed *Smilodon*, deer (Cervidae), tapirs (Tapiridae) and new camelid species. The anteaters (Myrmecophagidae) seem to be the only South American group that crossed the isthmus towards North America. GABI 3 took place between 1 and 0.8 Ma, marking the arrival of opossums (Didelphidae) to North America and new varieties of cats, deer and peccaries to South America. Finally, GABI 4 at 125 ka is the last migration large wave towards South America with the arrival of otters, wolves, ocelots, coatis and rabbits, among others.

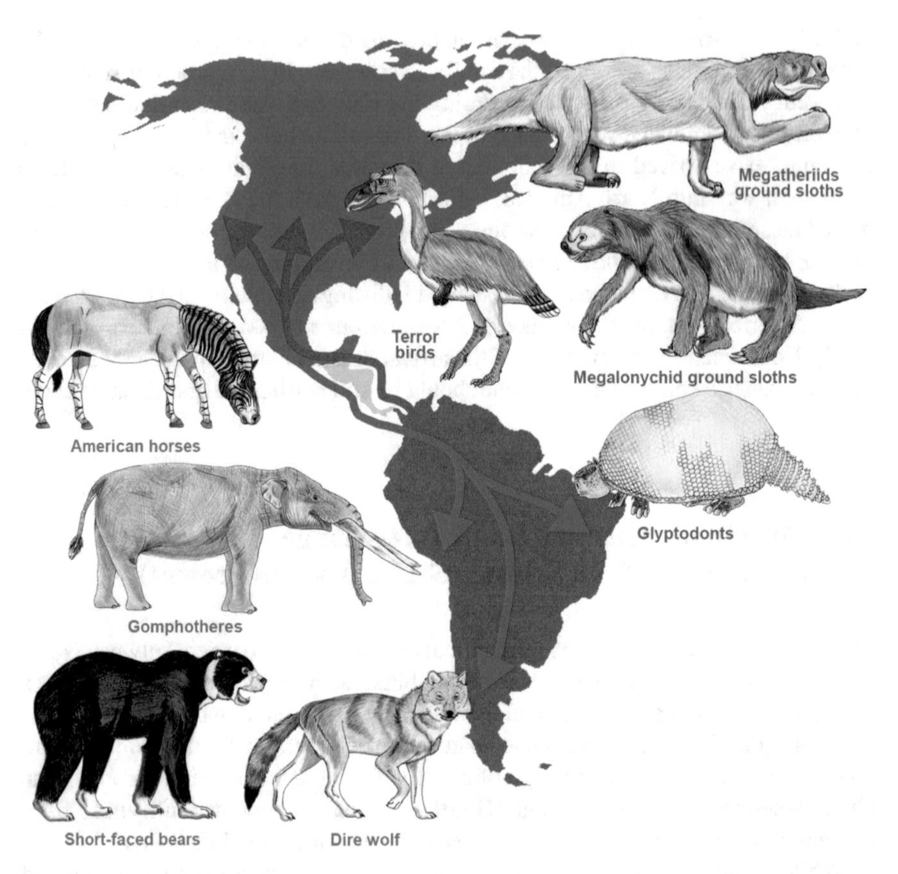

Fig. 2.4 American Biotic Interchange. The Neogene dispersion of some species across South and North America during the Early and Great American Biotic Interchange (EABI and GABI, respectively)

These migrations correspond only to the main events reconstructed from the fossil record, however, a constant flow of mammals and other animals was most likely crossing the isthmus during and between each of these pulses. In fact, it is important to understand that although most of this exchange occurred between the Late Pliocene and the Pleistocene, it is a multidirectional and still-active process. The fossil record shows that the modern cougar *Puma concolor* originated in North America during the Late Miocene and arrived in South America during GABI 2. However, genetic analyses demonstrate that all current populations of cougars in North America originated 100 ka ago from South American populations (Culver et al. 2000). Hence, this species completely disappeared from the Northern Hemisphere by the end of the Pleistocene, only to recolonize North America later on from the south. This example clearly demonstrates the multidirectionality of the exchange between North and South America, as well as its role on the preservation and wide distribution of some species.

An even more recent example can be seen on flying animals. At the end of the nineteenth century wandering individuals of the western cattle egret *Bubulcus ibis*, originally from Africa, started to appear in Suriname and French Guiana in northern South America (Krebs et al. 1994), from where the species began its expansion in the Americas. By the 1970s, the egrets were already spread from Canada to Tierra del Fuego. In 1957 the first African bees were brought to Brazil in order to improve honey production. However, its accidental escape allowed them to mingle with native bees and thus originating a new variety known as Africanized bee or ‘killer bee’. By 1990 this new variety had crossed the Isthmus of Panama and reached the north of Mexico and the south of the United States. These examples demonstrate that far from being a unique and inactive process, the American Exchange continues even today and it will continue to operate as long as the connection between the two continents exists.

The group of South American animals that successfully colonized and still inhabit the Northern Hemisphere comprises opossums, armadillos, porcupines, tyrant flycatchers (Tyrannidae), parrots and hummingbirds, among others. Some animals like vampire bats (Desmodontinae) and capybaras managed to settle in the Northern Hemisphere during the Pleistocene but became extinct later on. Other animal such as megatheres, glyptodonts and pampatheriids did not only settle in North America but evolved into new forms that subsequently recolonized South America, such as the genera *Glyptotherium* and *Megalonyx*. Nevertheless, they finally became extinct in both continents as was also the case with the mylodons, megalonychids, terror birds and other large animals that comprised the American megafauna.

The most successful participants in this process were the North American immigrants that settled in the southern continent. Many groups of invasive mammals that arrived before the formation of the isthmus, such as monkeys and rodents, were successfully integrated into the local fauna. Instead of replacing the original inhabitants, many of them seem to have filled specific niches that were not explored by the largest groups of native mammals (e.g. arboreal frugivores and omnivores, fossorial microgranivorous). Nevertheless, the connection with North America enabled a massive invasion of species competing for key roles covered by native forms (e.g. large herbivores and top predators). This finally led to the gradual replacement of many South American animals in favour of those coming from North America. Although the reasons why northern species were more successful are still debated, the result is evident: almost half of the mammal species currently present in South America are descendants of Northern Hemisphere immigrants. In fact, many of the animals that today are considered typical of South America are actually the descendants of recent invaders. Foxes, cats, skunks, otters, rabbits, mice, deer and rattlesnakes (Crotalinae) are some of the successful immigrants that survived until the present; although some immigrants became extinct in the Americas (e.g. horses) or globally (e.g. gomphotheres).

The closure of the Central American isthmus not only altered the continental faunas, but also caused a simultaneous divergence event between the tropical marine biota known as the Great American Schism. This event resulted from the closure of the Atrato channel which connected the Pacific and the Atlantic Ocean, causing changes in the nutrient transport towards the Atlantic and vicariance. This leads to a great faunal replacement by the Late Pliocene—Early Pleistocene in invertebrates such as molluscs and corals. Studies on molecular divergence in 115 marine species with populations in both oceans in the Central American region, show that the age of divergence for 30% of these populations coincides with the final stages of the closure of the isthmus, while 63% would have diverged prior to that (Lessios 2008). This high-level divergence predating closure of the isthmus appears to be a consequence of environmental changes in the region since the Miocene. Studies on foraminifera suggest that marine ventilation and productivity conditions changed prior to the Late Miocene isthmus closure (7.6–4.2 Ma), coincident with an increase in organic debris and inhibited deep-water ventilation (Jain and Collins 2007). The final closure of the terrestrial corridor at 4.2–2.5 Ma has been associated with a decrease in productivity, an improvement in deep-water ventilation, and a reduced ocean currents speed. Many of the populations separated by the isthmus started to develop mechanisms of reproductive isolation, which is an important precursor for their divergence into new species. Nevertheless, a complete isolation has been identified only for sea urchins of the genus *Diadema*. It is also noted that 7% of the populations seem to maintain some degree of connection postdating the closure of the Central American isthmus, which could be at least partially a recent phenomenon resulting from the opening of the Panama Channel.

2.7 A Brief Discussion

In summary, the fauna that inhabits South America today seems to have its origin in a long series of events and geographical changes. At the time of the first glaciations, South America was populated by numerous descendants from North American immigrants that had arrived during the GABI and some native forms that had survived the impact of this invasion. Glacial climate led to the emergence of large open areas with abundant vegetation, which provided suitable environments for the development of many herbivorous mammals, which at the same time were prey for many predators. During the Pleistocene large mammals and the so-called megamammals were particularly abundant, defining this period as the reign of the megafauna. Thus, scenes and animal communities found in the South American plains and prairies resembled the modern African savannas. Large gomphotheres or 'mastodons' wandered across the continent along with giant ground sloths, armoured glyptodonts and pampatheres, the peculiar toxodonts and macrauchenia. These were the last large native mammals that dominated the plains along with a number of other recently arrived large herbivores, including horses, llamas, and deer. The saber-toothed cats, pumas, jaguars, wolves and bears from the plains had replaced the marsupial sparassodonts and the terror birds as the main predators of the continent. Towards the end of the Pleistocene, the vast majority of these large mammals disappeared, leaving the tapir as the largest mammal on the continent and the anteater bears as the only large xenarthrans in the grasslands formerly populated by many species of giant sloths.

However, these large animals were not the only ones crossing the Isthmus of Panama. The modern humans evolved in northern Africa, rapidly colonized the Old World and reached North America across the Bering Strait. Towards the end of the Pleistocene humans had already reached the south of South America and their arrival coincides with the gradual extinction of the megafauna. While it is likely that factors such as climate change and even the introduction of new diseases would have play an important role in the extinction of large mammals, there is no doubt that the human presence influenced this phenomenon. The end of the ice age coincides with the extinction of the megamammals and most of the large mammals in general. This highlights the importance of the Pilauco site in order to study the coexistence of the last megafauna with the first South American settlers in southern Chile, as well as the origin of modern wildlife in this region.

References

Alvarenga H, Jones W, Rinderknecht A (2010) The youngest record of phorusrhacid birds (Aves, Phorusrhacidae) from the late Pleistocene of Uruguay. Neues Jahrb Geol Paläontol. Abh 256:229–234

Alvarenga H (1985) Um novo Psilopteridae (Aves: Gruiformes) dos sedimentos Terciários de Itaboraí, Rio de Janeiro, Brasil. In: Congresso Brasileiro de Paleontologia, 8. Anais. Rio de Janeiro, NME-DNPM, 1983. p 1720

Angst D, Buffetaut E, Lécuyer C, Amiot R (2013) "Terror Birds" (Phorusrhacidae) from the Eocene of Europe Imply Trans-Tethys Dispersal. PLoS ONE 8(11):e80357

Antoine PO, Marivaux L, Croft DA, Billet G, Ganerod M, Jaramillo C, Martin T, Orliac MJ et al (2011) Middle Eocene rodents from Peruvian Amazonia reveal the pattern and timing of caviomorph origins and biogeography. Proc R Soc Biol Sci Ser B 279:1319–1326

Beck RMD, Godthelp H, Weisbecker V, Archer M, Hand SJ (2008) Australia's oldest marsupial fossils and their biogeographical implications. PLoS ONE 3(3):e1858

Bertrand OC, Flynn JJ, Croft DA, Wyss AR (2012) Two new taxa (Caviomorpha, Rodentia) from the early Oligocene Tinguiririca Fauna, (Chile). Am Mus Novit 3750:1–36

Bochenski Z, Bochenski ZM (2008) An old world hummingbird from the Oligocene: a new fossil from Polish Carpathians. Int J Ornithol 149:211–216

Bond M, Tejedor MF, Campbell KE, Chornogubsky L, Novo N, Goin F (2015) Eocene primates of South America and the African origins of New World monkeys. Nature 520(7548):538–541

Brochu CA (2011) Phylogenetic relationships of Necrosuchus ionensis Simpson, 1937 and the early history of caimanines. Zool J Linn Soc 163:S228–S256

Brochu C, Jiménez O (2014) Enigmatic crocodyliforms from the early Miocene of Cuba. J Vertebr Paleontol 34(5):1094–1101

Campbell K, Prothero D, Romero-Pittman L, Hertel F, Rivera N (2010) Amazonian magnetostratigraphy: dating the first pulse of the Great American Faunal Interchange. J S Am Earth Sci 29:619–626

Carlini, AA, Pascual, R, Reguero, MA, Scillato Yané, GJ, Tonni, EP, Vizcaíno, SF (1990) The first Paleogene land placental mammal from Antarctica: its paleoclimatic and paleobiogeographical bearings. Paper presented at IV international congress of systematic and evolutionary biology, University of Maryland, 1–7 July 1990

Carroll RL (1988) Vertebrate paleontology and evolution. W. H. Freeman and Company

Cawood P, Buchan C (2007) Linking accretionary orogenesis with supercontinent assembly. Earth-Sci Rev 82:217–256

Cenizo M (2012) Review of the putative Phorusrhacidae from the Cretaceous and Paleogene of Antarctica: new records of ratites and pelagornithid birds. Pol Polar Res 33:225–244
Cione AL, Tonni EP, Bargo S, Bond M, Candela AM, Carlini AA, Deschamps CM, Dozo MT, Esteban G, Goin FJ, Montalvo CI, Nasif N, Noriega JI, Ortiz Jaureguizar E, Pascual R, Prado JL, Reguero MA, Scillato-Yané GJ, Soibelzon L, Verzi DH, Vieytes EC, Vizcaíno SF, Vucetich MA (2007) Mamiferos continentales del Mioceno tardío a la actualidad en Argentina: cincuenta años de estudios. Asoc Paleontol Arg Pub Espec (11, Ameghiniana 50th aniversario: pp 257–278)
Coates AG, Stallard RF (2013) How old is the Isthmus of Panama? Bull Mar Sci 89:801–813
Cody S, Richardson J, Rull V, Ellis C, Pennington RT (2010) The Great American Biotic Interchange revisited. Ecography 33:326–332
Culver M, Johnson WE, Pecon-Slattery J, O'Brien SJ (2000) Genomic Ancestry of the American Puma. J Hered 91(3):186–197
Ding S (1979) A new edentate from the Paleocene of Guandong. Vertebrat PalAsiatic 17:57–64
Fratantoni D, Johns W, Townsend T, Hurlburt H (2000) Low-latitude circulation and mass transport pathways in a model of the tropical Atlantic Ocean. J Phys Oceanogr 30(8):1944–1966
Gandolfo MA, Hermsen EJ, Zamaloa MC, Nixon KC, González CC et al (2011) Oldest known Eucalyptus Macrofossils are from South America. PLoS ONE 6(6):e21084
Gaudin TJ (1999) The morphology of xenarthrous vertebrae (Mammalia: Xenarthra) (No. 41). Field Museum of Natural History
Gelfo JN, Goin FJ, Woodburne MO, Muizon CD (2009) Biochronological relationships of the earliest South American Paleogene mammalian faunas. Palaeontol 52:251–269
Goin F, Candela A, Muizon C (2003) The Affinities of Roberthoffstetteria nationalgeographica (Marsupialia) and the Origin of the Polydolopine Molar Pattern. J Vertebr Paleontol 23(4):869–876
Hackett SJ, Kimball RT, Reddy S, Bowie RCK, Braun RC, Bowie EL, Braun MJ, Chojnowski JL (2008) A phylogenomic study of birds reveals their evolutionary history. Science 320(5884):1763–1768
Jain S, Collins LS (2007) Trends in Caribbean Paleoproductivity related to the Neogene closure of the Central American Seaway. Mar Micropaleontol 63:57–74
Kiesling N, Yi SXK, Sperone G, Wildman D (2015) The tempo and mode of New World monkey evolution and biogeography in the context of phylogenomic analysis. Mol Phylogenet Evol 82B:386–399
Kondrashov P, Agadjanian A (2012) A nearly complete skeleton of Ernanodon (Mammalia, Palaeanodonta) from Mongolia: morphofunctional analysis. J Vertebr Paleontol 32(5):983–1001
Krebs E, Riven-Ramsey D, Hunte W (1994) The Colonization of Barbados by Cattle Egrets (Bubulcus ibis) (1956–1990. Colon Waterbirds 17:86–90
Le M, Raxworthy CJ, McCord WP, Mertz L (2006) A molecular phylogeny of tortoises (Testudines: Testudinidae) based on mitochondrial and nuclear genes. Mol Phylogenet Evol 40(2):517–531
Lessios H (2008) The Great American Schism: divergence of marine organisms after the rise of the Central American Isthmus. Annu Rev Ecol Evol Syst 39:63–91
Lim BK (2009) Review of the origins and biogeography of bats in South America. Chiropt Neotrop 15(1):391–410
Luo Z-X, Ji Q, Wible JR, Yuan C-X (2003) An early Cretaceous tribosphenic mammal and metatherian evolution. Science 302(5652):1934–1940
MacPhee R, Iturralde-Vinent MA, Gaffney ES (2003) Domo de Zaza, an Early Miocene vertebrate locality in south-central Cuba, with notes on the tectonic evolution of Puerto Rico and the Mona Passage. Am Mus Novit 3394:1–42
Marshall LG, Berta A, Hoffstetter R, Pascual R, Reig OA, Bombin M, Mones A (1984) Mammals and stratigraphy: geochronology of the continental mammal-bearing quaternary of South America. Palaeovertebr Mém Extraord:1–76
Mayr G (2006) First fossil skull of a Palaeogene representative of the Pici (woodpeckers and allies) and its evolutionary implications. Ibis 148:824–827
Mayr G (2009) Paleogene fossil birds. Germany, Springer, Heidelberg

Mayr G, Pietri VLD (2014) Earliest and first Northern Hemispheric hoatzin fossils substantiate Old World origin of a "Neotropic endemic". Naturwissenschaften 101(2):143–148
McGuire JA, Witt CC, Remsen JV Jr, Corl A, Rabosky DL, Altshuler DL, Dudley R (2014) Molecular phylogenetics and the diversification of hummingbirds. Curr Biol 24(8):910–916
McLoughlin S (2001) The breakup history of Gondwana and its impact on pre-cenozoic floristic provincialism. Aust J Bot 49:271–300
Mitchell KJ, Llamas B, Soubrier J, Rawlence NJ, Worthy TH, Wood J, Lee MSY, Cooper A (2014) Ancient DNA reveals elephant birds and kiwi are sister taxa and clarifies ratite bird evolution. Science 344(6186):898–900
Montes C, Cardona A, Jaramillo C, Pardo A, Silva JC, Valencia V, Ayala C, Perez-Angel LC, Rodriguez-Parra LA, Ramirez V, Niño H (2015) Middle Miocene closure of the Central American Seaway. Science 348:226–229
Moore WS, Migliab KJ (2009) Woodpeckers, toucans, barnets and allies (Piciformes). In: Hedges B, Kumar S (eds) The timetree of life. Oxford University Press, pp 445–450
Mothe D, Avilla L (2015) Mythbusting evolutionary issues on South American Gomphotheriidae (Mammalia: Proboscidea). Quat Sci Rev 110:23–35
Mourer-Chauvire C, Tabuce R, Mahboubi M, Adaci M, Bensalah M (2011) A Phororhacoid bird from the Eocene of Africa. Naturwissenschaften 98:815–823
Nilsson MA, Churakov G, Sommer M, Tran NV, Zemann A, Brosius J, Schmitz J (2010) Tracking marsupial evolution using archaic genomic retroposon insertions. PLoS Biol 8(7):e1000436
O'Leary MA, Bloch JI, Flynn JJ, Gaudin TJ, Giallombardo A, Giannini NP, Goldberg SL, Kraatz BP, Luo Z-X Meng J Ni X, Novacek MJ, Perini FA, Randall ZS, Rougier GW, Sargis EJ, Silcox MT, Simmons NB, Spaulding M, Velazco PM, Weksler M, Wible JR, Cirranello, AL (2013) The Placental Mammal Ancestor and the Post-K-Pg Radiation of Placentals. Science 339(6120):662–667
Oliveira F, Cassola E, Marroig G (2009) Paleogeography of the South Atlantic: a Route for Primates and Rodents into the New World? In: Garber et al (eds) South American Primates, developments in primatology: progress and prospects. Springer, pp 55–68
Patel S, Weckstein J, Patané J, Bates J y Aleixo A (2011) Temporal and spatial diversification of Pteroglossus araçaris (AVES: Ramphastidae) in the neotropics: Constant rate of diversification does not support an increase in radiation during the Pleistocene. Mol Phylogenet Evol 58:105–115
Perez SI, Tejedor MF, Novo NM Aristide L (2013) Divergence times and the evolutionary radiation of New World Monkeys (Platyrrhini, Primates): an analysis of fossil and molecular data. PLoS ONE 8(6):e68029
Pol D, Powell J (2011) A new sebecid mesoeucrocodylian from the Rio Loro formation (Palaeocene) of north-western Argentina. Zool J Linn Soc 163:S7–S36
Poole I, Hunt R, Cantrill D (2001) A Fossil Wood Flora from King George Island: ecological Implications for an Antarctic Eocene Vegetation. Ann Bot (Lond) 88:33–54
Reguero MA, Marenssi SA, Santillana N (2002) Antarctic Peninsula and Patagonia Paleogene terrestrial environments: biotic and biogeographic relationships. Palaeogeogr Palaeoclimatol Palaeoecol 179:189–210
Reguero MA, Gelfo JN, López GM, Bond M, Abello A, Santillana SN, Marenssi SA (2014) Final Gondwana breakup: the paleogene South American native ungulates and the demise of the South America-Antarctica land connection. Glob Planet Change 123B:400–413
Simpson GG (1980) Splendid isolation: the curious history of South American Mammals. Yale University Press, New Haven
Stehli FG, Webb SD (1985) The great American biotic interchange. Topics in geobiology. Plenum Press, New York
Takai M, Anaya F, Shigehara N, Setoguchi T (2000) New fossil materials of the earliest new world monkey, Branisella boliviana, and the problem of platyrrhine origins. Am J Phys Anthropol 111:263–281
Veevers J (2004) Gondwanaland from 650–500 Ma assembly through 320 ma merger in Pangea to (185–100 Ma breakup: supercontinental tectonics via stratigraphy and radiometric dating. Earth Sci Rev 68:1–132

Velez-Juarbe J, Martin T, Macphee RDE, Ortega-Ariza D (2014) The earliest Caribbean rodents: Oligocene caviomorphs from Puerto Rico. J Vertebr Paleontol 34(1):157–163
Vidal N, Azvolinsky A, Cruaud C, Hedges SB (2007) Origin of tropical American burrowing reptiles by transatlantic rafting. Biol Lett 4(1):115–118
Vidal N, Marin J, Morini M, Donnellan S, Branch WR, Thomas R, Vences M, Wynn A, Cruaud C, Hedges SB (2010) Blindsnake evolutionary tree reveals long history on Gondwana. Biol Lett 6(4):558–561
Voloch C, Vilela J, Loss-Oliveira L, Schrago C (2013) Phylogeny and chronology of the major lineages of New World hystricognath rodents: insights on the biogeography of the Eocene/Oligocene arrival of mammals in South America. BMC Res Notes 6:160
Woodburne MO, Cione AL, Tonni EP (2006) Central American provincialism and the Great American Biotic Interchange. In: Carranza-Castañeda O, Lindsay EH (eds) Advances in late tertiary vertebrate paleontology in Mexico and the Great American Biotic Interchange, Univ Nac Aut Mex. Inst Geol Centro Geoscien Publ Espec, vol 4, pp 73–101
Woodburne MO (2010) The great American biotic interchange: dispersals, tectonics, climate, sea level and holding pens. J Mamm Evol 17:245–264

Chapter 3
Geology, Stratigraphy, and Chronology of the Pilauco Site

Mario Pino, Alejandra Martel-Cea, Rodrigo M. Vega, Daniel Fritte and Karin Soto-Bollmann

Abstract The Pilauco site is located in north-western Chilean Patagonia, Region de Los Lagos (40°34′12″S, 73°06′12″W). The stratigraphy of the site comprises of two distinct Pleistocene units. The lower unit corresponds to the sequence San Pablo of MIS5e age and consists of clastic and volcanoclastic sediments (PB-1 to PB-5). The upper unit includes the strata PB-6 to PB-9 with age constraints between 17,370 and 4,340 cal. year BP. The strata bearing fossil mammals and archaeological evidence (PB-7 and PB-8; 16,400–12,800 cal. year BP) correspond to layers of sand derived by means of colluvial transport from the nearby hills to the north of the site, in a fine-grained matrix of organic anoxic material and dispersed gravel clasts deposited in a seasonal wetland. The contact between the layers PB-8 and PB-9 is characterized by an erosional unconformity, a drastic increase in charcoal particles coeval with the nearly complete disappearance of pollen and other plant remains. Moreover, layer PB-9 lacks paleontological and archaeological remains. The discordance has been dated on ~12,800 cal. year BP, which coincides with the Younger Dryas Northern Hemisphere climatic oscillation. The local geological characteristics are concordant with regional geomorphological interpretations that have identified several terraces formed during the last glaciation and the Holocene.

Keywords Geology · Stratigraphy · Pilauco

M. Pino (✉) · R. M. Vega · K. Soto-Bollmann
Instituto de Ciencias de la Tierra, Universidad Austral de Chile, Valdivia, Chile
e-mail: mariopinoquivira@gmail.com

M. Pino · A. Martel-Cea · D. Fritte
Transdisciplinary Center for Quaternary Research in the South of Chile, Universidad Austral de Chile, Valdivia, Chile

A. Martel-Cea
Graduate School Facultad de Ciencias Forestales y Recursos Naturales, Universidad Austral de Chile, Valdivia, Chile

M. Pino and G. A. Astorga (eds.), *Pilauco: A Late Pleistocene Archaeo-paleontological Site*, The Latin American Studies Book Series,
https://doi.org/10.1007/978-3-030-23918-3_3

3.1 Background

The past 2.6 million years are characterized by the periodic alternation of cold episodes (glaciations) lasting about 100,000 years and shorter, warmer ~10,000 years periods with a climate similar to modern-day conditions or interglacials. In northwestern Chilean Patagonia, in the region that corresponds to the lakes district, the last glaciation has been referred as Llanquihue (Mercer 1976), and was preceded by the Valdivia Interglacial (c. 130,000 years ago; Brüggen 1945; Arenas et al. 2005; Latorre et al. 2007; International Unión of Geological Sciences 2009). The onset of the Valdivia Interglacial was characterized by an important episode of explosive Andean volcanic activity. The event has been associated with the deposition of a series of volcanoclastic flows of 50 m of minimum thickness, grouped and referred as Lahar Puyehue (Corvalán 1974), and temporally constrained to the Late Pleistocene (Moreno and Varela 1985).

Later on, this sequence has been assigned to the Pyroclastic–Epiclastic San Pablo Sequence (Plsp unit is a formal denomination of the Chilean Geological Survey (Pérez et al. 2003) for the Pyroclastic-Epiclastic San Pablo Sequence). According to these authors, the sequence includes pyroclastic subaerial and subaquatic flows, lacustrine deposits, and other deposits reworked by fluvial activity. The pyroclastic materials include ash and lapilli composed mainly of lithic fragments and scoria, but also pumice, crystals, and glassy fragments. The Plsp unit is above the drift of the penultimate glaciation and is covered by the Llanquihue age glacial deposits (Mercer 1976; Porter 1981).

Furthermore, Pérez et al. (2003) identified three clastic units exposed in the area of Osorno. The first one preceding the above-described units corresponds to sand deposits of the pre-Llanquihue glaciations (Plgf2). These deposits are scarce; display a maximum thickness of 5 m and present intense weathering, up to 1 m depth, characterized by a mixture containing iron and manganese oxides on clast surfaces. The second unit (Pigf1) is composed of clastic sediments and represents glaciofluvial deposits from the Llanquihue glaciation. The deposition of this unit is controlled by the preexisting relief and follows the direction of the present-day fluvial channels of the Damas River, among others. The layered gravels reach maximum thickness up to 20 m and lack weathering crust. According to Pérez et al. (2003), these deposits correspond to material transported by dynamic runoff water flows, with migrating channels that reflect abrupt changes in the capacity of fluvial discharge during the Llanquihue glaciation (Mercer 1976).

The third unit PlHf is represented by fluvial deposits composed of unconsolidated gravel and sand, which form terraces of 1–15 m of elevation above the present-day riverbed. The reduced thickness of the overlying soil and the morphostratigraphic relationship with the Llanquihue glacial deposits indicate a post-Llanquihue depositional age of about 13,900 ^{14}C year BP. (Lowell et al. 1995). This age dating was later evaluated by Denton et al. (1999), who identify an abrupt increase in temperatures at the sites Huelmo and Canal de Puntilla based on the higher abundance of *Nothofagus* at 14,600 ^{14}C year BP (Moreno 1997; Moreno and León 2003; Moreno et al. 1999).

According to Denton et al. (1999), the most recent glacial advances did not entirely occupy the main plain area. In general, the drainage attached to the interior moraine systems developed a secondary plain carved into the principal one in the proglacial domain and merging with it downstream. In Osorno, the Tijeras River is not part of the drainage system directly linked with the glaciers during their most recent advances, because all its tributaries originate from the external moraines in the west.

In turn, the Damas River apparently maintained the connection with the glaciated domain over the stream Quebrada Honda. It was linked over two deep canyons with the terminal position of the Puyehue and Rupanco glacial lobes during a glacial advance dated back to c. 19 Ka ^{14}C and 16 Ka ^{14}C (Bentley 1997).

3.2 Materials and Methods

To understand the sedimentary environment of the Pilauco layers we used textural and granulometric characteristics and also simultaneity we define the lateral and horizontal relationships between them. Color of wet sediments was based on the Munsell Color Chart. We conducted a textural analysis of sediments from a stratigraphic column 5 cm width including 255 samples (14AD grid). The sampling covered the entire PB-7 to PB-9 sequence and the uppermost 2 cm of layer PB-6. The analyzed column extends from 326 to 581 cm of local elevation. The samples were wet-sieved to quantify the gravel (2 mm), sand (2 mm to 62.5 μm), and mud (<62.5 μm) fractions. Following, the sand and mud fractions were incinerated at 550 °C for the calculation of total combustible organic carbon. In Pilauco, like in any sedimentary sequence, the presence of gravel, pebbles or cobbles in a column only 5 cm wide is totally random; sampling several kilograms from a column at an archaeological site is not possible. In this way, the presence of even a single shingle causes great variations in the percentage calculations of the other components such as sand, mud or organic matter. For this reason, the relative abundance of gravel was not considered in the percentage mass calculations. Also, we evaluate the relationships between the abundance of gravel fraction and the mean grain size of the sand in samples from layer PB-7 as a method to define the source of that sand. With the same purpose, we performed a granulometric, textural, and multivariate statistical analyses of the sandy fraction from 17 samples from the layer PB-7. The results were compared to another 10 samples from layer PB-1 that outcrops in the hill to the north of the site.

To investigate the dynamics of local wildfires and characteristics of vegetation at Pilauco during the transition between layers PB-8 and PB-9, pollen and charcoal analyses were performed following standard techniques (Faegri and Iversen 1989; Stockmarr 1971; Whitlock and Larsen 2001). The analysis of pollen grains and spores was performed by sampling 1 cm^3 of sediment at 1 cm intervals (540–553 cm) from the stratigraphic column in grid AD10. To quantify the concentration of charcoal particles in the sediments (>0.125 mm), 2 ml of sediment at intervals of 1 cm were sampled from 530 to 564 cm from the same column and grid.

Around 50 radiocarbon dates have been obtained at the Pilauco site since 2007 in bones, charcoal, and plant remains. Some relevant ages mentioned in the text were calibrated using the software CALIB 7.1 and the SHcal13 curve[1] (Stuiver and Reimer 1993; Hogg et al. 2013). Radiocarbon dates were done in NSF-Arizona AMS Lab, Dept. of Physics at the University of Arizona; Keck Carbon Cycle AMS Facility at Earth System Science Dept. University of California Irvine; DirectAMS, Bothell WA 98011; Isotrace Radiocarbon Laboratory Accelerator Mass Spectrometry Facility at the University of Toronto; Center for Accelerator Mass Spectrometry at the Lawrence Livermore National Laboratory and Radiocarbon Laboratory of the Energy and Environmental Sustainability Laboratories at the Pennsylvania State University. To date plants, we try always to use seeds or leaves avoiding wood or charcoal.

3.3 Stratigraphy and Sedimentology

The stratigraphy of the Pilauco site was described for the first time during the initial excavations in 2007 (Pino and Miralles 2008). Two sedimentary sequences have been identified and assigned to distinct geomorphological units, characterized by different granulometric features and age constraints.

Towards the north, the Pilauco site is delimited by a hill chain of 40–70 m a.s.l., composed by deposits of volcanoclastic origin (ashes, lahars, pyroclastic flows, ignimbrites and tuff layers), as well as small lenses of fluvial sediments (sand and gravel). These units correspond to the outcrops of the San Pablo Sequence (Pérez et al. 2003). Several authors have investigated these deposits and provided different interpretations regarding their origin. For instance, Corvalán (1974) and Pérez et al. (2003) assigned them to the laharic and volcanoclastic complex San Pablo. Denton et al. (1999), relate this unit to the main sandur plain associated to glacial deposits from the mid-Pleistocene glacial episodes, also referred to as Casma and Colegual (Laugénie 1982) or Santa Maria glaciations (Porter 1981). Laugénie (1982) describes the sediments as of fluvial origin associated with volcanoclastic deposits in the interfluve zone.

Our investigation suggests that the northern part of the Pilauco site is dominated by the volcanoclastic domain, whereas abundant fluvial units outcrop approx. 3 km to the west of the site. Some of the volcanoclastic material has been reworked by fluvial activity, and in turn, the fluvial units have been incorporated in volcanoclastic processes (e.g., lahars). Both processes are difficult to identify and distinguished in the field, mainly because the relief unit provides material of fluvial as well as volcanoclastic origin.

At Pilauco (40°34′12″S, 73°06′12″W; Fig. 3.1; Chap. 1, this volume), the San Pablo Sequence (Pérez et al. 2003) can be observed towards the northern part of the site forming a hill of smooth surface. At the southern flank of the hill, a discordant contact is observed between the volcanoclastic sediments (tuffs

[1] http://calib.qub.ac.uk/calib/, consulted in July 2015.

and lapilli tuffs) from the San Pablo Sequence and a surface of alluvial origin (Pino and Miralles 2008). The alluvial plain is composed of conglomerates and peaty layers of variable, but rather low, pebble fraction supported by a sandy matrix. Ten meters to the south, these layers gradually turn into fluvial material composed of gravels and sand. Below the first terrace (~10 m) and south from the Damas River, another terrace composed of pebbles and boulders supported by a sandy matrix can be observed. The clasts are characterized by presenting a typical weathering crust with a thickness of several millimeters (Fig. 3.2) of possibly pre- or Early-Llanquihue age as described by Lauer (1968) and Illies (1970). This terrace is defined by a well-developed surface uniting the drainages of different glacial advances during the late Pleistocene (Denton et al. 1999). It follows the pre-Llanquihue relief eventually merging with the modern flood plains. At Pilauco, the stratigraphy of the terrace is represented by PB-6, PB-7 and PB-8 layers (Pino et al. 2013). An assembly of three to five different levels is embedded within the main plain towards the level of the present-day flood plains.

A schematic representation of the site stratigraphy is shown in Fig. 3.3. Between the base of the excavation site and the strata of the San Pablo Sequence at the northern hill, five stratigraphic layers can be identified.

(A) San Pablo Unit (Plsp)

- **PB-1**: This lapilli tuff corresponds to the base of the entire sequence and outcrops near the base of the hill located to north of the site, as well as discordantly below layer PB-6 and sometimes PB-7. It is composed by angular clasts with different coloring (orange, grey, green, black), generally smaller than 15 mm, supported by a light bluish-gray ash matrix (between 5BG/7/1 and 10BG/7/1). The thickness of this layer is undetermined.
- **PB-2**: Layer of 15 cm thickness, composed mainly by a grayish-white (N/7) volcanic ash matrix, bearing some white, orange, and greenish pyroclastic grains, less than 1 mm in size. Layers PB-2 to PB-5 are lenticular in shape and present lengths of several meters.
- **PB-3**: Light bluish-gray lapilli (between 5BG/7/1 and 10BG/7/1) reaching up to 5 cm thickness. The grain size corresponds to medium-grained sand, occasionally including some orange-colored and greenish clasts, less than 5 mm in size.
- **PB-4**: This layer corresponds to a light bluish-gray colored ash (between 5BG/7/1 and 10G/7/1) and displays some areas of highly weathered sediments. It presents variable thickness from 2 to 15 cm and is the least consolidated layer in the sequence. Between the layers PB-4 and PB-5, there is an erosional unconformity.
- **PB-5**: The layer corresponds to a grayish-white (N8) tuff of medium-grained sand containing few dispersed angular pebbles. It presents a variable grade of consolidation.

(B) Layers of the Pilauco site

- **PB-6**: Forms the base of the sequence at the site. The layer corresponds to a conglomerate containing well-rounded clasts of diameters between 1 and 15 cm,

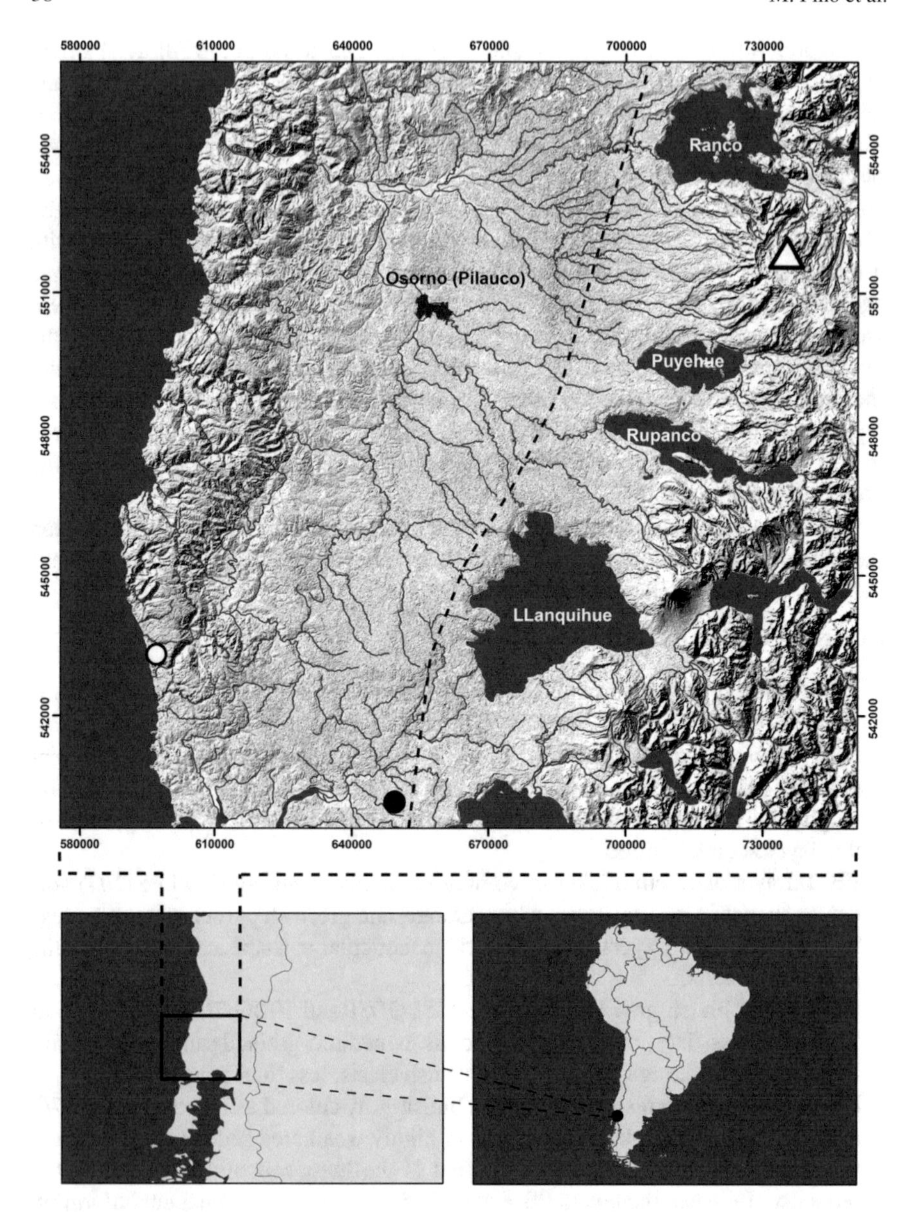

Fig. 3.1 Location of the Pilauco site in the city of Osorno. The dashed line indicates the ice margin during the last glacial maximum. The black circle shows the location of Monte Verde site. The coordinates correspond to the Universal Transverse Mercator 2-dimensional Cartesian coordinate system (UTM)

Fig. 3.2 Andesitic and basaltic gravel clasts showing differential weathering. Left: gravel clasts without weathering crust from the PB-6 layer; Right: Weathered pebbles characteristics of pre-Llanquihue deposits

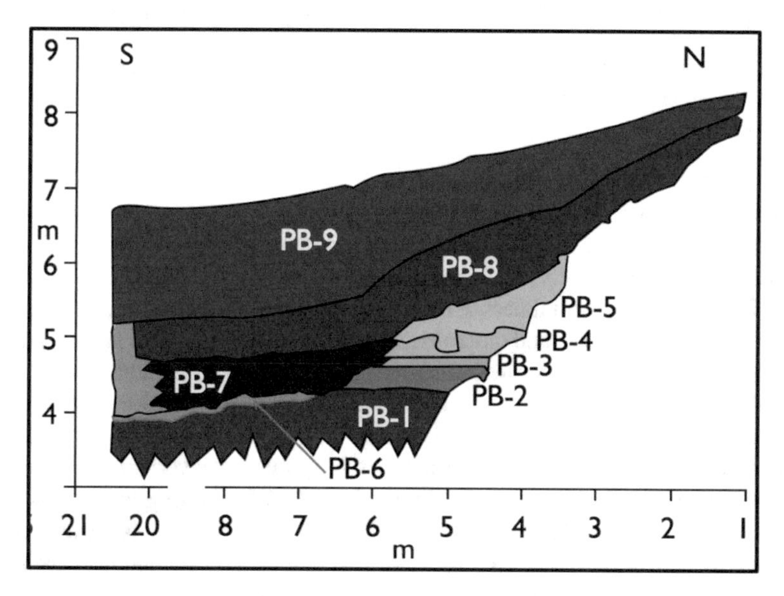

Fig. 3.3 Schematic representation of the Pilauco stratigraphy. Layers PB-7 to PB-9 correspond to sediments with high organic anoxic content. Layers PB-7 and PB-8 interfinger with fluvial deposits to the south

Fig. 3.4 Unconformity between layers PB-1 and PB-6 at the northern limit of the Pilauco site. The fossil-bearing layer PB-7 overlies PB-6. The cobbles under the PB-6 label measure 10 cm in diameter

composed of igneous rocks of Andean origin that exhibit fresh non-weathered surfaces (Fig. 3.2). The olive gray (2.5GY 5/1) sand matrix conforms 60% of the layer. It lies discordantly over fluvial channels cutting the top of layers PB-1 or PB-2. The maximum thickness of this stratum generally does not exceed the diameter of the largest clasts, and outcrops as lenticular bodies that bear large fragments of wood. This layer lacks fossil remains of megafauna and artifacts.

- **PB-7**: This layer is discordantly overlying PB-1, PB-2, or PB-6 (Figs. 3.3 and 3.4). It contains most of the fossil bones of extinct megafauna and extant micromammals, stone artifacts, a large number of plant and invertebrate remains. It corresponds to a dark-brown organic matrix (10YR 3/1) filling the porous space of the sand that supports the sediment. The layer contains isolated and poorly selected clasts of igneous rocks, up to 7 cm of diameter embedded in coarser sand (Fig. 3.5). This layer contains also some intraclasts derived from PB-2 to PB-3 of less than 2 cm in size. Towards the northern part of the site, layer PB-7 lies in unconformable contact with layer PB-5. In the eastern area of the site, a thin yellowish lamina marks the contact with the overlying PB-8 layer (Fig. 3.6).
- **PB-8**: This layer is very similar to PB-7 but presents a more intense brown color (2.5Y 4/2). Towards the north, it gradually changes into a yellowish, sandy sediment and finally into a gravel deposit. At the base of the layer in the northern part of the site, a light olive-green peat layer (10Y/6/2) can be observed.
- **PB-9**: This layer is characterized by vertical and horizontal heterogeneity and contains a number of lenticular bodies combining organic and inorganic components.

Fig. 3.5 Pebble cast in a PB-7 sample. That sedimentological feature is not possible in fluvial deposits

It corresponds to black peat (2.5 Y 2/0) bearing abundant (20–50%) intraclasts derived from the volcanoclastic units of the nearby northern hill, and numerous fragments of wood and charcoal. It laterally changes into organic soil. The contact with the underlying layer PB-8 is marked by an erosional unconformity. This layer overlies the volcanoclastic sediments that outcrop in the southern flank of the hill.

The percentage weight of the gravel fractions decreases towards the top of the sequence, averaging 21.5% (±11.6 SD) for PB-7, 16.7% (±9.4 SD) for PB-8, and 11% (±8.7 SD) for PB-9. In layers PB-7 to PB-9, sand is the dominating fraction and conforms the grain fabric supporting the sediments (Table 3.1). Layer PB-9 is characterized by the highest average values of sand content. The average mud content varies between 16.2% in PB-9/PB-7 and 23.6% in PB-8, reaching occasionally a maximum of 61.6% in layer PB-9.

In every terrigenous sediment, the percentage weight of organic matter is not representative of its actual abundance. This is mainly because its density around 0.7 g/cm^3 is low compared to the density of the mineral fraction (2.6 g/cm^3). Thus, the actual content of organic matter is approximately three times higher considering the volume of the sample. In Pilauco layers, the average organic matter weight fluctuates between 9.4 and 12.9% (Table 3.1). Although there are maximum values ~40% in

Fig. 3.6 Brown-yellowish colored lamina separating layers PB-7 and PB-8. The pebble in the lower right measures 2 cm

Table 3.1 Basic statistical analysis of grain size parameters derived from column AD14 including layers PB-7 to PB-9 at Pilauco

		PB-7	PB-8	PB-9
% Sand	Mean	70.7	63.9	74.4
	S.D.	10.6	13.2	11.6
	Maximum	93.6	84.6	90.3
	Minimum	25.6	34.9	21.4
% Mud	Mean	16.3	23.6	16.2
	S.D.	8.2	12.3	10.3
	Maximum	55.0	50.5	61.6
	Minimum	1.0	3.1	3.6
% Org. matter	Mean	12.9	12.6	9.4
	S.D.	5.7	3.2	2.4
	Maximum	62.3	21.7	17.1
	Minimum	1.9	2.7	4.1

few of PB7 samples, they likely represent local accumulations of macroscopic plant material. Therefore, in the Pilauco layers, the organic matter represents an average of ~30 vol%, which explains the plastic consistency of the deposits.

The sandy and gravel sediments of the layers PB-7 and PB-8 were initially interpreted as fluvial plain deposits similar to an oxbow lake or swamp associated with an abandoned river meander (Pino and Miralles 2008; Pino et al. 2013). The erosional outline of the hill suggests a meandering course for the ancient Damas River, probably some 20,000 years ago during the Last Glacial Maximum. The layer PB-6 corresponds to a high-energy fluvial deposit whereas the layers PB-7 and PB-8 lack sedimentary structures that might be associated with this type of depositional environment. Additionally, isolated shingles included in a muddy sand fabric cannot be deposited in sediment of fluvial origin (Fig. 3.5).

In a fluvial setting, gravel and coarse-grained sand are well related. When correlating the gravel and mud fractions of layer PB-7 through linear regression, it was observed a non-statistically significant relationship between the parameters. Even more, the results suggest that the abundance of gravel is not related to the energy of the colluvial processes that transported and deposited simultaneously the sand from the northern hill but represent a gravitational stochastic transport.

Tello et al. (2017) and Tello (Chap. 12, this volume) analyzed fossil Coleoptera from layer PB-7 assigning them to different taxonomic levels. They document the existence of six ecological groups, including dung beetles. These findings suggest that at least part of the surrounding area of Pilauco was free of water bodies and therefore not related to a wetland. Later on, Soto-Bollmann (2014) performed and multivariate statistical analyses using granulometric data of the sandy fraction from the layer PB-7 and from layer PB-1 that outcrops in the hill to the north of the site. The results indicate that the terrigenous sediments from the layer PB-7 correspond to moderately sorted, coarse-grained sand. A Discriminant Analyses demonstrated that the sand fractions in layer PB-7 and the deposits from the San Pablo unit are statistically indistinguishable (Fig. 3.7). Thus, the sediments dominated by coarse-grained sand and occasionally gravel may not necessarily relate to high-energy fluvial environments, but rather to transport and deposition by local colluvial processes (gravitational erosion and pluvial transport from the northern hills). Therefore, fluvial transport can be discarded for sediments from the PB-7 layer. Additionally, the coexistence of dry and wet environments at Pilauco is supported by the presence of ferns and aquatic plants (Abarzúa and Gajardo-Pichincura 2008; Abarzúa et al. Chap. 9 this volume).

3.4 The Radiocarbon Ages

Table 3.2 include the radiocarbon dates obtained at the Pilauco site since 2007 in bones, charcoal, and plant remains (CALIB 7.1 and the SHcal13 curve[2]; Stuiver and Reimer 1993; Hogg et al. 2013). Because the orientation of the layers is not always

[2]http://calib.qub.ac.uk/calib/, consulted in July 2015.

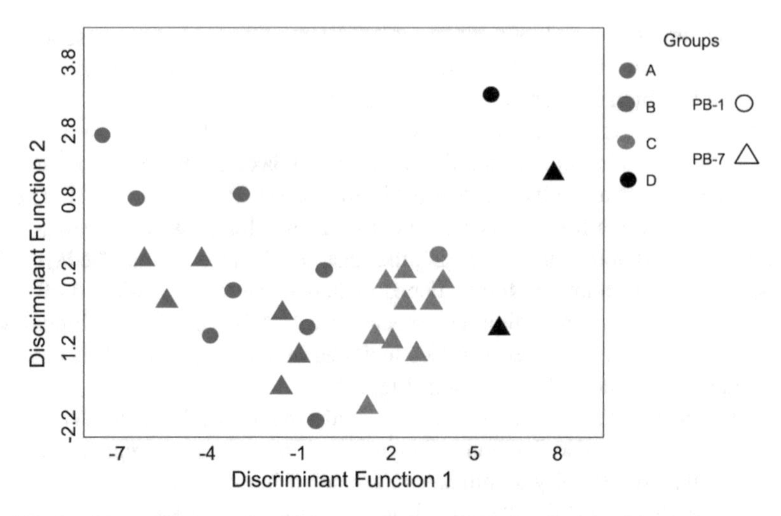

Fig. 3.7 Grouping of the samples in 4 assemblages (**a–d**) classified by Discriminant Analysis (Functions 1 and 2). The circles and triangles correspond to samples of the PB-1 and PB-7 layers, respectively. The Discriminant Functions 1 (84%) and 2 (14%) have values of $P < 0.05$ which are significant on the 95% confidence

horizontal, the elevations indicated in Table 3.2 are referential. Therefore, different ages can be found at the same elevation across the site. The oldest radiocarbon age (17,369 cal. year BP; Table 3.2) obtained at Pilauco was derived from rounded wood fragments deposited within the gravel layer of PB-6 (Fig. 3.4) that correspond to the base of the strata that confirm the Pilauco site. The age constraints for layer PB-7 are consistently younger than PB6, and they have been obtained using different animal and plant material. The first age dating at Pilauco was performed on a gomphothere jawbone rescued by workers in 1986, during the building of the residential area "Villa Los Notros". This dating yielded an age of 14,700 cal. year BP and is around 700 years older than the youngest date obtained for this layer (14,010 cal. year BP). However, the exact elevation of the piece rescued in 1986 (TO11477) is approximated (Table 3.2). The oldest remain of megafauna from layer PB-7 (16,300 cal. year BP) corresponds to a gomphothere tusk from the nearby site Los Notros (see Chap. 15). Additionally, a gomphothere skull has been dated directly, as well as using seeds and sediment that filled the skull, yielding ages between 15,840 and 15,550 cal. year BP (Table 3.2). Additionally, a human footprint (Macías 2012; Moreno et al. 2019) from the same layer was dated using seeds from the depressed sediment predating the footprint (15,727 cal. year BP) and from the infilling sediment postdating the footprint (16,156 cal. year BP, Table 3.2)

The layer PB-8 has also been dated using a variety of fossil materials. Remains of megafauna (a horse tooth and an antler) from this layer yielded ages of 13,850 and 12,720 cal. year BP, respectively. The youngest age in this layer was obtained from charcoal (12,600 cal. year BP). The PB-9 layer lacks bone remains and the dating was performed on plant material yielding ages between 12,658 and 4,340 cal. year BP.

Table 3.2 Radiocarbon ages from Pilauco site. Layer PB-6 (grey), PB-7 (violet), PB-8 (yellow) and PB-9 (green). TO: Isotrace Radiocarbon Laboratory Accelerator Mass Spectrometry Facility at the University of Toronto; AA: NSF-Arizona AMS Laboratory; UCIAMS: Keck Carbon Cycle Accelerator Mass Spectrometry Laboratory Irvine California; D-AMS: DirectAMS; PSUAMS: Penn State AMS 14C Facility; CAMS: Center for Accelerator Mass Spectrometry at the Lawrence Livermore National Laboratory. Dates have been published by 1: Pino et al. (2013), 2: González-Guarda et al. (2017), 3: Pino et al. (2019), 4: Natalia Villavicencio, personal communication 2017, 5: This chapter and Chap. 15, this volume

Lab. code	^{14}C age	1σ error	med.prob.	2σ range	elevation	grid	layer	source	material
D-AMS 024772	3944	30	4340	4231 - 4427	623	7AD	PB-9	5	charcoal
AA108169	4671	31	5400	5293 - 5469	625	10AC	PB-9	5	seed
UCIAMS110206	9135	20	10240	10199 - 10275	639	14AD	PB-9	3	charcoal
AA108171	10059	36	11500	11306 - 11716	602	8BA	LN-4	5	charcoal
AA108164	10214	37	11850	11700 - 12019	602	8BA	LN-4	5	charcoal
AA108167	10416	38	12220	12038 - 12412	542	10AD	PB-9	3	charcoal
D-AMS 024773	10456	44	12260	12039 - 12432	550	10AD	PB-9	5	charcoal
AA81809	10517	150	12320	11930 - 12702	560	6E	PB-9	1	charcoal
D-AMS 024774	10506	42	12400	12364 - 12554	554	10AD	PB-9	5	charcoal
AA81805	10630	124	12500	12358 - 12721	550	9E	PB-9	1	wood
UCIAMS101684	10710	30	12660	12625 - 12706	576	14AD	PB-9	3	charcoal
UCIAMS101668	10660	30	12600	12551 - 12672	571	14AD	PB-8	3	charcoal
AA81806	10739	128	12610	12370 - 12837	545	9E	PB-8	1	wood
AA109502	10860	60	12720	12660 - 12806	596	BH7/BI7	LN-2	5	antler
UCIAMS101669	10950	30	12760	12707 - 12825	560	14AD	PB-8	3	charcoal
AA108165	10916	67	12760	12680 - 12924	589	8BA	LN-2	5	charcoal
AA81812	11004	186	12870	12552 - 13219	505	7AC	PB-8	1	coprolite
AA108168	11079	40	12900	12762 - 13037	505	14AD	PB-8	3	seed
AA81804	11122	178	12950	12694 - 13274	467	9E	PB-8	1	wood
CAMS175049	11320	90	13140	12940 - 13325	429	12G	PB-8	2	bone
AA81810	11457	140	13260	12940 - 13325	446	7G	PB-8	1	tooth
AA81807	11665	136	13450	13190 - 13745	430	9E	PB-8	1	wood
AA81808	11834	186	13640	13264 - 14079	426	9E	PB-8	1	wood
PSUAMS2417	12035	50	13850	13726 - 14026	433	SD	PB-8	4	tooth
AA108170	12173	42	14010	13815 - 14148	420	14AD	PB-7	3	seed
AA108166	12449	52	14490	14155 - 14860	526	7BA	LN-1	5	seed
TO11477	12540	90	14700	14232 - 15107	357	11H	PB-7	1	bone
UCIAMS102670	12665	35	15010	14757 - 15183	401	15AD	PB-7	3	wood
UCIAMS101670	12725	40	15110	14865 - 15278	416	15AD	PB-7	3	bone
UCIAMS101771	12735	40	15120	14892 - 15289	352	15AD	PB-7	3	bulk
UCIAMS101830	12760	50	15150	14911 - 15330	416	15AD	PB-7	3	bone
UCIAMS101672	12860	35	15270	15112 - 15471	358	14AD	PB-7	3	wood
UCIAMS101685	12905	40	15340	15169 - 15574	401	15AD	PB-7	3	seed
UCIAMS110204	12935	25	15380	15223 - 15590	403	14AD	PB-7	3	wood
UCIAMS101768	13010	35	15500	15296 - 15711	401	15AD	PB-7	3	bulk
PSUAMS2416	13040	60	15550	15292 - 15785	381	15AD	PB-7	4	bone
UCIAMS101770	13045	30	15560	15332 - 15749	353	14AD	PB-7	3	leaf
PSUAMS2421	13135	50	15710	15426 - 15942	365	16AC	PB-7	4	bone
UCIAMS101673	13145	35	15730	15505 - 15945	352	14AD	PB-7	3	leaf
UCIAMS101832	13165	55	15750	15490 - 16001	318	15AC	PB-7	3	osteoderm
UCIAMS101769	13175	40	15770	15573 - 15987	352	14AD	PB-7	3	seed
UCIAMS101674	13195	35	15800	15627 - 16000	352	14AD	PB-7	3	wood
PSUAMS2418	13210	60	15820	15587 - 16056	369	14AC	PB-7	4	bone
UCIAMS102087	13220	40	15840	15659 - 16029	318	15AC	PB-7	3	osteoderm
UCIAMS101831	13220	60	15840	15608 - 16065	400	15AD	PB-7	3	bone
PSUAMS2420	13240	60	15860	15644 - 16084	351	15AC	PB-7	4	bone
PSUAMS2415	13240	60	15860	15644 - 16084	377	17AC	PB-7	4	tooth
PSUAMS2419	13260	70	15890	15645 - 16128	387	14AB	PB-7	4	bone
AA91450	13332	72	15980	15744 - 16217	340	12F	PB-7	3	wood
UCIAMS101671	13470	35	16160	15977 - 16320	353	14AD	PB-7	3	bulk
UCIAMS110203	13570	70	16290	16047 - 16560	339	14AD	PB-7	3	seed
AA109501	13585	81	16310	16048 - 16614	504	BI5/BI6	LN-1	5	tusk
UCIAMS110205	13650	70	16400	16155 - 16681	403	14AD	PB-7	3	seed
UCIAMS101675	14195	40	17230	17042 - 17441	330	19AC	PB-6	3	wood
UCIAMS101676	14300	40	17370	17161 - 17543	328	18AC	PB-6	3	wood

3.5 The Stratigraphic Relation Between PB-8 and PB-9 Layers

The interpretation of the erosional unconformity between layers PB-8 and PB-9 requires a particular discussion. Until 2013, both layers were described as deposited in an abandoned oxbow lake with sporadic fluvial flooding deposits corresponding to a mixture of mud and peat in a sandy matrix with some gravel content (Pino et al. 2013). Later on, a multivariate analysis was applied demonstrating that the sand fraction, and probably most of the gravel were derived from local colluvial sources (i.e., gravitational erosion from the southern flank of the northern hill during rainy periods). Moreover, the sands have been identified as corresponding to the fluvial and volcanoclastic deposits of the San Pablo Sequence that form the northern hill.

Layer PB-9, on the other hand, has quite different characteristics. Whereas PB-7 and PB-8 are mainly composed of a muddy peat in a dark-brown sandy matrix, layer PB-9 corresponds to a black peat lacking fossil animal remains and knapped/polished artifacts (see Chap. 16, this volume). The most distinctive feature of PB-9 is the occurrence of a yellow-orange zone (10YR 8/8) containing intraclasts and volcanic ash (tuffs) also derived from the northern hill.

The two layers are marked by a distinct erosional unconformity. In some stratigraphic columns, this erosional unconformity is characterized by a significant accumulation of angular wood fragments and charcoal pieces centimetric in size (Fig. 3.8). A large amount of locally derived angular intraclasts likely indicate a period of intense precipitation promoting the erosion and transportation of sediments over short distances from the nearby hill located on the northern part of the site (Fig. 3.9). This erosional–depositional scenario implies that the hill was completely exposed, or covered by scarce vegetation, and subjected to heavy rainfalls. The unconformity has been dated in two stratigraphic profiles from grids B9 and AD14. The first profile (grid B9) yielded ages of 12,600 and 12,500 cal. year BP derived from wood fragments from the upper section of PB-8 and base of PB-9 (Stuiver and Reimer 1993; Hogg et al. 2013). In the second stratigraphic profile (grid AD14, ~120 m to the SW), the age of seeds collected above and below the erosional unconformity correspond to 12,760 and 12,660 cal. year BP. This age constraints are concordant with the Northern Hemisphere Younger Dryas event (see Chap. 15, this volume, for the Bayesian model of this event).

According to Björk (2007), the Younger Dryas event was a reversal to glacial conditions occurred during the final phase of the last glaciation. The event was originally described based on paleobotanic and litho-stratigraphic studies from peats and lakes in Sweden and Denmark (Andersson 1896; Hartz and Milthers 1901). The term "Dryas" refers to the abundant fossil findings of *Dryas octopetala* (Rosaceae) in siliciclastic lake sediments. The characterization "Younger" is derived from the idea that the original "Dryas" period was preceded by a warmer period (Allerød), which in turn overlies sediments indicating at least one older cold period. Currently, it is commonly accepted that the Younger Dryas chronozone occurred between 12,800 and 11,600 cal. year BP, whereas the age of the onset has been estimated with particular precision (Björck 2007).

Fig. 3.8 Erosional unconformity between layers PB-8 (below) and PB-9 (above) in grid 9B. Differences in texture and color are evident between the two layers. The discordance surface is denoted by an accumulation of coarse-grained sediments, representing a lag deposit (i.e., that could not be moved by a transport agent)

Fig. 3.9 Accumulation of irregularly shaped wood fragments above the erosional unconformity between layer PB-8 and PB-9 in the stratigraphic profile from grid AD14. The unconformity is marked by the change in color from grey to dark-olive (image corresponds to humid sediments)

In the Southern Hemisphere, there is also abundant evidence of a cold period close to the Pleistocene–Holocene transition. According to Barnosky and Lindsey (2010), palynological and glaciological data from South America indicate a cold oscillation contemporary with the Younger Dryas, although there is no agreement about the precise timing and intensity of this event. In the mid-latitudes of Chile and Argentina close to the Andes, this period is well recorded in lacustrine sediments and is called Huelmo/Mascardi Cold Reversal (HMCR; Moreno 2000; Hajdas et al. 2003; Boës and Fagel 2008; Massaferro et al. 2009). Additionally, in a study of climate, wildfires, and vegetation reconstruction near Puerto Montt (41°45′S, 73°07E), Moreno (2004) recognized a cold period between c. 12,200 ^{14}C year BP and 9,900 ^{14}C year BP. Using a high-resolution radiocarbon chronology, Hajdas et al. (2003) demonstrated that this cold event occurred between 11,400 ^{14}C and 10,150 ^{14}C year BP (Fig. 3.10). Based on sedimentological proxies from lacustrine sediments, Boës and Fagel (2008) placed the cold reversal between 13,300 and 12,200 cal. year BP and proposed an interruption by a dry period between 12,800 and 12,600 cal. year BP (Fig. 3.9). Recently, Massaferro et al. (2009) integrated information from chironomids and distinguished a humid period during the HMCR (13,500–12,800 cal. year BP), followed by a prominent dry period (12,800–11,500 cal. year BP; Fig. 3.10).

The erosional discordance between PB-8 and PB-9 in Pilauco is constrained within the HMCR (12,700 to 12,500 cal. yr BP; Table 3.2 and Fig. 3.10). It provides evidence of a short duration period, characterized by significant precipitation that may have caused the erosion of the top section of PB-8, as well as of the hill formed by volcaniclastic sediments to finally deposit layer PB-9. These observations partially coincide with the end of the HMCR humid period postulated by Massaferro et al. (2009), although their conclusions are contrary to the dry phase proposed by Boës and Fagel (2008).

The dynamics of local wildfires and characteristics of vegetation at Pilauco during the transition between layers PB-8 and PB-9, will be discussed in detail in Chap. 9, and only a brief overview of the main findings will be provided here. The results of a pollen analysis show that during the final episode of deposition of layer PB-8, the forests (Myrtaceae and *Maytenus*) and herbaceous taxa (Poaceae) presented similar relative abundance, and very low plant charcoal concentration (Chap. 9, this volume). This suggests that preceding the erosional unconformity between layers PB-8 and PB-9 at c. 12,600–12,761 cal. year BP (Fig. 3.11), the area of Pilauco was characterized by an open landscape associated with the initial stages of forest succession at the end of the glaciation. Low temperatures and high precipitation may have prevented the occurrence of wildfires during this period. Additionally, the high concentration of pollen suggests dense vegetation cover indicating a rather productive environment. Importantly, a marked change was observed at the base of layer PB-9, immediately above the erosional unconformity (12,658 cal. yr BP; Table 3.2). A sharp reduction of arboreal taxa, displaying minimum abundance values at 550 cm in layer PB-9, indicates the occurrence of an open landscape with scarce arboreal vegetation (see Chap. 9, this volume.) At the same time, a peak in charcoal concentration indicates intense wildfire activity, gradually reducing towards the top of column AD10 (Fig. 3.11). Based on the increased occurrence of the shrub *Aristotelia*

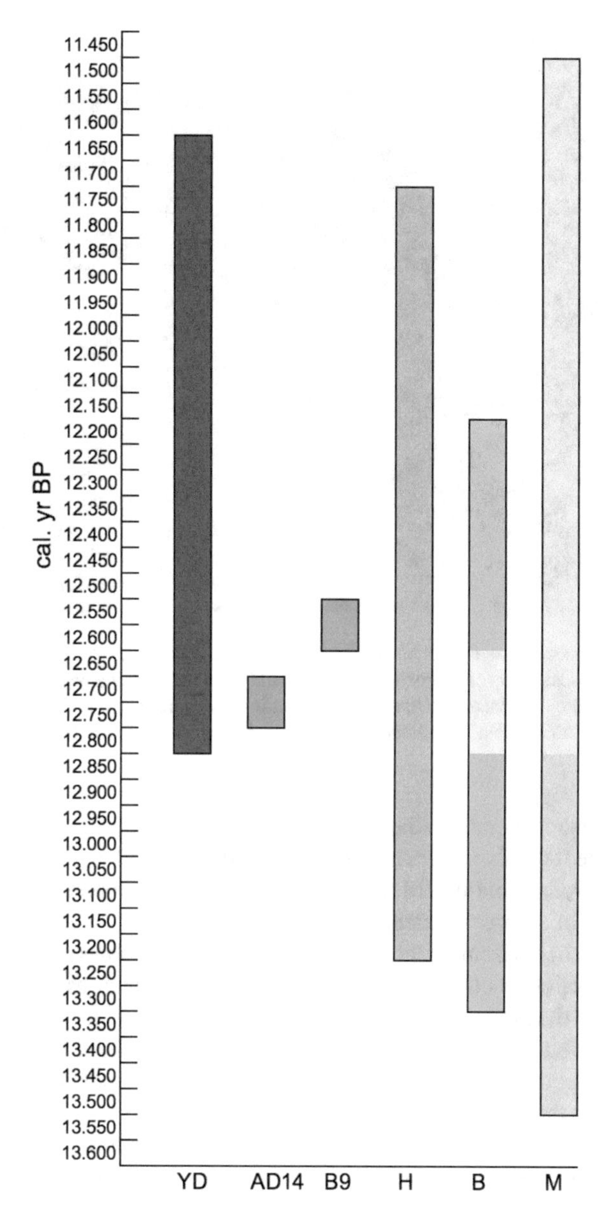

Fig. 3.10 Age ranges proposed for the Younger Dryas and Huelmo/Mascardi Cold Reversal (HMCR). YD: Younger Dryas age constraints in the Northern Hemisphere (Björck 2007). AD14: age ranges of the erosional unconformity between layers PB-8 and PB-9 in grid AD14 at Pilauco. B9: age constraints of the unconformity in grid B9. H: HMCR sensu Hajdas et al. (2003). B: HMCR sensu Boës and Fagel (2008) (lighter color indicates an abrupt dry event). M: HMCR, Massaferro et al. (2009), the lighter color indicates drier conditions. All ages from South America were calibrated following Stuiver and Reimer (1993) and Hogg et al. (2013) using the Southern hemisphere curve (SHcal13)

Fig. 3.11 Profile section analyzed for pollen and charcoal concentration across layers PB-8 and PB-9. The yellow scale shows the erosional unconformity between the two layers at 542 cm. The red scale indicates the occurrence of charcoal concentration (particles per cm^{-1}). The yellow arrows indicate volcanic tuff intraclasts embedded in PB-9 that originate from the San Pablo Sequence

chilensis and its regenerative adaptation after disturbances, we can confidently infer the establishment of this species near Pilauco during a period of post-perturbation by the fire (Chap. 9, this volume). Moreover, the low pollen concentration would support the occurrence of scarce vegetation and low productivity in the area of Pilauco.

The significant environmental changes identified by the pollen and charcoal analyses at the transition between layers PB-8 and PB9, as well as by other regional records and the dynamics of local wildfires (Moreno et al. 2015), are coherent with erosion by gravitational and pluvial processes along the flanks of the hills formed by the San Pablo Sequence due to low forest cover, intense precipitation, and significant wildfires during the deposition of PB-9 (responsible for the black color of this layer). The scarcity of vegetation left the San Pablo Sequence exposed to climatic events reflected in the occurrence of numerous intraclasts derived from the hill to the north of the site.

3.6 Discussion and Conclusions

The Pilauco site has a clear and simple stratigraphy that is well supported by tens of concordant radiocarbon dates and can be related to the dominant environmental conditions during the late Pleistocene. The layers and erosional surfaces described can be grouped into three associations. The first group corresponds to a pre-Llanquihue relief, carved by the Damas River during the earlier or rather later (~19.000 ^{14}C year BP) stages of the Llanquihue glaciation. This erosional surface outcrops both at the base of the excavation and in the pit located in the northernmost sector of the excavation site (units 2H–5H, see Chap. 1), and cut the fine-grained volcano-sedimentary sequence represented by the layers PB-1 to PB-5.

The second group represents deposits from the later stages of the Llanquihue Glaciation and is located at least in a depression directly below the excavation site and over the recently described surface. It corresponds to the channel-type boulder deposits of layer PB-6. This layer is included in the (Pigf1) unit by Pérez et al. (2003). At this time the Damas River was nourished by water from the melting glaciers, and its base was about 100 m below modern level. Thus, is very probable that the stream had a large capacity and competence. Based on this scenario and the age of layer PB-6 ($\sim$17.400 cal. year BP), is possible to assign the formation of subsidiary terraces to this time. The Damas River must have been incised into the main plain, forming the largest of the subsidiary plains, which was abandoned towards the end of this period when the Damas River was disconnected of the glaciers. This scenario explains the Andean origin of clasts in layers PB-6 at Pilauco (intrusive and volcanic rocks), some of which (aphanitic basalts) were probably used for the fabrication of the material artifacts by early human settlers (Navarro et al. 2019).

The third stratigraphic group is composed of layers PB-7 and PB-8. These layers may have originated during a period in which the Damas River left the upper plain of the fluvial system, moving towards its modern location, south of the site. At that time, both layers may have received material from the pre-Llanquihue relief through hillside erosion processes, and from the Damas River through flooding events. This group is the chronostratigraphic equivalent to the unit PlHf (Pérez et al. 2003) represented by fluvial deposits and terraces between 1 and 15 m above the modern-day riverbed. The layers PB-7 and PB-8 do not include fluvial deposits, although the possibility of occasional flooding reaching the swamps cannot be ruled out.

Finally, the hillside removal processes characterize the PB-9 layer, which was deposited at Pilauco without fluvial contribution, simultaneously with large forest fires. The particular deposition of layer PB-9 will be discussed in Chap. 15.

References

Abarzúa A, Gajardo-Pichincura A (2008) ¿Y qué nos cuentan los polen?: Reconstruyendo la historia climática y vegetacional del sitio Pilauco. in: Pino M (ed.) Pilauco: Un sitio complejo del Pleistoceno tardío. Osorno, Norpatagonia chilena. Universidad Austral de Chile. Imprenta América, Valdivia, Chile, pp 49–53

Andersson G (1896) Svenska växtvärldens historia. P. A. Norstedt & Söner, Stockholm

Arenas M, Milovic J, Pérez Y, Troncoso R, Behlau J, Hanisch J, Helms F (2005). Geología para el ordenamiento territorial: área de Valdivia, Región de Los Lagos. Servicio Nacional de Geología y Minería, Carta Geológica de Chile, Serie Geología Ambiental, No. 8, 71 p., 6 mapas escala 1:100.000 y 1 mapa escala 1:25.000

Bentley JM (1997) Relative and radiocarbon chronology of two former glaciers in the Chilean Lake District. J Quat Sci 12(1):25–33

Björck S (2007) Younger Dryas oscillation, global evidence. In: Mock CJ, Scott E (eds) The encyclopedia of quaternary sciences. Elsevier, pp 1995–1983

Boës X, Fagel N (2008) Timing of the late glacial and Younger Dryas cold reversal in southern Chile varved sediments. J Paleolimnol 39(2):267–281

Barnosky AD, Lindsey EL (2010) Timing of Quaternary megafaunal extinction in South America in relation to human arrival and climate change. Quat Int 217:10–29

Brüggen J (1945) Miscelánea geológica de las provincias de Valdivia y Llanquihue. Rev Chil Hist Geogr 2:90–113

Corvalán N (1974) A new lahar in the central south Chile at the latitude of the Osorno (41oS). In: Abstracts of the IAVCEI symposium, Santiago, 4–6 June 1974

Denton GH, Lowell TV, Heusser CJ, Schlüchter C, Andersen BG, Heusser LE, Moreno PI, Marchant DR (1999) Geomorphology, Stratigraphy, and radiocarbon chronology of Llanquihue driftin the area of the southern Lake District, Seno Reloncaví, and Isla Grande de Chiloé, Chile. Geogr Ann A 81(2):167–229

Faegri K, Iversen J (1989) Textbook of pollen analysis. Wiley, Londres

González-Guarda E, Domingo L, Tornero C, Pino M, Fernández MH, Sevilla P, Agustí J (2017) Late Pleistocene ecological, environmental and climatic reconstruction based on megafauna stable isotopes from northwestern Chilean Patagonia. Quat Sci Rev 170:188–202

Hajdas I, Bonani G, Moreno PI, Ariztegui D (2003) Precise radiocarbon dating of Late-Glacial cooling in mid-latitude South America. Quat Res (Orlando) 59:70–78

Hartz N, Milthers V (1901) Det senglacie ler i Allerød tegelværksgrav. Medd Dansk Geol Føren 8:31–60

Hogg AG, Hua Q, Blackwell PG, Niu M, Buck CE, Guilderson TP, Heaton TJ, Palmer JG, Reimer PJ, Reimer RW Turney CSM, Zimmerman SRH (2013) Shcal13 Southern Hemisphere calibration, 0–50,000 years cal BP. Radiocarbon 55(4):1889–1903

Illies H (1970) Geología de los alrededores de Valdivia y Volcanismo y Tectónica en márgenes del Pacífico en Chile Meridional. Universidad Austral de Chile, Valdivia

International Unión of Geological Sciences (2009) Request for IUGS ratification to establish the quaternary as a system/period of the cenozoic and revise the associated base of the Pleistocene Series. http://www.iugs.org/

Laugénie C (1982) La région des Lacs, Chili méridional, recherches sur l'évolution géomorphologique d'un piémont glaciaire quater-naire andin. Dissertation, University of Bordeaux III

Lauer W (1968) Die glaziallandschaft des südchilenischen Seengebietes. Acta Geogr 20:215–236

Latorre C, Moreno P, Vargas G, Maldonado A, Villa-Martínez R, Armesto J, Villagrán C, Pino M, Núñez L, Grosjean M (2007) Quaternary environments and landscape evolution. In: Moreno T, Gibbons W (eds) Geology of Chile. Geological Society Press, London, pp 309–328

Lowell TV, Heusser CJ, Anderson BG, Moreno PI, Heusser LE, Heusser C, Slucter S, Marchant DR, Denton GH (1995) Interhemispheric correlation of Late Pleistocene glacial events. Science 269:1541–1549

Macías C (2012) Análisis de posible icnita en el Sitio Pilauco, Pleistoceno tardío, Osorno, Centro-Sur de Chile, Undergraduate Dissertation. Universidad Austral de Chile

Massaferro JI, Moreno PI, Denton GH, Vandergoes M, Dieffenbacher-Krall A (2009) Chironomid and pollen evidence for climate fluctuations during the Last Glacial Termination in NW Patagonia. Quat Sci Rev 28:517–525

Mercer JH (1976) Glacial history of southernmost South America. Quat Res (Orlando) 6:125–166

Moreno PI (1997) Vegetation and climate change near Lago Llanquihue in the Chilean Lake District between 20,200 and 9500 ^{14}C yr BP. J Quat Sci 12:485–500

Moreno PI (2000) Climate, fire, and vegetation between about 13,000 and 9200 ^{14}C year BP in the Chilean Lake District. Quat Res (Orlando) 54:81–89

Moreno PI (2004) Millennial-scale climate variability in northwest Patagonia over the last 15 000 yr. J Quat Sci 19(1):35–47

Moreno PI, León AL (2003) Abrupt vegetation changes during the last glacial to Holocene transition in mid-latitude South America. J Quat Sci 18:787–800

Moreno H, Varela J (1985) Geología, volcanismo y sedimentos piroclásticos cuaternarios la región central y sur de Chile. In Tosso J (ed) Suelos volcánicos de Chile. Capítulo 6. Instituto de Investigaciones Agropecuarias (INIA), pp 491–526

Moreno PI, Jacobson GL, Andersen BG, Lowell TV, Denton GH (1999) Abrupt vegetation and climate changes during the last glacial maximum and the last Termination in the Chilean Lake District: a case study from Canal de la Puntilla (41°S). Geogr Ann A: Series A, Phy Geogr 81A:285–311

Moreno PI, Denton GH, Moreno H, Lowell TV, Putnam AE, Kaplan MR (2015) Radiocarbon chronology of the last glacial maximum and its termination in northwestern Patagonia. Quat Sci Rev 122:233–249

Moreno K, Bostelmann JE, Macías C, Navarro-Harris X, De Pol-Holz R, Pino M (2019). A late Pleistocene human footprint from the Pilauco archaeological site, Northern Patagonia, Chile. PLOS ONE PONE-D-18-03927R2 in press

Pérez Y, Milovic J, Troncoso R, Hanish F, Helms F, Toloczky M (2003) Geología para el ordenamiento territorial: Área de Osorno, Región de los Lagos. Servicio Nacional de Geología y Minería, Carta Geológica de Chile, Serie Geológica Ambiental No 6, 62 p, 7 mapas escala 1:100.000. Santiago

Pino M, Miralles C (2008) La geología cuaternaria de Pilauco. In: Pino M (ed.). Pilauco, un Sitio complejo del Pleistoceno tardío. Osorno, Norpatagonia Chilena. Universidad Austral de Chile. Imprenta América, Valdivia, Chile, p 37–42

Pino M, Chavez M, Navarro X, Labarca R (2013) The late Pleistocene Pilauco site, south central Chile. Quat Int 299:3–12

Pino, M., Martel-Cea, A., Astorga, G., Abarzúa, A.M., Cossio, N., Navarro, X., Lira, M.P., Labarca, R., Lecompte, M.A., Adedeji, V., Moore, C.R., Bunch, T.E., Mooney, C., Wolbach, W. S., West, A., Kennett, J.P. (2019). Sedimentary record from Patagonia, southern Chile supports cosmic-impact triggering of biomass burning, climate change, and megafaunal extinctions at 12.8 Ka. Sci Rep 9(1):4413

Porter S (1981) Pleistocene glaciation in the Southern District of Chile. Quat Res (Orlando) 16:262–292

Soto-Bollmann K (2014) Origen de la fracción arenosa de la capa portadora de megafauna del Sitio Pilauco, Osorno, Norpatagonia de Chile. Undergraduate Dissertation, Universidad Austral de Chile

Stockmarr J (1971) Tablets with spores used in absolute pollen analysis. Pollen Spores 13:615–621

Stuiver M, Reimer PJ (1993) Extended 14C database and revised CALIB radiocarbon calibration program. Radiocarbon 35:215–230

Tello F, Elgueta M, Abarzúa AM, Torres F, Pino M (2017) Fossil beetles from Pilauco, south-central Chile: An Upper Pleistocene paleoenvironmental reconstruction. Quat Int 449:58–66

Whitlock C, Larsen C (2001) Charcoal as a proxy fire. In: Smol JP, Birks HJB, Last WM (eds) Tracking environmental change using Lake Sediments: Terrestrial, Algal, and Siliceous indicators. Kluwer Academic Publishers, Dordrecht, The Netherlands, pp 75–97

Chapter 4
The Proboscidean Gomphotheres (Mammalia, Gomphotheriidae) from Southernmost South America

Omar P. Recabarren

Abstract The Gomphotheriidae family included a large number of taxa, and a fossil record with extensive geographic and stratigraphic distribution. In South America, the gomphotheres evolved the dibelodont and brevirostrian characteristics, with brachydont, bunodot and subhypsodont molar teeth. Among palaeontologists it is generally accepted the division of gomphotheres into two different groups: the lowland and Andean gomphotheres. There is scientific consensus to designate the species *Cuvieronius hyodon* as representative of the Andean-type gomphotheres, and in the last few decades, the genus *Stegomastodon* including its two species *S. platensis* and *S. waringi* was described as a lowland-type gomphothere. Recent studies included the genus *Stegomastodon* within the genus *Notiomastodon*, with the species *Notiomastodon platensis* as endemic for the region. In Chile, the most abundant fossils of megafauna from the late Pleistocene—early Holocene limit are gomphotheres, counting with one of the southernmost fossil records that extend to Castro, on the island of Chiloé. The most common fossil remains are molars that are the base of many scientific studies aiming to synthesize the complex history of these mammals. The investigation of the Pilauco site in south-central Chile has improved our paleoecological knowledge of these mammals, as well as the taphonomic processes that favoured their fossilization in southern South America.

Keywords Gomphotheriidae · Stegomastodon · Notiomastodon

4.1 Background

Proboscideans are a group of eutherian mammals characterized by possessing an upper lip and nose merged in a structure similar to those of modern elephants. Overall, proboscideans have massive bodies and a skull highly pneumatized comprising

---

O. P. Recabarren (✉)
Departamento de Geodinámica, Estratigrafía y Paleontología, Facultad de Ciencias Geológicas, Universidad Complutense de Madrid, Madrid, Spain
e-mail: o.recabarren@hotmail.com

TAQUACH, Transdisciplinary Center for Quaternary Research in the South of Chile, Universidad Austral de Chile, Valdivia, Chile

M. Pino and G. A. Astorga (eds.), *Pilauco: A Late Pleistocene Archaeo-paleontological Site*, The Latin American Studies Book Series,
https://doi.org/10.1007/978-3-030-23918-3_4

a big and complex brain. They also have large tusks, which correspond to the second incisors projected outwards. In some taxa, the tusks may be present in the jaw (Shoshani 1998; Shoshani and Tassy 1996). The nostril in these animals is significantly retracted (Shoshani 1998; Shoshani and Tassy 1996), whereas representatives of the group have no canines, nor first premolars, and the radio is in a pronated position. However, in the basal forms these characteristics may differ, because they are considered synapomorphies of the group (Shoshani and Tassy 1996; Lambert and Shoshani 1998). A particularity of the group is the way in which the molars emerge and are replaced. The molars are replaced one at a time reaching M3/m3 (final molars, two lower and two upper ones). The emerging molars push their predecessors in a way similar to a conveyor belt. The modern elephants have six molars on each hemi-maxilla and hemi-mandible (three premolars and three molars), which adding the two upper incisors (the tusks), totalling 26 teeth (Shoshani and Tassy 1996).

Throughout their evolutionary history, proboscideans colonized and occupied almost all continental habitats—including swamps, moorlands, forests, deserts, savannahs and high mountains—and they adapted morphologically and physiologically to the demands of the specific environments. Thus, their fossil remains are found in all continents except Antarctica and Australia (Shoshani 1998; Lambert and Shoshani 1998; Sukumar 2003). Systematic investigations have assigned proboscideans to the grandorder Paenungulata, a group that experienced rapid radiation at the beginning of the Paleogene. This group lies within the hierarchy of the superorder clade Afrotheria, a mammal group including the orders: Proboscidea (elephants), Sirenia (dugongs), Hyracoidea (dassies), Macroscelidea (elephant shrews), Tubulidentata (aardvarks) and Afrosoricida (tenrecs and golden moles) (Roca and O'Brien 2005). Molecular studies suggest the superorder originated in the Cretaceous about 145 Ma ago and that proboscideans diverged during the early Palaeocene (66 Ma), experiencing a rapid expansion in the early Eocene (c. 56 Ma) (Sukumar 2003; Sanders et al. 2010). During the Late Eocene in Africa, the ancestors of proboscideans had a semiaquatic life (Liu et al. 2008; Gheerbrant et al. 2012). Such is the case, for instance, of *Moeritherium*, a primitive genus of proboscideans (Liu et al. 2008). The size and appearance of the representatives of this group was similar to that of modern pigs with a small proboscis (Haynes 1991; Sukumar 2003; Liu et al. 2008). However, one of the most significant changes in the evolution of proboscideans is the considerable increase in body size. From their origin as small animals (e.g. *Moeritherium*), the group evolved towards modern elephants such as *Loxodonta africana* (African savannah elephant) or *Elephas maximus* (Asian elephant) that represent the largest terrestrial living mammals today (Shoshani 1998; Shoshani and Tassy 1996). The only exception is *Loxodonta cyclotys* (African forest elephant), which smaller size is favourable for dwelling in forests (Sanders et al. 2010).

The evolutionary history of the proboscideans indicates that the Gomphotheriidae family was the group with the largest number of taxa and fossil records over a wide geographical and stratigraphic area (Shoshani 1998). This family experienced a great radiation—the second largest within the order Proboscidea—from eastern Africa and Eurasia during the early Miocene, some 20 million years ago. The gomphotheres (common name for members of the family Gomphotheriidae) have an extensive evolutionary history, and their fossil remains are recorded all over the world including South America. Thus, gomphotheres persisted in Africa until the Pliocene–Pleistocene (c. 2.5 Ma), whereas their fossil records in Eurasia extend back to the late Miocene, approximately 7 Ma ago (Shoshani and Tassy 1996). To reach North America, gomphotheres probably used the corridor formed by the Bering Strait. In this region of America, gomphotheres fossil ages fluctuate between the late Miocene (c. 13 Ma) and the Pleistocene–Holocene (11 ka) (Shoshani and Tass 1996). From here, gomphotheres dispersed towards South America, reaching the southernmost regions of the continent.

The main evolutionary novelties of gomphotheres were the increase in size of the molars, as well as the increase in the number of lophos/lophids in bunodont and zygodont molars (Shoshani and Tassy 1996; Sukumar 2003; Sanders et al. 2010). Additionally, gomphotheres developed cement and conules in areas near the middle line of the molars (Lambert 1996; Sanders et al. 2010). Thus, when the occlusal surface (the lophos/lophids and the conules) is worn down through the process of chewing, a figure similar to a clover emerges (Lambert and Shoshani 1998) (Fig. 4.1). The tusks also increased in length reaching a differentiation in each taxon (Shoshani and Tassy 1996; Sanders et al. 2010). In groups with mandible tusks, present in basal gomphotheres (e.g. *Gomphotherium angustidens*), these are rounded or oval in transversal cross-section (Lambert and Shoshani 1998). The nasal cavity was also widened in comparison to the other proboscidean families (Shoshani and Tassy 1996; Lambert and Shoshani 1998).

4.2 The Family Gomphotheriidae in South America

The arrival of proboscideans to South America occurred as a consequence of the establishment of a connection between the northern and southern part of the American continent through the Isthmus of Panama some 3 Ma ago during the middle Pliocene (Leigh et al. 2014). This natural bridge enabled the migration of plants and animals in both directions in a process of migration known as the Great American Biotic Interchange (GABI) (Leigh et al. 2014; see Chap. 2, this volume). The family Gomphotheriidae is the only one out of three-proboscidean families from North America (Elephantidae, Mammutida and Gomphotheriidae) with fossil records in South America (Cabrera 1930; Simpson and Paula-Couto 1957; Haynes 1991; Alberdi et al. 2002).

The South American gomphotheres are dibelodonts, which means that they only have upper tusks in adult age (Hoffstetter 1950, 1952). The tusks can reach one meter in length, and in some cases, they are almost straight, curved, or displaying a slight or more pronounced torsion. The tusks can present a band of enamel in adults or only in juvenile specimens. Additionally, South American gomphotheres are brevirostrian species, meaning that the mandibular symphysis is short and display a downward curvature. The intermediary molars (Dp4/dp4, M1/m1 and M2/m2) are trilophodonts

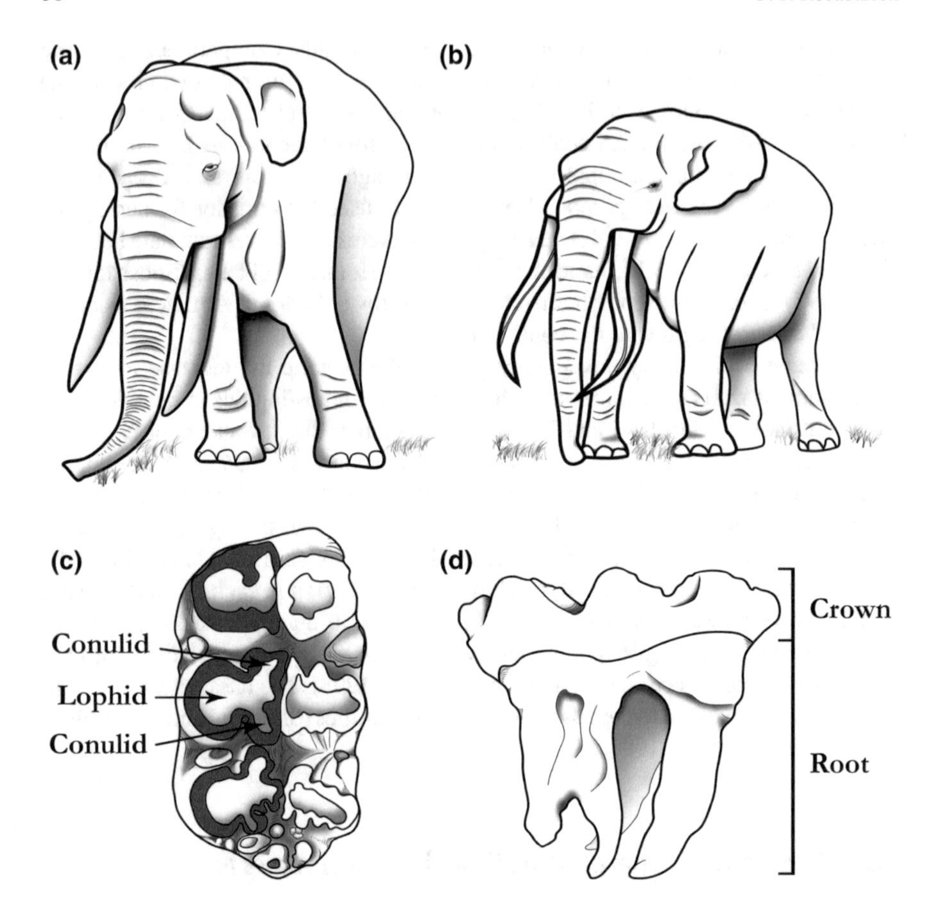

Fig. 4.1 Gomphotheres from South America. **a** Adult lowland-type gomphothere. The tusks might exhibit a band of enamel. **b** Adult Andean-type gomphothere. The tusks show a spiral curvature and a band of enamel. **c** Scheme of m2 molar (lower left molar). The red areas outline the clover figure (abrasion of lophid and conulid). **d** Dental crown and root of m2 molar (labial view)

(three successive lophos). The molars are brachydonts, bunodonts or subhypsodonts and exhibit a small internal or external curvature of the terminal lophos (Alberdi et al. 2002; Hoffstetter 1950, 1952; Lambert and Shoshani 1998). The molars M3/m3 have four to six successive lophos followed by a robust talon/talonid (Lambert and Shoshani 1998; Alberdi et al. 2002). The skull of the gomphotheres is brachycephalic similar to that of modern elephants (Alberdi et al. 2002).

Based on their fossil record Gomphotheres are assigned to a mid-Pleistocene to Holocene age (781–11 Ka) with fossil findings located mainly along the eastern and western margins of the American continent. These sites are located either in high-altitude domains close to the Andean Range and associated to temperate-cold climate, or in lower elevation areas that correspond to plains of rather temperate-warm climate conditions (Prado et al. 2005).

However, the taxonomy of gomphotheres in South America has been complicated since early findings by a number of problematic issues that originated from the use of multiple synonyms and the assignment of diagnostic value to anatomic elements lacking taxonomic relevance. This gave origin to a number incorrectly defined or questioned genera and species since the first studies of South American gomphotheres until the present causing a lack of consensus about the definition of certain groups. Since the 1990s, efforts have been made in order to synthesize the synonymy of the group and provide structure to the taxonomic classification of the family in South American. However, there is general agreement regarding the distribution of two types of gomphotheres in South America: the high-altitude Andean domain and the low-altitude plains gomphotheres (Alberdi et al. 2002; Fig. 4.1).

The fossil records from the high-altitude Andean region are consistent with the genus *Cuvieronius*. This genus is endemic to South America where it is recorded since the late Pleistocene, whereas in North America it persisted until the end of the Pliocene migrating to South America during the GABI (Lambert and Shoshani 1998; Leigh et al. 2014). The species *Cuvieronius hyodon* is widely accepted in South America as representative of the genus *Cuvieronius*. The species is characterized by having an elongated short cranium, upper tusks with subcircular cross-section and spiral inward or outward torsion (Cabrera 1930; Simpson and Paula-Couto 1957; Parodi 1962; Frassinetti and Alberdi 2000; Prado et al. 2005). The tusks exhibit a band of enamel along the spiral (Cabrera 1930; Prado et al. 2005) (Fig. 4.1). Additionally, the alveoli in these animals—sockets in the mandible where the upper tusks are rooted and projected outward from the middle line of the skull (Prado et al. 2005). The juvenile individuals exhibit tusks in the mandible that disappear in adult age. The fossil findings of *Cuvieronius hyodon* are distributed along the entire western margin of South America, in northern Colombia, Ecuador, Peru, northern Chile, Argentina and southern Bolivia (Tarija Region) (Cabrera 1930; Frassinetti and Alberdi 2000; Prado et al. 2005). The oldest fossil stratigraphic occurrence corresponds to the mid-Pleistocene of Imbabura in northwestern Ecuador, whereas the youngest fossils are from the late Pleistocene to early Holocene from Tibitó in Colombia and Tagua-Tagua in Chile (Frassinetti and Alberdi 2000; Prado et al. 2005).

However, there is still no clear consensus regarding the classification of the lowland gomphotheres. For instance, the genus *Haplomastodon*, defined based on the single species *Haplomastodon chimborazi*—presenting a high cranium and curved tusks without enamel band—was later integrated into the genus *Stegomastodon* (Parodi 1962; Prado et al. 2005; Lucas and Alvarado 2010). However, there are still authors arguing for the validity of the genus *Haplomastodon* (Ferretti 2010; Lucas and Alvarado 2010). The fossil remains assigned to *H. chimborazi* are present in Peru, Brazil, Ecuador, Venezuela and Colombia (Ferretti 2010).

The genus *Stegomastodon* dates back in North America to the Pliocene—early Pleistocene (Lucas et al. 2011) and includes three recognized species. *S. primitivus* considered the most primitive species, is characterized by curved upper tusks, trilophodont M2/m2 molars, pentalophodont M3/m3 molars and in some cases a sixth underdeveloped lopho (Lambert and Shoshani 1998). *S. mirificus*, characterized by upward curved upper tusks without enamel band, trilophodont M2/m2 molars, and in some cases a fourth-underdeveloped lopho (Lambert and Shoshani 1998). The molars M3/m3 exhibit six lophos, and sometimes even seven and a half. Finally, *S. aftoniae*, a gomphothere described based on an M3 of seven lophos and a talon (Lambert and Shoshani 1998).

In South America, two species have been assigned to the genus *Stegomastodon*—*S. platensis* and *S. waringi* (Alberdi et al. 2002; Prado et al. 2005). Their fossil remains date back to the middle-to-late Pleistocene (Alberdi et al. 2002; Prado et al. 2005). *Stegomastodon platensis* is characterized by having a tall, short skull and body-size similar to modern Asian elephants. The tusks are long and massive, and sometimes exhibit a slight upwards-directed curvature. The tusks of the adults ʃ lack enamel and torsion. The tusk alveoli are located close to the middle line of the cranium (Prado et al. 2005), and the molars M2/m2 are trilophodont, whereas the molars M3/m3 are tetra- or pentalophodont (Prado et al. 2005). The occlusal morphology of the upper and lower molars is complicated due to the presence or absence of central conules and conuletes (Cabrera 1930; Prado et al. 2005). Fossils representatives of *Stegomastodon platensis* are typical of the Pampean region of Argentina and particularly frequent in Buenos Aires, Córdoba, Santa Fe, and the Entre Ríos Province. There are also findings of *S. platensis* in Uruguay and Paraguay, as well as in Monte Verde and Pilauco in southern Chile (Cabrera 1930; Simpson and Paula-Couto 1957; Prado et al. 2005; Frassinetti and Alberdi 2005; Recabarren 2008; Recabarren et al. 2014).

The skull of *Stegomastodon waringi* is short and tall, resembling the elephantoid type, but not as much as *S. platensis*, and it is less depressed compared to *Cuvieronius* (Alberdi et al. 2002). The tusks of adult specimens are relatively straight or with slight upward curvature and exhibit a slight torsion (Alberdi et al. 2002). A band of enamel is present only in juvenile individuals and it has not been observed in adults (Alberdi et al. 2002; Prado et al. 2005). Similar to *S. platensis*, the molars M2/m2 are trilophodont, whereas the occlusal morphology of the molars M3/m3 is complex and characterized by a clear tendency to the displacement of the internal cones compared to the external ones (Alberdi et al. 2002; Prado et al. 2005). Fossil remains of *Stegomastodon waringi* have been recorded in Peru, Ecuador (Santa Elena Peninsula), Venezuela (Taima) and Brazil (Lagoa Santa, Minas Gerais; Bonito, Mato Grosso and Rio Grande do Sul) (Simpson and Paula-Couto 1957; Alberdi et al. 2002; Prado et al. 2005).

However, later studies have challenged the diagnostic features separating the two species of *Stegomastodon* (Mothé et al. 2012). This is mainly because the characteristic distinguishing the lowland-type gomphotheres are minimal and are thought not to be important to separate specimens up to the species-level. Furthermore, the occurrence of the genus *Stegomastodon* in South America has been questioned (Ferretti 2010; Lucas and Alvarado 2010; Mothé et al. 2012, 2013), partly because of the lack of unquestionable fossil records in Central America (Lucas and Alvarado 2010). The fossil remains in question have been assigned instead to the genus *Notiomastodon* (Hoffstetter 1950; Mothé et al. 2012, 2013), an endemic South American genus described by Cabrera (1930), who considered the species *Notiomastodon ornatus* as holotype (Cabrera 1930). The main characteristics of this species are a mandible with a short symphysis without tusks. The upper tusks exhibit an upward curvature and a lateral enamel band. The molars are bunodont and feature lophos with accessory conules, which are characterized by a double clover figure resulting from the abrasion. Recent studies have described the new species *Notiomastodon platensis* (Lucas and Alvarado 2010; Mothé et al. 2013), which encompasses all the characteristics attributed to the genus *Stegomastodon* in South America, considering also, in adult individuals, the presence of an enamel band, which was not considered before as a diagnostic feature for *Stegomastodon* (Mothé et al. 2012, 2013).

In addition to the taxonomic complexity, the age of the oldest gomphotheres in South America has been also controversial. For instance, the species *Amahuacatherium peruvium* was considered the oldest gomphothere of South America (Campbell et al. 2000). Its fossil remains are associated to late Miocene sediments (c. 9.5 Ma) from the Peruvian Amazon region (Campbell et al. 2000, 2009). It is described as a tetrabelodon (four-tusked) gomphothere with tusks embedded in a brevirostrian mandible. The occlusal view of the molars exhibits a complex pattern (Campbell et al. 2000, 2009). However, the taxonomic validity and age of this species have been strongly questioned (Lucas 2013). Some authors assign *A. peruvium* to the genus *Stegomastodon*, but this needs also careful revision (Prado et al. 2005). One of the reasons for challenging the proposed age is the lack of proboscidean fossils in South America until the mid-Pleistocene (Prado et al. 2005).

4.3 The Gomphotheres in Chile

The gomphotheres represent one of the most common types of fossils from the late Pleistocene-early Holocene in Chile. Fossil remains of the group have been found in Tarapacá, Tongoy, Los Vilos, Llampaico, Chacabuco, Pudahuel, Padre Hurtado, La Ligua, Casablanca, San Vicente, Tagua–Tagua, Talca (Cerro del Chivato), Cauquenes, Parral, Chillán, Galvarino, Carahue, San Pablo de Tramalhue, Máfil, Los Lagos, Paillaco (El Trébol), Futrono (La Plata), Osorno (Lomas Blancas, El Caracol, Los Pinos, Mulpulmo, Pilauco, Nochaco and Huilma), Frutillar, Monte Verde, and Chiloé (Castro) (Latcham 1929; Montané 1968; Paskoff 1971; Frassinetti and Alberdi 2000; Recabarren et al. 2014). The reasons for the greater proportion of fossil findings can perhaps be explained by the exceptional conditions of preservation of the sites and the hunting of gomphotheres by early human settlers, which may have favoured the accumulation of preserved elements (Fernández-López 2000). The most commonly found anatomical elements are molars, usually found isolated, near rivers or in peatland related sediments. The isolated occurrence of molars is likely the results of allochthonous accumulation, resedimentation or human-related transport (Fernández-López 2000).

In Chile, the geographic distribution of gomphothere fossils encompasses the Central Depression and both sides of the Coastal Cordillera (Frassinetti and Alberdi 2000). Statistical analyses based on molar analyses characterize the Chilean gomphotheres as high-altitude Andean type. These gomphotheres migrated along the entire western margin of the continent reaching Chile (Frassinetti and Alberdi 2000). This classification probably needs a revision given new studies that assign originally high-altitude type features to the lowland plain type gomphotheres (e.g. the enamel ban of the tusks that might be present) (Mothé et al. 2012, 2013). Along these lines, it appears noteworthy to mention that the Tagua-Tagua deposit revealed both types of gomphotheres at the same site (Frassinetti and Alberdi 2000, 2005), which arise from the reconstruction of alveoli from a gomphothere premaxilla. Later studies of fossil gomphotheres between 39°39′ and 42°49′ S recognized low-altitude plain type gomphotheres as far south as the island of Chiloé (Recabarren et al. 2014). In northern Patagonia, the Andes would not have represented a significant geographical barrier due to the deep glacial incision during the Pleistocene providing a number of low-elevation migration corridors (Moreno et al. 1994; Recabarren et al. 2014).

4.4 Gomphotheres from the Pilauco Site

Thirty gomphothere fossils have been found since the first fieldwork campaign at Pilauco. They are non-articulated fossils and present a variable degree of preservation (Labarca et al. 2014). The first fossil findings date back to 1986 when the site was discovered. They were rescued by construction workers and correspond to a left hemi-mandible with m2 and m3 molars, a right hemi-mandible with an m3 molar in anatomic position, an isolated m2 molar, a right humerus and three thoracic vertebrae with variable degree of conservation (Fig. 4.2).

These fossils can be associated with at least two individuals based on the record of two atlas vertebrae (the first cervical vertebra), two m2 molars (one on the left hemi-mandible (MHMOP/PI/14) and another one in isolated position (MHMOP/PI/16)) (Figs. 4.2 and 4.3). However, none of the atlas vertebrae seems to match the size of the skull that was found later. Four of the fossil elements reveal marks from carrion eaters and include a proximal epiphysis of the humerus (MHMOP/PI/610), a talus (MHMOP/PI/612), a heel bone (MHMOP/PI/613), and a tibia (MHMOP/PI/614). Some of the marks are very deep exposing the spongy bone (Labarca et al. 2014) (Fig. 4.4). The fossils discovered during more recent field campaigns between 2010 and 2014 include a skull (Fig. 4.5) lacking the alveolar zone and the tusks (Labarca et al. 2014).

The distribution of the fossils suggests the presence of at least one individual on the west and other on the eastern part of the excavation site. The position of the fossil remains may have been affected by taphonomic processes, such as: human or carnivo-

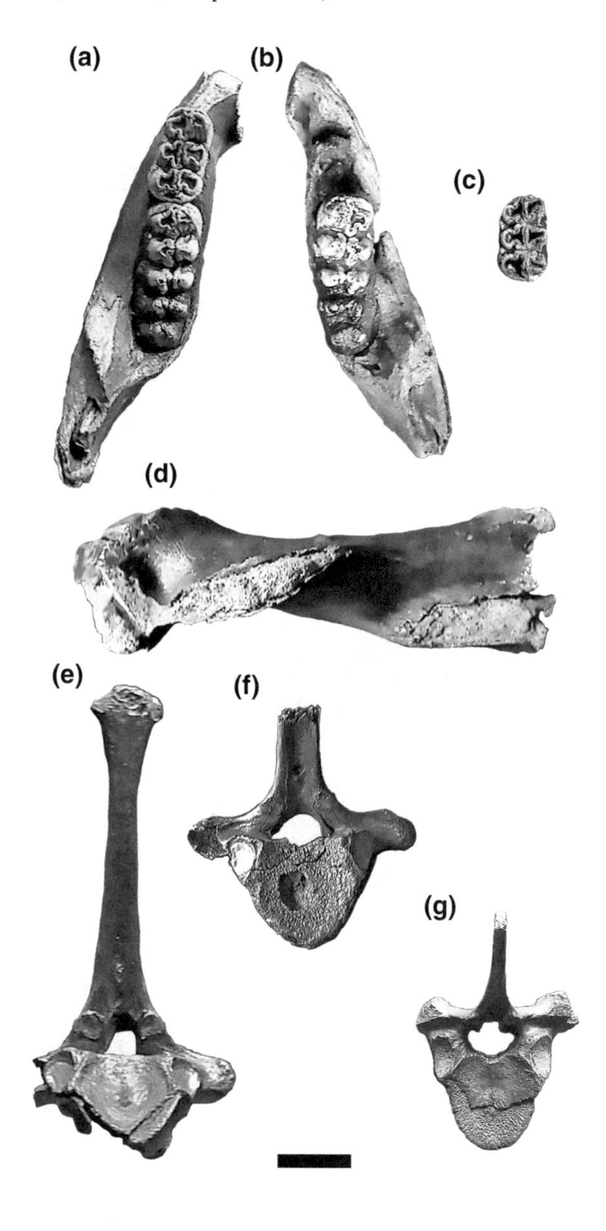

Fig. 4.2 Gomphothere fossil remains from the excavation in 1986. **a** MHMOP/PI/14, left hemi-mandible. **b** MHMOP/PI/15, right hemi-mandible. **c** MHMOP/PI/16, M2 molar in occlusal view. **d** 1986-1, right humerus. **e–g** 1986-2, 1986-3 and 1986-4, thoracic vertebrae in frontal view. Scale bar: 10 cm

rous activities, colluvial landslides and trampling. The different levels of preservation might be due to asynchronous burial meaning that some bones were buried faster than others and exposed during less time to climatic effects. The teeth marks on the bones can be attributed to a large carnivore, but further identification is complicated. However, the action of canines (dogs) and ursids (bears) can be discarded, mainly because the former leaves smaller marks and those and bear teeth marks are often more superficial, shorter and parallel. The best candidate is a feline, so far unrec-

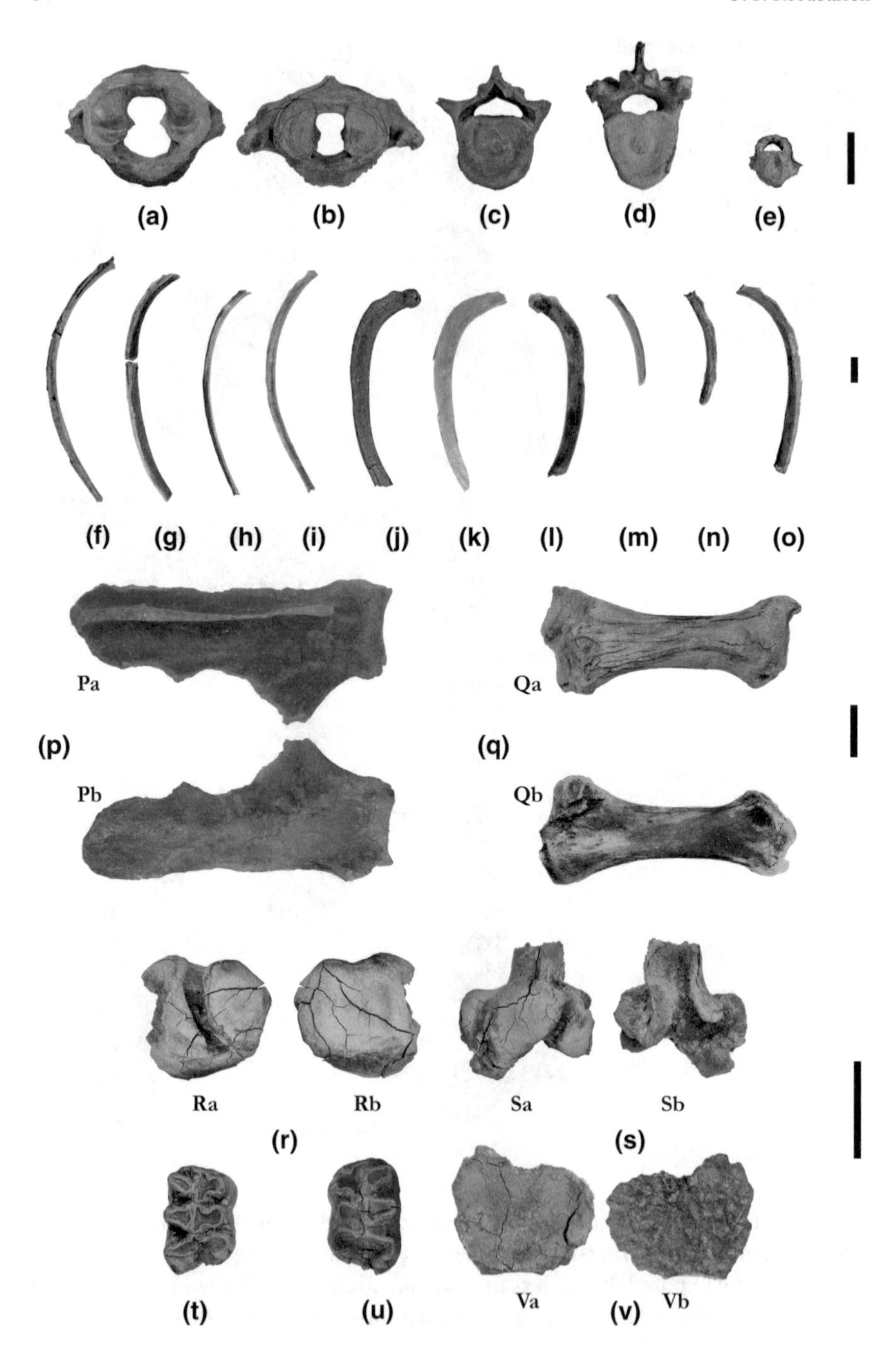
(a)
(b)
(c)
(d)
(e)
(f)
(g)
(h)
(i)
(j)
(k)
(l)
(m)
(n)
(o)
Pa
Pb
(p)
Qa
Qb
(q)
Ra
Rb
(r)
Sa
Sb
(s)
(t)
(u)
Va
Vb
(v)

◂**Fig. 4.3** Gomphothere fossil remains from Pilauco recovered during recent field campaigns. **a–b** MHMOP/PI/40 and MHMOP/PI/629, Atlas vertebrae in frontal view. **c** MHMOP/PI/626, Cervical. **d** MHMOP/PI/632, thoracic, and **e** MHMOP/PI/625, caudal vertebrae in frontal views. **f–k** MHMOP/PI/619, MHMOP/PI/621, MHMOP/PI/616, MHMOP/PI/617, MHMOP/PI/618, MHMOP/PI/611, left side ribs. **l–o** MHMOP/PI/615, MHMOP/PI/624, MHMOP/PI/622, MHMOP/PI/620, right side ribs. **p** MHMOP/PI/43, right scapula in side (Pa) and in medial (Pb) view. **q** MHMOP/PI/614, right tibia in frontal (Qa) and posterior (Qb) view. **r** MHMOP/PI/612, right-side talus in plantar (Ra) and dorsal (Rb) view. **s** MHMOP/PI/613, right calcaneus in dorsal (Sa) and plantar (Sb) view. **t** MHMOP/PI/628, M2 right molar. **u** MHMOP/PI/627, M2 left molar. **v** MHMOP/PI/610, fragment from a proximal femoral epiphysis in external (Va) and internal (Vb) views. Scale bar: 10 cm

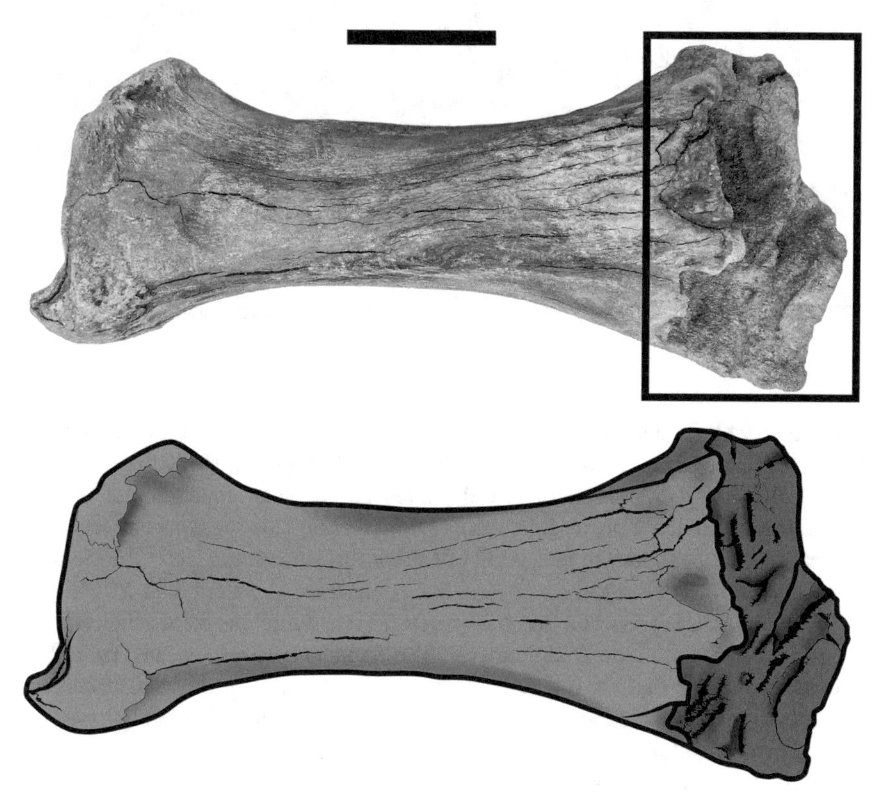

Fig. 4.4 Gomphothere right side tibia in frontal view (MHMOP/PI/614). Teeth marks can be observed on the frontal epiphysis (rectangle) and schematically indicated in the lower image. Scale bar: 10 cm

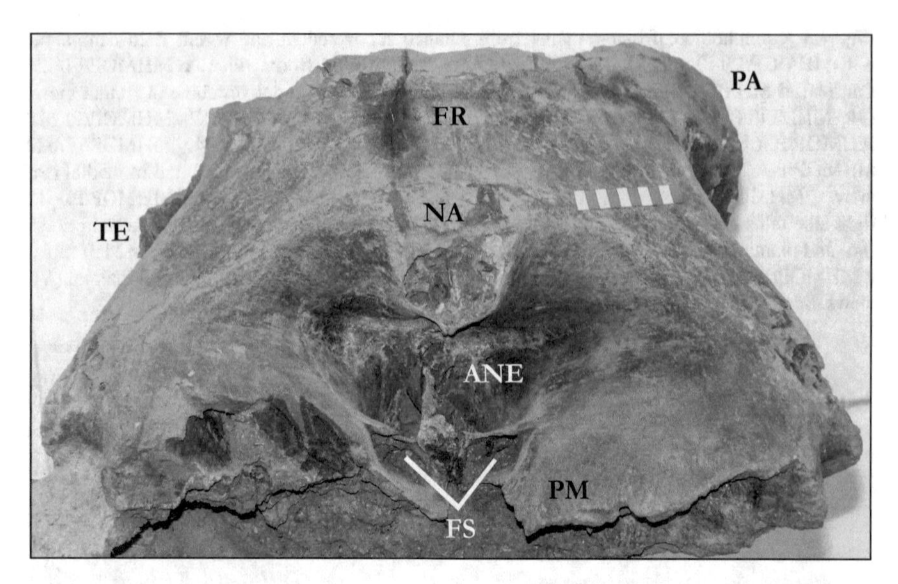

Fig. 4.5 Gomphothere skull in frontal view (specimen MHMOP/PI/630). Parietal (PA), temporal (TE), frontal (FR), nasal (NA), external nasal aperture (ANE), subnasal cavity (FS), premaxilla (PM). Scale bar: 10 cm

ognized from this region. Scavenging occurred at least in one adult animal and is evidenced in the marks left on a tibia with totally fused epiphysis (Labarca et al. 2014) (Fig. 4.4). Among cranial (skull), axial (vertebrae, ribs) and appendicular skeleton elements only the latter presented carrion-eating marks. This suggests that the skeleton was partially exposed to the action of carnivores, since statistically the elements of the appendicular skeleton are the last to be consumed (Labarca et al. 2014).

In summary, it is proposed that gomphotheres fossil remains from Pilauco correspond to the lowland type gomphotheres. The remains originate at least from three individual animals that are not fully articulated in the site. Additionally, there are four fossil bones that display scavenging marks, the cause of which might have been an unknown large feline.

Future field campaigns at Pilauco will likely reveal more proboscideans fossils that might also increase the number of identified individuals. Continuing the investigation of the site will contribute to gain a better understanding of the site's formation processes and to provide an improved taxonomy of proboscideans in South America.

References

Alberdi MT, Prado JL, Cartelle C (2002) El registro de Stegomastodon (Mammalia, Gomphotheriidae) en el Pleistoceno superior de Brasil. Rev Esp Paleontol 17:217–235

Cabrera A (1930) Una revisión de los mastodontes argentinos. Revista Mus La Plata 32:61–144

Campbell KE Jr, Frailey CD, Romero-Pittman L (2000) The late Miocene gomphothere Amahuacatherium peruvium (Proboscidea: Gomphotheriidae) from Amazonian Peru: Implications for

the Great American Faunal Interchange. Instituto Geológico, Minero y Metalúrgico del Perú, Serie D: Estudios Regionales, Boletín 23:1–152
Campbell KE Jr, Frailey CD, Romero-Pittman L (2009) In defense of Amahuacatherium (Proboscidea: Gomphotheriidae). Neues Jahrb Geol Paläontol 252:113–128
Fernández-López SR (2000) Temas de Tafonomía. Departamento de Paleontología, Universidad Complutense de Madrid
Ferretti MP (2010) Anatomy of Haplomastodon chimborazi (Mammalia, Proboscidea) from the late pleistocene of ecuador and its bearing on the phylogeny and systematics of South American gomphotheres. Geodiversitas 32:663–721
Frassinetti D, Alberdi MT (2000) Revision y estudio de los restos fosiles de mastodontes de Chile (Gomphotheriidae): cuvieronius hyodon, pleistoceno superior. Estud Geol (Madr) 56:197–208
Frassinetti D, Alberdi MT (2005) Presencia del género Stegomastodon entre los restos fósiles de mastodonte de Chile (Gomphotheriidae), pleistoceno superior. Estud Geol (Madr) 61:101–107
Gheerbrant E, Bouya B, Amaghzaz M (2012) Dental and cranial anatomy of Eritherium azouzorum from the Paleocene of Morocco, earliest known proboscidean mammal. Palaeontogr Abt A: Palaeozoology-Strat 297:151–183
Haynes G (1991) Mammoths, mastodonts, and elephants: biology, behavior, and the fossil record. Cambridge University Press, Cambridge
Hoffstetter R (1950) Observaciones sobre los mastodontes de Sud América y especialmente del Ecuador. Haplomastodon, subgn. nov. de Stegomastodon. Rev Politéc 1:1–51
Hoffstetter R (1952) Les mammifères pléistocènes de la République de l´Équater. Mémoires, Bull Soc Géol Fr 66:1–391
Labarca R, Recabarren O, Canales-Brellenthin P, Pino M (2014) The Gomphotheres (Proboscidea: Gomphotheriidae) from Pilauco site: scavenging evidence in the late pleistocene of the chilean patagonia. Quat Int 352:75–84
Lambert WD (1996) The biogeography of the gomphotheriid proboscideans of North America. In: Shoshani J, Tassy P (eds) The Proboscidea: evolution and palaeoecology of elephants and their relatives. Oxford University Press, New York, pp 143–148
Lambert WD, Shoshani J (1998) Proboscidea. In: Janis CM, Scott KM, Jacobs LL (eds) Evolution of tertiary mammals of North America. Volume 1: terrestrial carnivores, ungulates, and ungulatelike mammals. Cambridge University Press, pp 606–621
Latcham R (1929) Los mastodontes chilenos. Rev Educ (Santiago) 6:423–432
Leigh EG, O'Dea A, Vermeij GJ (2014) Historical biogeography of the Isthmus of Panama. Biol Rev 89:148–172
Liu AGS, Seiffert ER, Somons EL (2008) Stable isotope evidence for an amphibious phase in early proboscidean evolution. Proc Natl Acad Sci USA 105:5786–5791
Lucas SG (2013) The palaeobiogeography of South American gomphotheres. J Palaeogeog 2:19–40
Lucas SG, Alvarado GE (2010) Fossil proboscidea from the upper cenozoic of Central America: taxonomy, evolutionary and paleobiogeographic significance. Rev Geol Am Central 42:9–42
Lucas SG, Aguilar RH, Spielmann JA (2011) Stegomastodon (Mammalia, Proboscidea) from the Pliocene of Jalisco, Mexico and the species level taxonomy of Stegomastodon. In: Sullivan RM, Lucas SG, Spielmann JA (eds) Fossil record 3, vol 53. Bulletin—New Mexico Museum of Natural History and Science, pp 517–553
Montané J (1968) Primera fecha radiocarbónica de Tagua–Tagua. Not Mens Mus Nac Hist Nat 139:11
Moreno PI, Villagran C, Marquet PA, Marshall LG (1994) Quaternary paleobiogeography of northern and central Chile. Rev Chil Hist Nat 67:487–502
Mothé D, Avilla L, Cozzuol M (2013) The South American gomphotheres (Mammalia, Proboscidea, Gomphotheriidae): taxonomy, phylogeny, and biogeography. J Mamm Evol 20:23–32
Mothé D, Avilla L, Cozzuol M, Winck G (2012) Taxonomic revision of the quaternary gomphotheres (Mammalia: Proboscidea: Gomphotheriidae) from the South American lowlands. Quat Int 276–277:2–7

Parodi R (1962) Los mastodontes sudamericanos y su clasificación. Universidad Nacional de Tucumán, Facultad de Ciencias Naturales, Cuaderno de la Revista de la Facultad de Ciencias Naturales, no 2

Paskoff R (1971) Edad radiométrica del mastodonte de Los Vilos: 9.100 ± 300 años B. P. Bol Mus Nac Hist Nat 15:11

Prado JL, Alberdi MT, Azanza B, Sánchez B, Frassinetti D (2005) The pleistocene gomphotheriidae (Proboscidea) from South America. Quat Int 126–128:21–30

Recabarren O (2008) El gonfoterio de Pilauco. In: Pino M (ed) Pilauco: un sitio complejo del Pleistoceno Tardío. Norpatagonia chilena, Universidad Austral de Chile, Osorno, pp 148–150

Recabarren O, Pino M, Alberdi MT (2014) La familia Gomphotheriidae en América del Sur: evidencia de molares al norte de la Patagonia chilena. Estud Geol (Madr) 70:e001

Roca AL, O´Brien SJ (2005) Genomic inferences from Afrotheria and the evolution of elephants. Curr Opin Genet Dev 15:652–659

Sanders WJ, Gheerbrant E, Harris JM, Saegusa H, Delmer C (2010) Proboscidea. In: Werdelin L, Sanders W (eds) Cenozoic mammals of Africa. University of California Press, Berkeley

Shoshani J (1998) Understanding proboscidean evolution: a formidable task. Trends Ecol Evol 12:480–487

Shoshani J, Tassy P (1996) The proboscidea: evolution and palaeoecology of elephants and their relatives. Oxford University Press, New York

Simpson GG, Paula-Couto C (1957) The mastodonts of Brazil. Bull Am Mus Nat Hist 2:125–190

Sukumar R (2003) The living elephants: evolutionary ecology, behavior, and conservation. Oxford University Press, New York

Chapter 5
The Camelids (Artiodactyla: Camelidae) and Equids (Perissodactyla: Equidae) from the Pilauco Site, Northwestern Chilean Patagonia

Rafael Labarca

Abstract This chapter synthesizes the findings of fossil remains from equids and camelids at the Pilauco site, including the first discoveries from 1986, as well as the systematic excavations since 2007. The group of horses is exclusively represented by the genus *Equus*, as was identified on cranial and postcranial remnants. The findings from the eastern and western sectors display metric differences that might be indicative for the presence of two species or morphotypes. The horse remains are most abundant in stratigraphic layer PB-8. The metric analysis assigns most of the camelid remnants to cf. *Hemiauhenia paradoxa*, whereas the remaining, less diagnostic, material is classified merely on the family level. The camelid remnants have been recovered from stratigraphic layers PB-7 and PB-8. Based on the distribution of camelid and equid fossils, two occupational events can be spatially distinguished within layer PB-7. A preliminary taphonomic analysis indicates some differences in the depositional history between the eastern and western sector of the site.

Keywords Camelids · Equids · Pilauco · Taxonomy · Taphonomy

5.1 Background

The fossil findings of Pleistocene mammals in Chile are generally abundant (Labarca 2015). Nevertheless, besides some exception (i.e. Magallanes, Tierra del Fuego), the vast majority of these are chance findings without established stratigraphic and chronological context. The systematically analysed sites in south-central Chile (32°–42°S), such as Quereo or Taguatagua (Núñez et al. 1983, 1994a, b), feature a long list of taxa including small mammals, as well anurans and birds (i.e. Jiménez-Huidobro and Sallabaerry 2015; Sallaberry et al. 2015). In some cases, the osteo-morphological similarity between some extinct and living

R. Labarca (✉)
Escuela de Arqueología, TAQUACH, Transdisciplinary Center for Quaternary Research in the South of Chile, Universidad Austral de Chile, Puerto Montt, Chile
e-mail: r.labarca.e@gmail.com

M. Pino and G. A. Astorga (eds.), *Pilauco: A Late Pleistocene Archaeo-paleontological Site*, The Latin American Studies Book Series,
https://doi.org/10.1007/978-3-030-23918-3_5

species (as in the case of the genus *Equus*) has probably introduced some bias in their documentation (Oliver 1934). These circumstances apply also for Chilean northern Patagonia (approx. 40°–43°S), where occasional findings of Pleistocene mammals have been reported since at least 1883 (Oliver 1926). Almost a century later, towards the end of the 1970s, the studies at Monte Verde leaded by Tom Dillehay initiated the systematic paleontological investigation in this area (Dillehay 1989, 1997). The Pilauco site represents the second systematically studied Pleistocene site in northern Chilean Patagonia. It documents a wide assemblage of so far unknown taxa for this region, including nondescript birds, rodents (*Loxodontomys micropus* Waterhouse and *Myocastor* cf. *M. coypus* Molina), deers (cf. *Pudu* sp.), proboscideans (*Notiomastodon* aff. *N. platensis* Ameghino), equids (*Equus andium* Branco), and camelids (cf. *Hemiauchenia paradoxa* Gervais and Ameghino) (Recabarren et al. 2011, 2014; Labarca et al. 2013, 2014; González et al. 2014; see also Chap. 7, this volume). The results confirm that the systematic study of such randomly found sites significantly contributes to the knowledge about Pleistocene fauna in a specific area. This chapter briefly summarizes the available information about camelids and equids from the site of Pilauco. The fossil material is described within a regional and South American context referring to some taphonomic, evolutionary and biogeographic background aspects.

5.2 Camelids and Equids: Origin, Taxonomy and Dispersion

The families Equidae and Camelidae emerge for the first time in the early and middle Eocene of North America, respectively (McFadden 1992; Honey et al. 1998; see Chap. 2, this volume). Contemporaneous registers of equids in Europe and Asia suggest a rapid colonization processes (Lucas and Kondrashov 2004; Missiaen et al. 2013). The camelids, in turn, followed an isolated evolution on the subcontinent until the Miocene, when first radiations towards Europe and Africa have been recorded (Honey et al. 1998). Currently, there are no native representatives of these families in North America, given that the equids and camelids both disappear towards the end of the Pleistocene (McFadden 1992; Scherer 2009).

The arrival of both families on the subcontinent is part of the Great American Biotic Interchange between North and South America (GABI, Webb 1991; Woodburne 2010; Cione et al. 2015; see Chap. 2, this volume). It certainly represents one of the most important evolutionary processes in the Americas, as almost half of the South American terrestrial mammals originate from North America (Cione et al. 2015). This migration event was not continuous, but rather characterized by pulses of variable intensity over the course of at least 7 million years, likely triggered by tectonic and climatic events (Woodburne 2010; Cione et al. 2015). The GABI initiated with the uplift of the Isthmus of Panama towards the end of the Pliocene, some 2.6 Ma ago, and included at least four large migration pulses until the end of the Pleistocene (Woodburne et al. 2006; Woodburne 2010). Nevertheless, the fossil register presents foreign taxa in South America such as the procyonids (racoons), at least since the Late Miocene (about 7 Ma ago) (Prevosti et al. 2013). This first, more sporadic and less intense process of interchange has been denominated Proto-GABI (Cione et al. 2015) or Early American Biotic Interchange (see Chap. 2, this volume). Some authors (i.e. Prothero et al. 2014) place the presence of Holarctic forms as far back as 9.5 million years ago based on fossil registered in strata of probably Miocene age in the Amazonas (Campbell et al. 2000).

There are two Pleistocene lineages of equids in South America represented by the genera *Equus* and *Hippidon* (Alberdi and Prado 2002, 2004; Prado and Alberdi 2017). Traditionally, five species have been described and assigned to the first group: *E. andium* (Branco 1883), *E. neogeus* (Lund 1840), *E. insulatus* (Ameghino 1904), *E. santaeelenae* (Spillmann 1938) and *E. lasallei* (Daniel 1948) (Alberdi and Prado 2004). However, recently, Prado and Alberdi (2017) considered only the first three of these as valid species. Three species have been assigned to the genus *Hippidon*: *H. principale* (Lund 1846), *H. devillei* (Gervais 1855), and *H. saldiasi* (Roth 1899) (Alberdi and Prado 2004; Prado and Alberdi 2017; Orlando et al. 2009). Nevertheless, the taxonomy of the genus *Equus* has been recently discussed by Machado et al. (2017) suggesting that the metric variables used by Alberdi and Prado (2004) merely represent gradual variations between meta-populations, and therefore cannot be applied for the definition of species.

The oldest evidence of *Equus* in South America has been found in the Tolomosa formation (Tarija, Bolivia) and corresponds to the species *Equus insulatus* of approximately 0.99 Ma (McFadden 2013), which coincides with GABI 3 (Woodburne et al. 2006). The genus *Equus* was already differentiated in North America at least since the Pliocene (Orlando et al. 2013). In turn, the oldest record of the genus *Hippidon* in South America originates from the Uquía formation (Argentina) and has been dated with 2.5 Ma and assigned to the species *H. devillei* (Prado et al. 1998). Although traditionally *Hippidon* has been considered an endemic South American genus (Prado and Alberdi 2014), recent studies of fossil DNA demonstrate that the separation between *Equus* and *Hippidon* occurred between 5.6 and 6.5 Ma ago (Der Sarkissian et al. 2015), indicating a North American origin for *Hippidon,* thus predating the GABI.

The taxonomy of fossil South American camelids was recently revised by Scherer (2009, 2013). According to these studies, five genera can be distinguished within the Lamini tribe: *Hemiauchenia, Palaeolama, Eulamaops*, *Lama* and *Vicugna* distributed over the entire continent. The number of species is still under debate, as several forms are poorly known (Scherer 2009). Although the family is very diverse during the Late Pleistocene (SALMA Lujanense, Scherer 2009), only *Lama* and *Vicugna* managed to overcome the Pleistocene–Holocene transition towards a present-day record.

The arrival of the tribe to South America apparently occurred at the end of the Pliocene (SALMA Chapadmalense, ca. 3.55–2.59 Ma), as has been deduced from the record of *Hemiauchenia* sp. in Olavarría (Argentina) (Scherer 2013; Gasparini et al. 2017). This taxon has also an early Pliocene record in Central America (Montellanos 1989). The genus *Lama* possibly derived from *Hemiauchenia*, as suggested by the analysis of Scherer (2009, 2013), who situates *Hemiauchenia* sp. as a sister clade with *Lama* and *Vicugna*. Nevertheless, the first fossil record of the genus *Lama* is relatively contemporaneous with *Hemiauchenia*, and some Pampean formations from Mar de Plata in Argentina even document their co-existence (Scherer 2009). The latter indicates a rather rapid separation of the two genera on the subcontinent. On the other hand, the first clear evidence of *Palaeolama* dates back to 0.99–0.78 Ma ago and originates from the Tolomosa formation (Bolivia) (McFadden and Shockey 1997; Guérin and Faure 1999; McFadden 2013). This indicates a second migratory pulse towards South America during GABI 3 (Woodburne 2010), which is coeval with the arrival of *Equus*. Due to the scarce occurrence of fossil material, the emergence of the genus *Vicugna* is not clearly defined yet. The fossil material from the Pampean region has been tentatively assigned to the Late Pliocene (Scherer 2009) and Mid-Pleistocene (Scherer 2013), whereas the findings in Tarija are of Mid-Pleistocene age (Guérin and Faure 1999). *Vicugna* apparently derived from *Lama* at some time during the Pleistocene (Menegaz et al. 1989; Scherer 2009).

5.3 The Pleistocene Record of Camelidae and Equidae in Chile: Taxonomy and a Biogeographical Overview

The Pleistocene camelids in Chile are distributed over the entire country (Fig. 5.1). There is a significant concentration of findings from the intensively studied southern part of the semi-arid northern Chile and the Magallanes basin (i.e. Núñez et al. 1983, 1994a; Jackson et al. 2007; López 2007; Bird 1988; Massone 1988; Nami 1987; Prieto 1991, amongst others). The absence of findings from some regions is probably due to the absence of investigations. Although many of the localities lack a precise chronology, almost all findings have been assigned to the end of the Pleistocene or the Lujanian mammal age (Latorre 1998; Casamiquela 1999; Cartajena et al. 2013; Labarca 2015; Martin 2013).

Numerous absolute radiocarbon data on Camelidae bones from the Magallanes region almost exclusively coincide with the Last Glacial Maximum (LGM), i.e. post-dating 17.000 cal yr BP (Massone 2004; Martin 2013; Metcalf et al. 2016, and references therein). On the other hand, the GNL Quintero 1 site (central Chile) yielded absolute age data from *Palaeolama* and Camelidae indet. of approximately 24,100–23,700 cal yr BP. (López et al. 2016a, b). The fossil records from Marchigüe and Los Sauces have been tentatively assigned to the Mid-Pleistocene (López and Cartajena 2006; Casamiquela 1969–1970).

At least 5 different forms of Camelidae have been recognized in the Pleistocene record of Chile, heterogeneously distributed across 29 localities. The fossil record of *Vicugna* is basically concentrated in the high Andean region and Magallanes and represented by the species *Vicugna provicugna* (Gervais and Ameghino 1880) (Cartajena et al. 2010) and *Vicugna gracilis* (Gervais and Ameghino 1880) (=*Lama* (*Vicugna*) *gracilis*) (Menegaz 2000; Metcalf et al. 2016), respectively. Based on the systematic

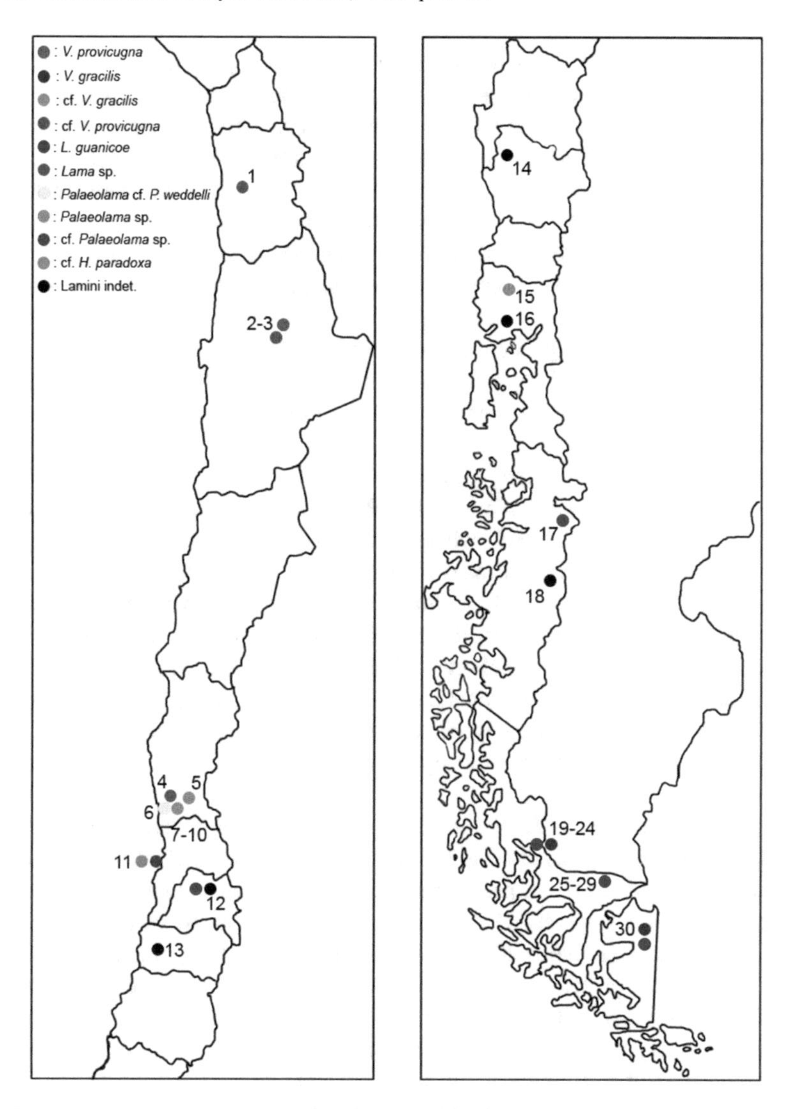

Fig. 5.1 Relative location of sites with fossil findings from Lamini in Chile. 1. Pampa del Tamarugal; 2. Kamac Mayu; 3. Ojo de Opache; 4. Quebrada Lazareto; 5. Santa Julia; 6. Quebrada Quereo; 7. Las Monedas; 8. Valle de los Caballos D; 9. El Membrillo; 10. El Avistadero; 11. GNL Quintero 1; 12. Chacabuco; 13. Marchigüe; 14: Los Sauces; 15. Pilauco; 16. Monte Verde; 17. Cueva Baño Nuevo 1; 18. Cueva Las Guanacas; 19: Cueva del Milodón; 20. Cueva del Medio; 21. Cueva Chica; 22. Cueva Escondida; 23. Cueva Lago Sofía 1; 24. Cueva Lago Sofía 4; 25. Cueva Fell; 26. Cueva Cerro Sota; 27. Cueva Pali Aike; 28. Cueva de los Chingues; 29. Cueva del Puma; 30. Tres Arroyos 1. Data from Cartajena et al. (2010, 2013), López et al. (2007, 2010, 2016b), Labarca (2015, 2016), López and Cartajena (2006), Casamiquela (1969, 1970, 1999), Labarca et al. (2008, 2013), Casamiquela and Dillehay (1989), López (2009), Roth (1902), Bird (1988), Latorre (1998), Martin (2013), Martin et al. (2015), Metcalf et al. (2016)

review of Scherer (2009), Labarca (2015) suggested the presence of *V. provicugna* in northern Chile considering the known geographic distribution for this species. In turn, recent studies of fossil DNA (Metcalf et al. 2016) recognize *V. gracilis* as a valid species, which is contrary to morphological (Scherer 2009), metric (Labarca and Prieto 2009), and previously performed fossil DNA studies (Weinstock et al. 2009) that assigned it to *Vicugna vicugna* (Molina 1782). Given that the genera *Lama* and *Vicugna* are clearly genetically differentiated (Marín et al. 2007; Labarca 2016) suggested the usage of the denomination *Vicugna gracilis* for this extinct taxon.

Apart from these areas, the occurrence of cf. *V. gracilis* was recently reported from the submerged site GNL Quintero 1 (López et al. 2016a). This evidence extends the Pleistocene distribution of the genus at least central Chile, which is not surprising given the wide Pleistocene dispersion of this form in Argentina (Menegaz 2000; Metcalf et al. 2016).

The guanaco (*Lama guanicoe* Müller 1776) has only one record in the Late Pleistocene of Central Chile, particularly in Las Pozas of Chacabuco (Santiago de Chile) (Oliver 1935; Fuenzalida 1936), whereas it is particularly abundant in Patagonia. There, the findings are concentrated mainly around rockshelters and caves (Roth 1899; López 2009; Latorre 1998; Nami and Menegaz 1991; Martin 2013; Labarca 2016) (Fig. 5.1). In several sites, there is a clear evidence for the exploitation of guanacos by groups of hunter-gatherers inhabiting the area (Bird 1988; Martin 2013; Labarca 2016). Notably, recent genetic studies (Metcalf et al. 2016) indicate that the Late Pleistocene guanaco of Patagonia constitutes a divergent clade from the actual form, and therefore represents a locally extinct population that was replaced during the Holocene.

The record of the genus *Palaeolama* extends south of 30°S across subsuperficial and superficial coastal sites close to the city of Los Vilos (Núñez et al. 1983; López 2007; Jackson et al. 2007). A preliminary metric study shows that the species from the Quebrada Quereo site and nearby findings might correspond to *Palaeolama weddelli* (Gervais 1855) (Labarca 2015). Towards the south (GNL Quintero 1), the fossil record confirms the presence of this genus along the coast of central Chile (Cartajena et al. 2013; López et al. 2016a, b). The possible presence of *Palaeolama* cf. *P. weddelli* around Los Vilos locality would represent the southernmost record of this species.

The evidences of cf. *Hemiauchenia paradoxa* from Pilauco (Región de Los Lagos) represent the first record west from the Andes (Labarca et al. 2013) for a taxon that was so far restricted to the Pampean region of Argentina (Scherer 2009).

Finally, the non-diagnostic fossil remains have been assigned to Lamini indet. by Labarca (2015). Fuenzalida (1936) reported fossil remnants from a large-sized camelid from one outcrop at Las Pozas de Chacabuco and assigned it to *Lama major* (=*Palaeolama major* Liais 1872). Later on, these remnants were included within *Paleolama* sp. (sic) by Casamiquela (1999) and Frassinetti and Alberdi (2001), however, without clear arguments for such assignation. To the south, in the locality of Los Sauces (37°58′S), Casamiquela (1969) reported about evidence for a large-sized metapodial from Camelidae. The same was observed in Monte Verde (41°30′S; Dillehay 1989) and in Cueva Las Guanacas in Región de Aysén (46°18′S; Labarca et al. 2008). The latter fossil finding might correspond to *Lama guanicoe*. Finally,

López and Cartajena (2006) documented some remnants from a camelid larger than the actual guanaco from the Marchigüe site (34°24′S).

Similar to the camelids, the fossil findings of equids are irregularly distributed between 49 localities across Chile (Fig. 5.2). There are some notable concentrations in the Calama basin (22°27′S), the Chilean Central Depression and the Austral Patagonia (Fig. 5.2). The fossil material indicates the presence of the two known genera from this family in South America: *Hippidon* and *Equus*, although there are only rare evidences for their co-existence. Due to the lack of contextual information, their stratigraphic distribution is challenging to define.

Nevertheless, it is possible to divide the fossil remains between three major time periods: Pliocene–Pleistocene, Mid-to-Upper Pleistocene, and Upper Pleistocene. The first age includes the ichnites described by Casamiquela and Chong (1974) in La Chimba (23°40′, Antofagasta, Fig. 5.2) and assigned to *Hippidion* sp. (Casamiquela 1999). The fossil remains assigned to *Equus* sp. from El Salado (22°19′S, Antofagasta) (Orellana 1965; Salinas et al. 1991) dated between the Lower-to-Mid-Pleistocene, whereas the mandibular and dental remnants from *Hippidon* sp. in Huallillinga (30°37′S, Ovalle) belong to the Lower Pleistocene (Hoffstetter and Paskoff 1966). The remaining findings have been attributed to the Upper Pleistocene with a particular concentration of radiocarbon ages obtained from southern Patagonia (Casamiquela 1999; Alberdi and Frassinetti 2000; Villavicencio et al. 2015; Metcalf et al. 2016).

Hippidon saldiasi has been unambiguously registered in three separate areas: the Calama basin at the Ojos de Opache, Kamac Mayu, and Betecsa 1 sites (Alberdi et al. 2007; López et al. 2010), in the Chilean Central Depression near Santiago (Santa Rosa de Chena and Chacabuco), and in the most austral part of Chile between the regions Aysén and Magallanes (Alberdi and Prieto 2000; Labarca et al. 2008). *Hippidon* is the common genus in the fossil record of the austral region and it was evidently prey for groups of hunters-gatherers, as well as big felines inhabiting this region (Martin 2013; Labarca 2016). According to Alberdi and Frassinetti (2000), this taxon, apparently derived from *H. lasallei*, took a dispersion route associated with the Andean Cordillera towards Patagonia, where the vast number of fossil findings suggests the prevalence of optimal conditions for its proliferation. Nevertheless, the findings in high altitude sites in northern Chile, as well as in the Intermediate Depression in central Chile, suggest an important environmental tolerance for these taxa.

Hippidon principale has been identified only in Laguna de Taguatagua (34°26′S, Fig. 5.2). Its closest record is from the Pampean region (Alberdi and Prado 2004). *Hippidon* sp. remains come from various localities (Fig. 5.2), whereas the material from Huallillinga and the ichnites from La Chimba represent particular highlights due to their Plio-Pleistocene origin (Hoffstetter and Paskoff 1966; Casamiquela and Chong 1974). *H. devillei* is so far the only *Hippidon* form with such an early record. On the other hand, *H. principale* has been documented for the first time in the Mid-Pleistocene of Bolivia (Alberdi and Prado 2004; McFadden 2013; Prado and Alberdi 2017).

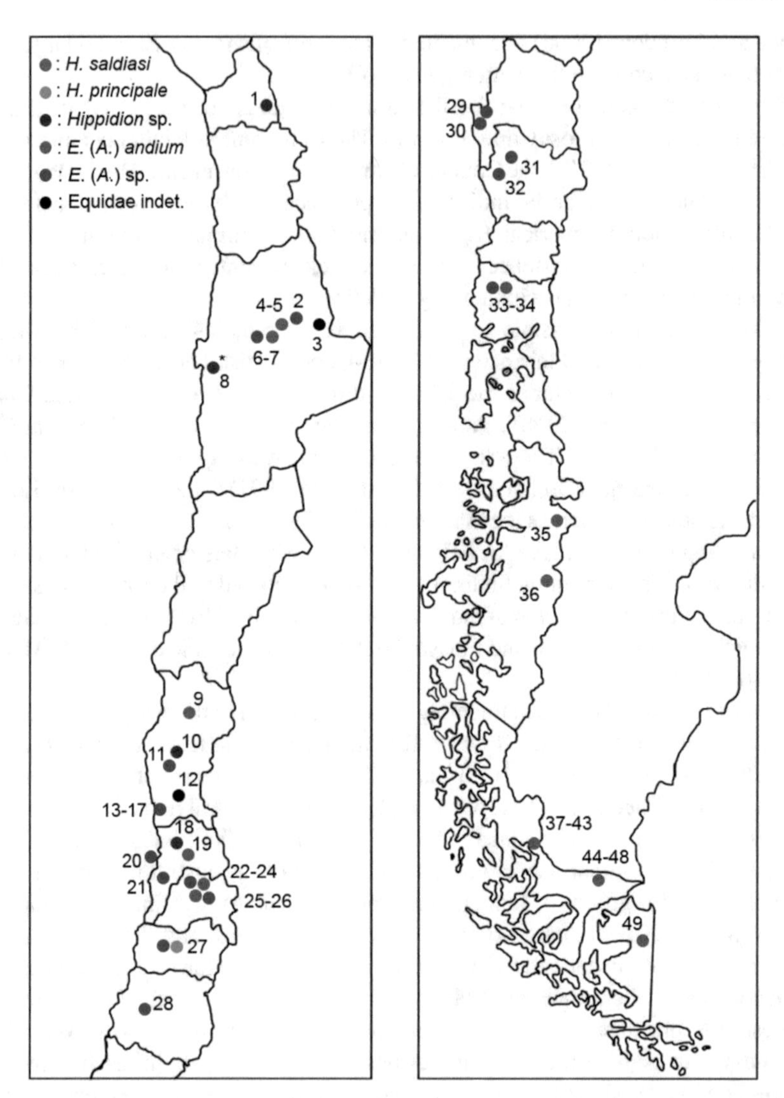

Fig. 5.2 Localities with Equidae remnants in Chile. 1. Salar de Surire; 2. Río Salado; 3. Tuina; 4. Kamac Mayu; 5. Betecsa 1; 6. La Vega; 7. Ojo de Opache; 8. La Chimba (* = ichnites); 9. Valle del Elqui; 10. Huallillinga; 11. Placilla; 12. Santa Julia; 13. Quereo; 14. Las Monedas; 15. El Avistadero; 16. El Membrillo; 17. Valle de los Caballos D; 18. Tierras Blancas; 19. Lo Aguirre; 20. GNL Quintero 1; 21. Lagunillas; 22. Las Pozas de Chacabuco; 23. Colina; 24. Conchalí; 25. Santa Rosa de Chena; 26. Macul; 27. Taguatagua; 28. Nirvilo; 29. Cerro Caracol; 30. San Pedro; 31. Carahue; 32. Humpil; 33. Pilauco; 34. Los Notros; 35. Baño Nuevo 1; 36. Cueva Las Guanacas; 37. Cueva del Milodón; 38. Cueva del Medio; 39. Cueva de la Ventana; 40. Cueva Nordenskjöld; 41. Cueva Escondida; 42. Cueva Lago Sofía 1; 43. Cueva Lago Sofía 4; 44. Cueva Fell; 45. Cueva Cerro Sota; 46. Cueva Pali Aike; 47. Cueva de los Chingues; 48. Cueva del Puma; 49. Tres Arroyos 1. Data from Santoro pers.com, Orellana (1965), Cartajena (2001), Alberdi et al. (2007), López et al. (2010, 2016a), Casamiquela and Chong (1974), Alberdi and Frassinetti (2000), Hoffstetter and Paskoff (1966), Jackson et al. (2007), López (2007, 2009), Cartajena et al. (2013), Casamiquela (1999), Fuenzalida (1936), Oliver (1934), Recabarren et al. (2011), Lira et al. (2016), Labarca et al. (2008), Alberdi and Prieto (2000), Der Sarkissian et al. (2015)

Regarding the genus *Equus,* only *E. andium* was recognized on the species level in La Serena (30°S), La Calera (32°47′S, and Pilauco (Fig. 5.2) (Alberdi and Frassinetti 2000; Recabarren et al. 2011). This assignation is based on the small size of the recovered dental pieces and the fact that *E. andium* corresponds to the smallest-size taxon within the genus (Alberdi and Prado 2004; Prado and Alberdi 2017). Outside Chile, this form has been registered in Ecuador (Alberdi and Prado 2004), which led Prado and Alberdi (2014) to suggest an Andean dispersion route. The remaining Chilean evidences are distributed from Calama to Osorno (Fig. 5.2) and have been classified only on the genus level due to the absence of diagnostic material. Oliver (1934) and Fuenzalida (1936) suggested the south-central Chilean findings correspond to *Equus curvidens* (= *E. neogeus*), which is very common for the Late Pleistocene in the Pampean region (Alberdi and Prado 2004). On the other hand, Frassinetti and Alberdi (2000) emphasized that the materials from Chacabuco, Taguatagua and Huimpil have dimensions comparable with *E. insulatus* from northern South America and Bolivia (Alberdi and Prado 2004; McFadden 2013). Labarca and López (2006) reached similar conclusions with materials from Quebrada Quereo. Recently, Lira et al. (2016) (see Chap. 14, this volume) reported *Equus* dental fragments at Los Notros site with greater dimensions compared to those known for *E. andium.*

Therefore, the presence of only two forms of *Equus* can be confirmed for the Pleistocene in Chile: one small-sized compatible with *E. andium,* and another, which is relatively larger. The specific classification remains open, even more so considering that some large-sized forms of *Equus* present overlapping characteristics of the metapodial and phalangeal dimensions (Alberdi and Prado 2004; Prado and Alberdi 2017).

5.4 The Record of Camelids and Equids from Pilauco

Up to date, a total of 17 fossil camelid and 21 equid specimens have been recovered from the PB-7 and PB-8 layer in the two studied areas at the Pilauco site, (number of identifiable specimens (NISP) = 38) (Table 5.1). The taxonomy of both families has been partially addressed by Recabarren et al. (2011) for the equids and Labarca et al. (2013) for the camelids. The first study is focused only on material from the eastern area, whereas the second includes specimens from both eastern and western areas that have been recovered until 2010.

Based on the analysis of a sample of 15 specimens from the eastern area, Recabarren et al. (2011) reported the presence of the genus *Equus* and assigned at least three specimens (MHMOP/PI/063, 067 and 520) to *Equus andium* (Fig. 5.3). The classification on the species level of this taxon is based on the small size of the specimens (Recabarren et al. 2011). Although scarce, the equids materials from the western area confirm the presence of the genus *Equus* in this place. The specimen MHMOP/PI/605 (Fig. 5.3) corresponds to a proximal portion of a first phalanx, which exhibits the articulate cartilage for the third metapodial. The specimen MHMOP/PI/634 (Fig. 5.3) is a fragment from a left coxal bone retaining the ilium,

Table 5.1 Summary of fossil specimens from camelids and equids from the site of Pilauco (1986–2016)

Code	Taxon	Anatomic Unit	Piece	Unit	Sector	Layer	Observations
MHMOP/PI/061*	Camelidae indet.	Molar	Enamel fragment	Trench no 5	–	ND	
MHMOP/PI/501*	Camelidae indet.	Molar	Loph fragment	Trench no 5	–	ND	
MHMOP/PI/580**	cf. *H. paradoxa*	Talus	Complete	Trench no 7	–	ND	
MHMOP/PI/600**	cf. *H. paradoxa*	Metacarpal	Proximal and diaphysis	ND	East	ND	Excavation in 1986
MHMOP/PI/072**	cf. *H. paradoxa*	Tibia	Distal	12G	East	PB-8	
MHMOP/PI/543*	Camelidae indet.	Molar	Fragment	11F	East	PB-8	
MHMOP/PI/605*	Camelidae indet.	Tibia	Proximal diaphysis	14A	West	PB-8	
MHMOP/PI/644*	cf. *H. paradoxa*	Radius–ulna	Proximal and diaphysis	19AA	West	PB-8	Vertical
MHMOP/PI/073**	cf. *H. paradoxa*	Metatarsal	Proximal and diaphysis	9G	East	PB-7	
MHMOP/PI/540**	cf. *H. paradoxa*	Incisor/canine	Complete	11I	East	PB-7	
MHMOP/PI/601**	cf. *H. paradoxa*	Coxal	Acetabulum, part of an ischium, pubis and ilium	14AD–14AC	West	PB-7	Above gomphothere bones
MHMOP/PI/602**	cf. *H. paradoxa*	Scapula	Distal and blade fragment	14AC	West	PB-7	Above gomphothere bones
MHMOP/PI/636*	cf. *H. paradoxa*	Talus	Complete	16AD	West	PB-7	Associated with gomphotheres

(continued)

Table 5.1 (continued)

Code	Taxon	Anatomic Unit	Piece	Unit	Sector	Layer	Observations
MHMOP/PI/641*	cf. *H. paradoxa*	Mandible	Body with incomplete molar series	14A	West	PB-7	Above gomphothere bones
MHMOP/PI/642*	Camelidae indet.	Humerus	Diaphysis fragment	18A	West	PB-7	Associated with gomphotheres
MHMOP/PI/643*	cf. *H. paradoxa*	Lumbar vertebra	Complete	14AB	West	PB-7	Above gomphothere bones
MHMOP/PI/645*	Camelidae indet.	Incisor/canine	Complete	17AA	West	PB-7	Associated with gomphotheres
MHMOP/PI/021***	*Equus* sp.	Coxal	Acetabulum, part of an ischium, pubis and ilium	PB-7, PB-8 and PB-9	East	ND	Excavation in 1986
MHMOP/PI/065***	*Equus* sp.	Incisor	Complete	8H	East	ND	
MHMOP/PI/069***	*Equus* sp.	Incisor	Complete	ND	–	ND	
MHMOP/PI/070***	*Equus* sp.	Upper molar	Fragment	ND	–	ND	
MHMOP/PI/063***	*E. andium*	m1–2	Complete	11D	East	PB-8	
MHMOP/PI/067***	*E. andium*	Left PM3–4	Proximal	7G	East	PB-8	Radiocarbon date
MHMOP/PI/071***	*Equus* sp.	Incisor	Complete	ND	–	PB-8	
MHMOP/PI/520***	*E. andium*	Phalanx 1	Complete	12E	East	PB-8	
MHMOP/PI/541***	*Equus* sp.	Upper molar	Fragment	11E	East	PB-8	
MHMOP/PI/060***	*Equus* sp.	Long bone undeterm.	Diaphysis	11D	East	PB-7	
MHMOP/PI/062***	*Equus* sp.	Upper molar	Fragment	8F	East	PB-7	

(continued)

Table 5.1 (continued)

Code	Taxon	Anatomic Unit	Piece	Unit	Sector	Layer	Observations
MHMOP/PI/066***	*Equus* sp.	Right M1-2	Complete	11I	East	PB-7	
MHMOP/PI/068***	*Equus* sp.	Upper molar	Fragment	ND	–	PB-7	
MHMOP/PI/355***	*Equus* sp.	Incisor	Fragment	ND	–	PB-7	
MHMOP/PI/544***	*Equus* sp.	Scapula	Distal fragment	G9	East	PB-7	
MHMOP/PI/492*	*Equus* sp.	Molar	Fragment	12E	East	PB-7	
MHMOP/PI/605*	*Equus* sp.	Phalanx 1	Proximal	17AC	West	PB-8	
MHMOP/PI/634*	*Equus* sp.	Coxal	Acetabulum fragment with ilium	15AD	West	PB-7	Above gomphothere bones
MHMOP/PI/635*	*Equus* sp.	Mandible	Ramus	15AD	West	PB-7	Above gomphothere bones
MHMOP/PI/649*	*Equus* sp.	pm3–4	Complete	16AA	West	PB-8	
MHMOP/PI/651*	*Equus* sp.	Talus	Complete	13AA	West	PB-7	Associated with gomphotheres

ND No data; *assigned in this study; **assigned in Labarca et al. (2013); ***assigned in Recabarren et al. (2011)

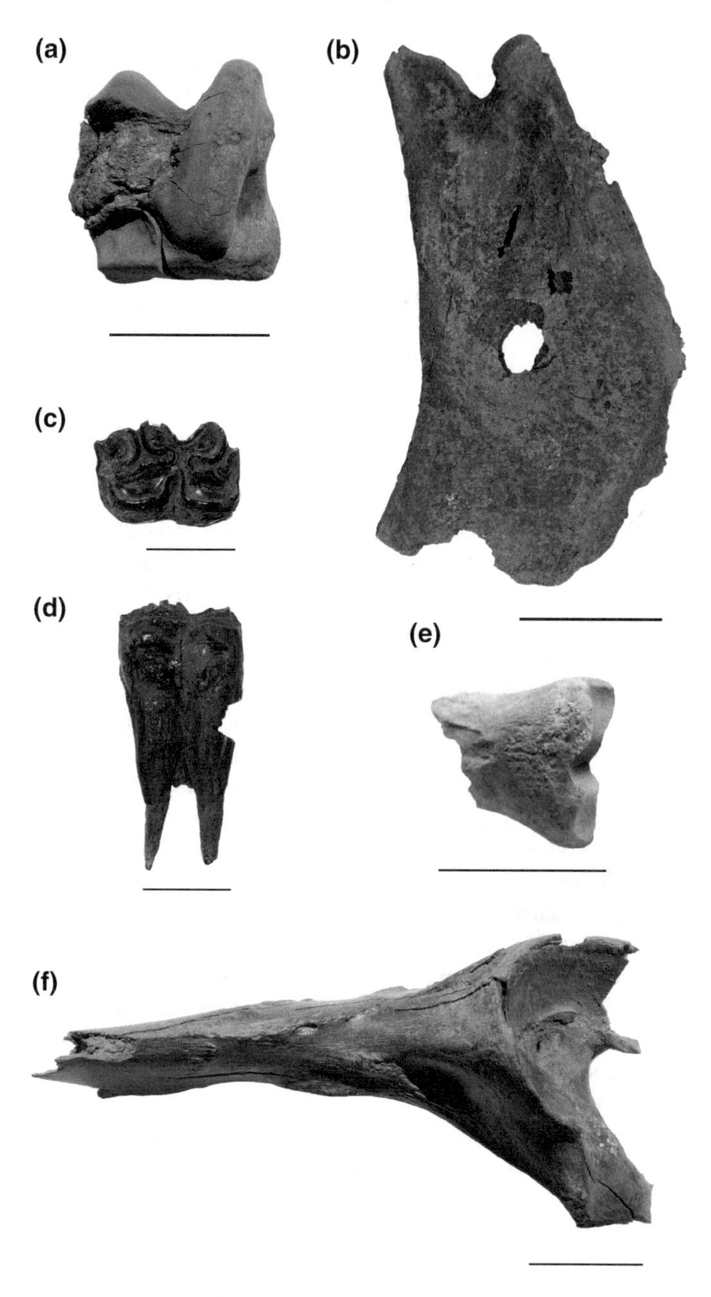

Fig. 5.3 *Equus* sp. remains from Pilauco. **a** MHMOP/PI/651 right talus, dorsal view; **b** MHMOP/PI/635 left mandibular ramus, lateral view; **c** MHMOP/PI/649 pm3–4, occlusal view; **d** MHMOP/PI/649 pm3–4 lingual view; **e** MHMOP/PI/605 first phalanx of third digit, frontal view; **f** MHMOP/PI/634 left coxal bone, frontal view. Length of scale bar is 5 cm (**a**, **b**, **e**, and **f**) and 2 cm (**c** and **d**)

Table 5.2 Measurements in mm from pm3–4 MHMOP/PI/649 after Eisenmann et al. (1988)

2	3	4	5	6
28	16	16	12	7

Table 5.3 Measurements in mm of talus MHMOP/PI/651. Numbers according to Eisenmann et al. (1988)

1	2	3	4	5	6	7
65	66	Fractured	62	54	39	54

with an almost intact acetabulum (except for a segment of the semilunar surface of the ischium), and the dorsal part of the pubis. On this specimen, the iliopubic eminence and the dorsal pubic tubercle can be observed. On the ilium, the nutrient foramen located in the lateral side and the two insertions for the *rectus femoris* muscle can be distinguished. Both are rather deep and oval-shaped, whereas the ventral insertion has a larger extension. The acetabulum is characterized by a wide acetabular fossa, continuing into a wide acetabular notch. The fossil specimen MHMOP/PI/635 (Fig. 5.3) represents a completely preserved mandibular ramus. The coronoid process is fractured, but it is still possible to recognize a narrow mandibular fossa. The condylar process is robust. The mandibular foramen can be observed in medial view and it is located in the middle-basal portion of the ramus. In its basal-towards-oral segment, a portion from the m3 alveolus can be observed. The specimen MHMOP/PI/649 (Fig. 5.3) corresponds to a right-side pm3–4. The occlusal surface displays the typical characteristics of the genus: a rounded metaconid and an angular metastilid, as well as a deep ectoflexid without contact to the linguaflexid. The measurements of this specimen follow Eisenmann et al. (1988) and are shown in Table 5.2.

Finally, the specimen MHMOP/PI/651 (Fig. 5.3) corresponds to a right talus, which is almost complete except for a missing lateral trochlea segment. The dorsal view is dominated by this feature, which has an oblique orientation with respect to the main axis. The four articular surfaces for the calcaneus can be distinguished on the plantar face, whereas from a distal view the surface for the central tarsus is particularly evident. The measurements on the specimen according to Eisenmann et al. (1988) are shown in Table 5.3.

A comparison of the dimensions of the pm3–4 MHMOP/PI/649 specimen with other fossil dental remains from the same genus (McFadden and Azzaroli 1987; Alberdi and Frassinetti 2000; Alberdi et al. 2003; Rincón et al. 2006; Prado and Alberdi 2008), suggest the presence of a large-size specimen, similar to *E. insulatus* or *E. neogeus* (Fig. 5.4). The size of the talus is conforming with that, although it could only be compared to *E. neogeus* (Alberdi et al. 2003). Hence, the fossil equid materials from the western sector have been assigned merely on the genus level. These results, together with the conclusions of Recabarren et al. (2011), would indicate the existence of two horse morphotypes at Pilauco, which following the proposal of Machado et al. (2017) may not correspond to two different species, would rather represent some metric variability within the same population. Recently, Lira et al.

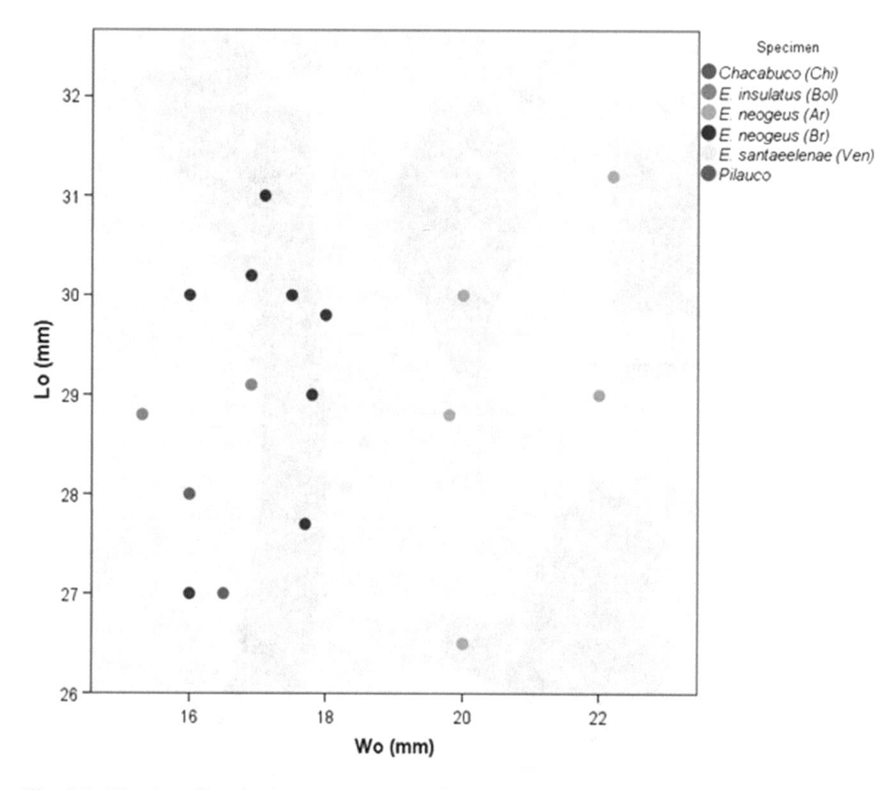

Fig. 5.4 Bivariate diagram for measurements of pm3–4. Lo: Occlusal length; Wo: Occlusal width (2 and 6 after Eisenmann et al. 1988). Data from Alberdi and Frassinetti (2000), Alberdi et al. (2003), McFadden and Azzaroli 1987, Prado and Alberdi (2008, 2012), Rincón et al. (2006)

(2016) reported the presence of a large-sized equid from the neighbouring site Los Notros, which shares the stratigraphy and chronology of the Pilauco site.

Regarding the camelids, ten new specimens have been added to the materials already described and assigned by Labarca et al. (2013). Three of them correspond to molar fragments (MHMOP/PI/61, 501 y 543) with low diagnostic value. The same is valid for a very fragmented incisor/canine (MHMOP/PI/645) that could not be described or measured. The above mentionated fossils were assigned to Camelidae indet. The remainig six specimens were included in cf. *Hemiauchenia paradoxa*. The specimen MHMOP/PI/603 corresponds to a proximal diaphysis fragment from a tibia with a fresh spiral fracture. Specimen MHMOP/PI/636 is a complete left talus (Fig. 5.5a). Its morphology is rectangular and the trochlea is located in a proximal position and well defined by the lateral and medial crests. The sustentacular facet can be distinguished in plantar view and shows a rectangular morphology, convex cross section and a small cavity in its distal portion, which merges with the sustentacular fossa. Two well-defined condylus can be distinguished in the distal trochlea. In dorsal view, the edges of the proximal trochlea reach almost halfway across the

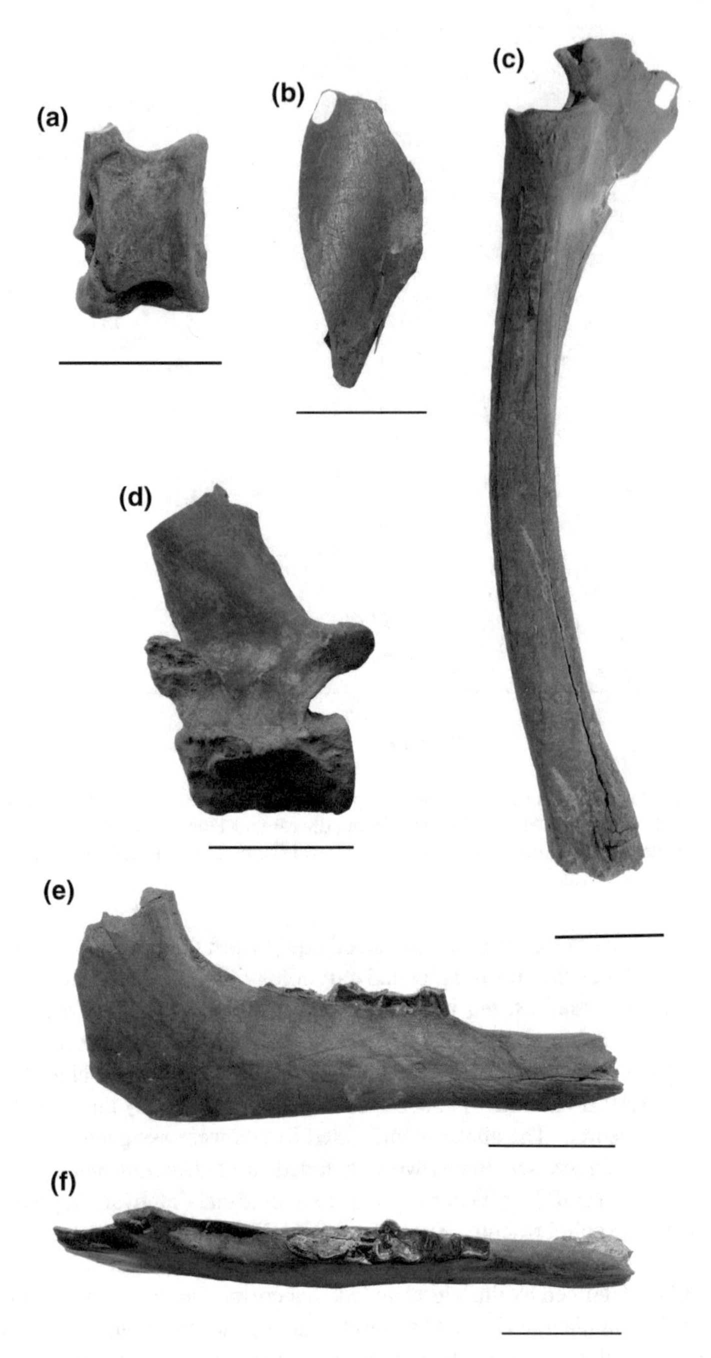

Fig. 5.5 Fossil camelid materials from Pilauco. **a** MHMOP/PI/636 left talus, plantar view; **b** MHMOP/PI/642 fragment from a humerus diaphysis; **c** MHMOP/PI/644 left radius–ulna, lateral view; **d** MHMOP/PI/643 lumbar vertebra, lateral view; **e** MHMOP/PI/641 right mandible, lateral view, **f** MHMOP/PI/641 right mandible, occlusal view. The length of the scale bar is 5 cm

Table 5.4 Measurements from talus specimen MHMOP/PI/636 in mm. Abbreviations according to L'Heureux (2008) and Scherer (2009). The bold numbers have been used for the Principal Component Analysis

Gl	Gb	LmT	GLm	Dl	SDl	Dm	Bd	Bp	CTT
56	**38**	32	51	**33**	**29**	30	**38**	**25**	**34**

specimen and coincide with the dorsal fossa. The specimen MHMOP/PI/641 (Fig. 5.5) corresponds to a right mandible consisting of a diastema, a body and the basal part of the ramus. It bears the complete dental series (p4–m3), which is significantly worn out and basically lacks dental figures. The m3 is fractured. The age estimation for tooth eruption and dental wearing proposed by Kaufmann (2009) suggests an age at death of this individual between 7 and 9 years. The wearing of the teeth is not homogeneous, as it is evident from a lateral perspective. The p4 is elevated relative to the remaining teeth, which suggests life loss of P4 or a certain malocclusion. The mandible body exhibits the caudal mental foramen between p4 and m1, whereas the mental foramen can be distinguished in the oral region close to the fracture. In medial view, the mandibular foramen aligns with the dental series. The MHMOP/PI/642 specimen (Fig. 5.5) corresponds to a humerus diaphysis fragment with a portion of the deltoid tuberosity. The specimen exhibits a fresh spiral fracture. MHMOP/PI/643 (Fig. 5.5) is a lumbar vertebra, possibly from the upper third portion. The dorsal part of the spine and the two transverse processes are missing. Finally, specimen MHMOP/PI/644 (Fig. 5.5) is an almost complete left radius–ulna. The olecranal tubercle, the anconeal process, and the distal articulate surface are missing. The diaphysis has a plano-convex section and the nutrient foramen can be observed in the proximal portion on the posteromedial side. The proximal articulation is characterized by a semicircular trochlea, which separates the lateral and medial surfaces through the capitulum of the radius located in its distal portion.

Based on the metric guidelines proposed by L'Heureux (2008) and Scherer (2009), a total of 10 measurements were obtained from the talus specimen MHMOP/PI/636 (Table 5.4). A Principal Component analysis on seven of them based on reference measurements from Scherer (2005), Scherer et al. (2007) and Labarca et al. (2013), suggest a slightly larger size for this specimen (Fig. 5.6).

5.5 Stratigraphic and Spatial Distribution

The Pilauco site can be spatially divided into two areas: an eastern sector excavated between 2007 and 2008 after the findings in 1986 and a western sector excavated between 2010 and 2017 (Chap. 16, this volume). They share similar stratigraphy with fossil remains restricted to layers PB-8 (13,813–12,725 cal yr BP) and PB-7 (16,552–14,458 cal years BP).

Table 5.5 Stratigraphic and spatial distribution of identified camelids and equids specimens (NISP) from the Pilauco site. In each column, the first number corresponds to camelids (Camelidae and cf. *H. paradoxa*) and the second number to equids (*Equus* sp. and *E. andium*)

Layer/Sector	East	West	ND	Total
PB-7	2/4	7/3	0/2	9/9
PB-8	2/4	2/3	0/1	4/8
ND	1/2	0/0	3/2	4/4
Total	5/10	9/6	3/5	17/21

ND No data

The distribution of fossil samples from each taxon across sectors and layers indicates some important variations that might be related to different depositional events (Table 5.5). The number of fossil equids versus camelids is higher in the east compared to the west and vice versa (Table 5.5). In terms of stratigraphy, the majority of fossil remains were recovered from layer PB-7 (NISP PB-7 = 18; NISP PB-8 = 12). The taxon distribution indicates higher abundance of equids versus camelids in layer PB-8, whereas their occurrence in layer PB-7 is rather similar (Table 5.5). Notably, *E. andium* was recovered only from layer PB-8 in the eastern sector (Table 5.1), whereas a larger size form assigned to *Equus* sp. has been found in layers PB-7 and PB-8 in the west (Table 5.1). This distribution might indicate a spatial and temporal segregation for these two morphotypes, however, such conclusion would be difficult to reconcile with the apparent stability of the geomorphological and palaeoenvironmental conditions of the site for this time period (see Chap. 9, this volume). According to Machado et al. (2017), these differences might rather reflect a variety within the population, without necessarily corresponding to two separated species.

More detailed observations reveal that within the PB-7 layer from the western sector a total of 10 camelid and equid specimens have been recovered, from which six are located 20–60 cm directly above gomphothere remains and four other specimens are associated with them (Table 5.1). This suggests at least two depositional events within the PB-7 layer—one coeval with and another one likely post-dating the main accumulation of gomphothere remains. The distribution of fossil equids and camelids within the PB-7 layer lacks spatial concentrations of bone remains. The fossil specimens are distributed across the entire studied area and do not reveal any anatomic coherence (Fig. 5.7).

5.6 Discussion and Conclusions

Some taphonomic traces have been reported mainly referring to fossil bones from cf. *H. paradoxa* (Labarca et al. 2013). These authors did not find any anthropic evidence on the studied fossils and restricted their studies exclusively on natural modifications. Labarca et al. (2013) report the presence of root etching, sedimentary abrasion, and

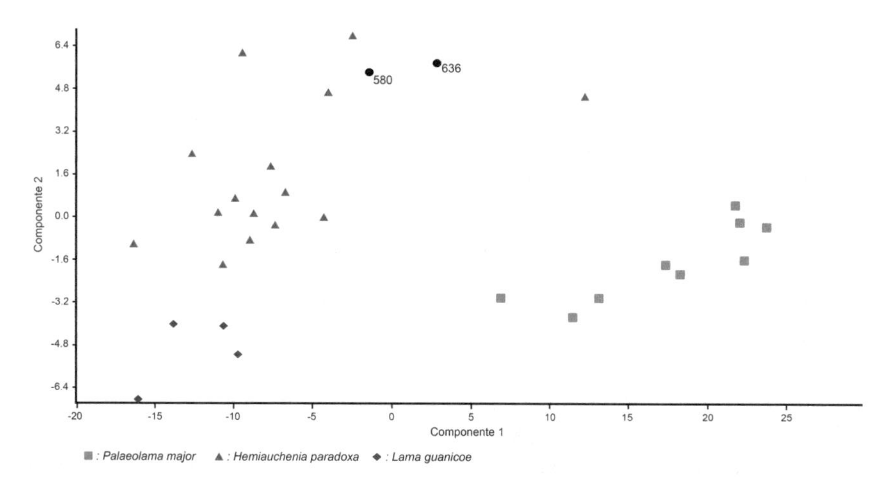

Fig. 5.6 Principal Component Analysis including the talus specimens MHMOP/PI/580 and/636. The first and second components accounts for 88% and 6% of the variance, respectively

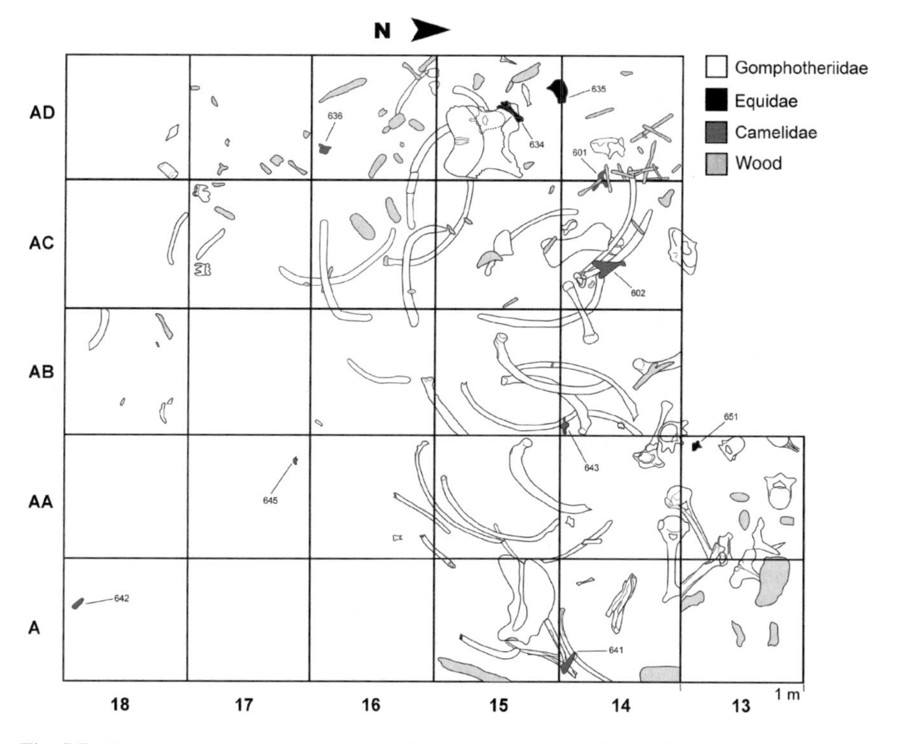

Fig. 5.7 Ground view of the eastern excavation sector at Pilauco, layer PB-7

trampling marks (Lyman 1994; Domínguez-Rodrigo et al. 2009). The abrasion marks might have been generated *in situ* due to a contact of the fossil specimens with abrasive material, or during its transport from other locations, probably the norther hill. The trampling would have been left by animals or humans (Moreno et al. 2019) inhabiting the area, pressing the fossil remains against the sedimentary matrix. This process might have contributed both to the horizontal and vertical displacement of fossil specimens. The identified marks are heterogeneously distributed between the samples, which is coherent with different taphonomic histories within the study site. The analysis of new samples from the western sector described here are partially coherent with previous studies and several samples show similar alterations. The trampling marks are visible on specimens MHMOP/PI/634/641/642/643/644 and /651, whereas the root imprints can be distinguished on specimens MHMOP/PI/634/635/641/644/643 and /651. Notably, in the western sector, no abrasion marks have been documented. The root marks and trampling can be found in both fossil-bearing layers given that the radius–ulna MHMOP/PI/644 from PB-8 shows similar modifications as specimens from PB-7. Labarca et al. (2014) described tooth marks on fossil gomphothere bones from a large extinct carnivore. Such marks have not been documented on camelid and equid bones, although the missing proximal and distal portions from radius–ulna MHMOP/PI/644 might have been left by a smaller sized carnivore preferring the parts of the bones, which are less dense (Binford 1981). This preliminary suggestion might also explain the missing distal portion from the metapodials MHMOP/PI/600 y/601 described by Labarca et al. (2013). A detailed study focusing on the analysis of the fracture surfaces of the fossil specimens might further test this hypothesis.

The fossil remains from equids in Pilauco constrain the presence of *E. andium* (Recabarren et al. 2011) to the PB-8 layer in the western sector of the site. In turn, the *Equus* materials from PB-7 and PB-8 layers in the eastern sector show larger dimensions, which points towards the presence of a large-sized form that has not been specifically assigned so far. If this hypothesis is confirmed, Pilauco would be the first South American site documenting the co-existence of two different *Equus* forms. Alternatively, and considering the review by Machado et al. (2017), these two morphotypes might also represent different metric expressions from the same population. On the other hand, almost all fossil remnants from camelids have been assigned to cf. *Hemiauchenia paradoxa*, and so far there are no materials that might indicate the presence of other forms. In terms of stratigraphy, the equid remnants are more abundant in the PB-8 layer, whereas within the PB-7 layer, the camelid and equid materials are rather equally represented. Within the PB-7 layer in the western sector, more than half of the materials assigned to these two families is located above the bulk gomphothere material suggesting at least two depositional events in this part of the site. The different taphonomic alterations and the lack of anatomic coherence are both in agreement with different depositional events for both families, including the possible transport of fossil material from the northern hill. Trampling have been registered in both layers and sectors, suggesting the indirect impact of animals and humans on the site might have contributed to the vertical and horizontal displacement of some fossil materials. The taphonomic analysis did not reveal any anthropic evidence on bones from camelids and equids.

References

Alberdi MT, Frassinetti D (2000) Presencia de Hippidion y Equus (Amerhippus) (Mammalia, Perissodactyla) y su distribución en el Pleistoceno superior de Chile. Est Geol 56(5–6):279–290

Alberdi MT, Prado JL (2004) Los caballos fósiles de América del Sur. Una historia de 3 millones de años. Universidad Nacional del Centro de la Provincia de Buenos Aires, INQUAPA: Serie monográfica 3

Alberdi MT, Prieto A (2000) Restos de Hippidion saldiasi en las cuevas de la Patagonia chilena. An del Inst de la Patagon 28:147–171

Alberdi MT, Cartelle C, Prado JL (2003) El registro Pleistoceno de Equus (Amerhippus) e Hippidion (Mammalia, Perissodactyla) de Brasil. Consideraciones paleoecológicas y biogeográficas. Ameghiniana 40:173–196

Alberdi MT, Prado JL, López P, Labarca R, Martínez I (2007) Hippidion saldiasi Roth 1899 (Mammalia Perissodactyla) en el Pleistoceno tardío de Calama. Rev Chil de Hist Nat 80:157–171

Ameghino F (1904) Recherches de Morphologie Phylognetique sur les molaires supèrieures des Ongulés. An Mus Nac 3:1–541

Binford R (1981) Bones: ancient men and modern myths. Academic Press, New York

Bird JB (1988) Travels and archaeology in South Chile. University of Iowa Press, Iowa City

Branco W (1883) Über eine fossile Säugetier-Fauna von Punin beo Riobamba in Ecuador. II: Beschreibung der Fauna. Pälaontologische Abhandlungen 1:39–204

Campbell KE, Frailey CD, Romero-Pittman L (2000) The late Miocene gomphothere Amahuacatherium peruvium (Proboscidea: Gomphotheriidae) from Amazonian Peru: implications for the Great American faunal interchange. Instituto de Geología, Minería y Metalurgia del Perú Serie D: Estudios Regionales, Boletín 23:1–152

Cartajena MI (2001) Los conjuntos arqueofaunísticos del Arcaico temprano en la Puna de Atacama, Norte de Chile. Dissertation, Berlin University, Germany

Cartajena MI, López P, Martínez I (2010) New camelid (Artiodactyla: Camelidae) record of the late Pleistocene of Calama (Second Region, Chile): a morphological and morphometric discussion. Rev Mex de Cien Geol 27(2):197–212

Cartajena MI, López P, Carabias D, Morales C, Vargas G, Ortega C (2013) First evidence of an underwater Final Pleistocene terrestrial extinct faunal bone assemblage from central Chile (South America): taxonomic and taphonomic analyses. Quat Int 305:45–55

Casamiquela R (1969) Enumeración crítica de los mamíferos continentales Pleistocenos de Chile. Rehue 2:143–172

Casamiquela R (1969–1970) Primeros documentos de la paleontología de vertebrados para un esquema estratigráfico y zoogeográfico del Pleistoceno de Chile. Bol Prehist Chile 2:65–73

Casamiquela R (1999) The Pleistocene vertebrate record of Chile. Quat S Am A 7:91–107

Casamiquela R, Chong G (1974) Icnitas (Mammalia, Equidae?) en rocas del Plio-Pleistoceno de la costa Provincia de Antofagasta (Chile). In: Abstracts of the Primer Congreso Argentino de Paleontología y Bioestratigrafía 2. Tucumán, 12–16 August 1974

Casamiquela R, Dillehay TD (1989) Vertebrate and invertebrate faunal analysis. In: Dillehay TD (ed) Monte Verde: a late pleistocene settlement in Chile. Smithsonian Institution Press, pp 205–210

Cione AL, Gasparini GM, Soibelzon L, Soibelzon LH, Tonni EP (2015) The great American biotic interchange: a South American perspective. Springer, Dordrecht

Daniel H (1948) Nociones de Geología y Prehistoria de Colombia. Medellín, p 360

Der Sarkissian C, Vilstrup JT, Schubert M, Seguinrlando A, Eme A, Weinstock J, Alberdi MT, Martin FM, López P, Prado JL, Prieto A, Douady C, Stafford TW, Willerslev E, Orlando L (2015) Mitochondrial genomes reveal the extinct Hippidion as an outgroup to all living equids. Biol Lett 11:20141058

Dillehay TD (1989) Monte Verde: a late pleistocene settlement in Chile. Palaeoenvironment and Site Context. Smithsonian Institution Press, Washington, DC

Dillehay TD (1997) Monte Verde: a late pleistocene settlement in Chile. The archaeological context and interpretation. Smithsonian Institution Press, Washington, DC

Domínguez-Rodrigo M, de Juana S, Galán AB (2009) A new protocol to differentiate trampling marks from butchery cut marks. J Archaeol Sci 36:2643–2654

Eisenmann V, Alberdi MT, De Giuli C, Staesche U (1988) Collected papers after the "New York International Hipparion Conference, 1981". In: Woodbrune M, Sondaar P (eds) Studying fossil horses 1, Methodology: 1–72. EJ Brill, Leiden, The Netherlands

Frassinetti D, Alberdi MT (2001) Los macromamíferos continentales del Pleistoceno superior de Chile: reseña histórica, localidades, restos fósiles, especies y dataciones conocidas. Estud Geol (Madr) 57(1–2):53–69

Fuenzalida H (1936) Noticia sobre los fósiles encontrados en la Hacienda Chacabuco, en Abril de 1929. Rev Chil Hist Nat 40:96–99

Gasparini GM, Reyes MD, Francia A, Scherer SC, Poire DG (2017) The oldest record of Hemiauchenia Gervais and Ameghino (Mammalia, Cetartiodactyla) in South America: comments about its paleobiogeographic and stratigraphic implications. Geobios 50(2):141–153

Gervais P (1855) Recherches sur les mammifères fossiles de l'Amérique Mèridionale. Chez P, Bertrand, Paris, p 63

Gervais P, Ameghino F (1880) Les Mammifères fossiles de l'Amerique du Sud. F. Savy-Igon Hnos. Paris-Buenos Aires, p 225

González E, Labarca R, Chávez-Hoffmeister M, Pino M (2014) First fossil record of the smallest deer cf. Pudu Molina, 1782 (Artiodactyla, Cervidae), in the Late Pleistocene of South America. J Vertebr Paleontol 34(2):483–488

Guérin C, Faure M (1999) Palaeolama (Hemiauchenia) niedae nov. sp., nouveau camelidae du nordeste brésilien et sa place parmi les lamini d´Amérique du Sud. Geobios 32(4):620–659

Hoffstetter R, Paskoff R (1966) Présence des genres Macrauchenia et Hippidion dans la faune Plêistocène du Chili. Bull Mus Nat d'Hist Nat, 2e Série 38:476–490

Honey JG, Harrison JA, Prothero DR, Stevens MS (1998) Camelidae. In: Janis CM, Scott KM, Jacobs LL (eds) Evolution of tertiary mammals of North America: Volume 1, Terrestrial carnivores, Ungulates, and Ungulatelike mammals, pp 439–462

Jackson D, Méndez C, Seguel R, Maldonado A, Vargas G (2007) Initial occupation of the Pacific coast of Chile during late pleistocene times. Curr Anthrop 48(5):725–731

Jiménez-Huidobro P, Sallabaerry M. (2015) Tetrápodos basales y anfibios fósiles de Chile. In: Rubilar-Rogers D, Otero R, Vargas A y Salaberry M (eds) Vertebrados fósiles de Chile. Publ ocas - Mus Nac Hist Nat 85–98

Kaufmann CA (2009) Estructura de Edad y Sexo en Lama guanicoe (guanaco). Estudios actualísticos y arqueológicos en pampa y Patagonia. Colección Tesis Doctorales, Sociedad Argentina de Antropología. Buenos Aires

L'Heureux GL (2008) El estudio arqueológico del proceso coevolutivo entre las poblaciones humanas y las poblaciones de guanaco en Patagonia meridional y norte de Tierra del Fuego. BAR International Series 1751, Oxford

Labarca R (2015) La meso y megafauna terrestre extinta del Pleistoceno de Chile. In: Rubilar-Rogers D, Otero R, Vargas A, Salaberry M (eds) Vertebrados fósiles de Chile. Publ ocas - Mus Nac Hist Nat 401–465

Labarca R (2016) La subsistencia de los cazadores recolectores de Patagonia meridional chilena durante la transición Pleistoceno – Holoceno: un enfoque integrador desde la zooarqueología. Dissertation, Universidad Nacional del Centro de la Provincia de Buenos Aires, Argentina

Labarca R, López P (2006) Los mamíferos finipleistocénicos de la Formación Quebrada Quereo: (IV Región-Chile): biogeografía, bioestratigrafía e inferencias paleoambientales. Mastoz Neotrop 13(1):89–101

Labarca R, Prieto A (2009) Osteometría de Vicugna vicugna Molina, 1782 en el Pleistoceno final de Patagonia meridional chilena: Implicancias paleoecológicas y biogeográficas. Rev del Mus de Antropol 2:127–140

Labarca R, Fuentes F, Mena F (2008) Los conjuntos faunísticos pleistocénicos de Cueva las Guanacas (Región de Aisén, Patagonia chilena): Alcances taxonómicos y tafonómicos. Magallania 36(2):123–142

Labarca R, Pino M, Recabarren O (2013) Los Lamini (Cetartiodactyla: Camelidae) extintos del yacimiento de Pilauco (Norpatagonia chilena): aspectos taxonómicos y tafonómicos preliminares. Estud Geol (Madr) 69(2):255–269

Labarca R, Recabarren O, Canales-Brellenthin P, Pino M (2014) The Gomphotheres (Proboscidea: Gomphotheriidae) from Pilauco site: scavenging evidence in the late Pleistocene of the chilean Patagonia. Quat Int 352:75–84

Latorre C (1998) Paleontología de mamíferos del Alero Tres Arroyos 1, Tierra del Fuego, XII Región, Chile. An Inst Patagonia, Serie Cs Nat 26:77–90

Liais E (1872) Climats, Geologie, Faune et Géographie Botanique du Brésil. Paris: Garnier Frères

Lira MP, Labarca R, Navarro-Harris X, Fritte D, Oyarce H, Pino M (2016) El Sitio Los Notros, Pleistoceno tardío, Osorno, norpatagonia occidental de Chile. In: Abstracts of the V Simposio de Paleontología en Chile, Concepción

López P (2007) Tafonomía de los mamíferos extintos del Pleistoceno tardío de la costa meridional del semiárido de Chile (IV Región-32° latitud S). Alcances culturales y paleoecológicos. Chungara Rev Antrop Chil 39(1):69–86

López P (2009) El mundo perdido de Patagonia central: Una aproximación tafonómica al estudio de los mamíferos extintos del sitio Baño Nuevo 1 (XI Región-Chile). In: López P, Cartajena MI, García C, Mena F (eds) Zooarqueología y tafonomía del confín del mundo. Monografías arqueológicas No 1, Universidad Internacional SEK-Chile, Facultad de estudios del patrimonio cultural, Área de Arqueología, pp 115–133

López P, Cartajena MI (2006) Hallazgos de mastodonte (Mammalia, Proboscidea) y un camélido extinto (Mammalia, Artiodactyla) en la comuna de Marchihue (VI Región). An Mus Hist Nat Valpso 25:72–78

López P, Rojas O, Mansilla P, Olivares L, Martínez I (2010) Mamíferos extintos del Pleistoceno de la Cuenca de Calama (Segunda Región, Chile). Viejas colecciones y nuevos hallazgos. Treb del Mus de Geol de Barc 17:11–25

López P, Cartajena MI, Carabias D, Prevosti FJ, Maldonado A, Flores-Aqueveque V (2016a) Reconstructing drowned terrestrial landscapes. Isotopic paleoecology of a late Pleistocene extinct faunal assemblage: Site GNL Quintero 1 (GNLQ1) (32° S, Central Chile). Quat Int 463:153–160

López P, Cartajena MI, Carabias D, Morales C, Letelier D Flores V (2016b) Terrestrial and maritime taphonomy: differential effects on spatial distribution of a late pleistocene continental drowned faunal bone assemblage from the Pacific coast of Chile. Archaeol Anthrop Sci 8:277–290

Lucas S, Kondrashov PE (2004) Early Eocene (Bumbanian) perissodactyls from Mongolia and their biochronological significance. New Mex Mus of Nat Hist and Sci Bull 26:215–220

Lund PW (1840) Nouvelles Recherches sur la Faune Fossile du Brésil. Ann Sci Nat 13:310–319

Lund PW (1846) Meddelelse af det Udbytte de 1844 Undersogte Knoglehuler Have Avgivet Til Kundskaben Om Brasiliens Dyreverden For Sidste Jordomvaeltning. Det Kongelige Danske Videnskabernes Selskabs Naturvidenskabelige Og Mathematisk Afhandlinger, Kjobenhavn 12:57–94

Lyman RL (1994) Vertebrate taphonomy. Cambridge University Press, Cambridge

Machado H, Grillo O, Scott E, Avilla L (2017) Following the footsteps of South American Equus: are the autopodia taxonomically informative? J Mamm Evol 25(3):397–405

Marín JC, Zapata B, González B, Bonacic C, Wheeler JC, Casey C, Bruford M, Palma E, Poulin E, Alliende MA, Spotorno A (2007) Sistemática, taxonomía y domesticación de alpacas y llamas: nueva evidencia cromosómica y molecular. Rev Chil Hist Nat 80:121–140

Martin FM (2013) Tafonomía de la Transición Pleistoceno-Holoceno en Fuego- Patagonia. Interacción entre humanos y carnívoros y su importancia como agentes en la formación del registro fósil. Ediciones de la Universidad de Magallanes, Punta Arenas

Martin FM, Tordisco D, Rodet J, San Román J, Morello F, Prevosti FJ, Stern Ch, Borrero LA (2015) Nuevas excavaciones en Cueva del Medio. Procesos de formación de la cueva y avances en los estudios de interacción entre cazadores recolectores y fauna extinta (Pleistoceno final, Patagonia meridional) Magallania 43(1):165–189

Massone M (1988) Artefactos óseos del yacimiento arqueológico Tres Arroyos (Tierra del Fuego). An del Instit de la Patagon, Ser Cienc Soc 18:107–112
Massone M (2004) Los Cazadores después del Hielo. Colección de Antropología 7, Centro de Investigación Barros Arana. Ediciones DIBAM
McFadden, BJ (1992) Fossil horses: systematics, paleobiology, and evolution of the family equidae. Cambridge University Press, Cambridge
McFadden BJ (2013) Dispersal of Pleistocene Equus (Family Equidae) into South America and calibration of GABI 3 Based on evidence from Tarija, Bolivia. PLoS ONE 8(3):e59277
McFadden BJ, Azzaroli A (1987) Cranium of Equus insulatus (Mammalia, Equidae) from the middle Pleistocene of Tarija, Bolivia. J Vert Paleontol 7:325–334
McFadden BJ, Shockey BJ (1997) Ancient feeding ecology and niche differentiation of pleistocene mammalian herbivores from Tarija, Bolivia: morphological and isotopic evidence. Paleobiology 23:77–100
Menegaz AN (2000) Los camélidos y cérvidos del cuaternario del sector bonaerense de la Región Pampeana. Unpublished, Dissertation, University La Plata
Menegaz AN, Goin, F. Ortíz-JaureguÍzar, E (1989) Análisis morfológico y morfométrico multivariado de los representantes fósiles y vivientes del género Lama (Artiodactyla, Camelidae). Sus implicancias sistemáticas, biogeográficas, ecológicas y biocronológicas. Ameghiniana 26(3–4):153–172
Metcalf L, Turney C, Barnett R, Martin FM, Bray SC, Vilstrup JT, Orlando L, Salas-Gismondi R, Loponte D, Medina M, De Nigris M, Civalero T, Fernández PM, Gasco A, Duran V, Seymour KL, Otaola C, Gil A, Paunero R, Prevosti FJ, Bradshaw CJA, Wheeler JC, Borrero LA, Austin JJ, Cooper A (2016) Synergistic roles of climate warming and human occupation in Patagonian megafaunal extinctions during the Last Deglaciation. Sci Adv 2(6):e1501682
Missiaen P, Quesnel F, Dupuis C, Storme J-Y, Smith T (2013) The earliest Eocene mammal fauna of the Erquelinnes Sand Member near the FrenchBelgian border. Geol Belg 16:262–273
Molina GI (1782) Saggio sulla storia naturale del Chili. Bologna. (Aquino)
Montellano M (1989) Pliocene Camelidae of Rancho El Ocote, Central Mexico. J Mamm 70:359–369
Moreno K, Bostelmann E, Macías C, Navarro-Harris X, De Pol-Holz R, Pino M (2019) A late pleistocene human footprint from the Pilauco archaeological site, Northern Patagonia, Chile. PLOS ONE in press
Müller, PLS (1776) Erste Classe, Säugende Thiere. Des Ritters Carl von Linné vollständiges Naturalsystem nach der zwölften Lateinischen Ausgabe
Nami HG (1987) Cueva del Medio: perspectivas arqueológicas para la Patagonia Austral. An Inst Patagonia 17:73–106
Nami HG, Menegaz AN (1991) Cueva del Medio: aportes para el conocimiento de la diversidad faunística hacia el Pleistoceno-Holoceno en Patagonia Austral. An Inst Patagonia 20:17–132
Núñez L, Varela J, Casamiquela R (1983) Ocupación Paleoindio en Quereo. Reconstrucción multidisciplinaria en el territorio semiárido de Chile. Universidad del Norte, Antofagasta
Núñez L, Varela J, Casamiquela R, Villagrán C (1994a) Reconstrucción multidisciplinaria de la ocupación prehistórica de Quereo, centro de Chile. Latin Am Antiq 5:99–118
Núñez L, Varela J, Casamiquela R, Schiappacasse V, Niemeyer H, Villagrán C (1994b) Cuenca de Taguatagua en Chile: el ambiente del Pleistoceno superior y ocupaciones humanas. Rev Chil Hist Nat 67:503–519
Oliver C (1926) Lista preliminar de los mamíferos fósiles de Chile. Rev Chil Hist Nat 30:144–156
Oliver C (1934) Los hallazgos de restos de caballos fósiles de Chile. Rev Universitaria 19(4):541–553
Oliver C (1935) Mamíferos fósiles de Chile. Adiciones y correcciones a una lista preliminar. Rev Chil Hist Nat 39:297–304
Orellana M (1965) Informe de la primera fase del proyecto arqueológico Río Salado. Antropología 3:81–117

Orlando L, Metcalf JL, Alberdi MT, Telles-Antunes M, Bonjean D, Otte M, Martin FM, Eisenmann V, Mashkour M, Morello F, Prado JL, Salas-Gismondi R, Shockey BJ, Wrinn PJ, Vasil'ev SK, Ovodov ND, Cherry MI, Hopwood B, Male D, Austin JJ, Hänni C, Cooper A (2009) Revising the recent evolutionary history of equids using ancient DNA. Proc Nat Acad of Sci 106(51):21754–21759

Orlando L, Ginolhac A, Zhang G, Froese D, Albrechtsen A, Stiller M, Schubert M, Cappellini E, Petersen B, Moltke I, Johnson PLF, Fumagalli M, Vilstrup JT, Raghavan M, Korneliussen T, Malaspinas AS, Vogt JF, Szklarczyk D, Kelstrup ChD, Vinther J, Dolocan A, Stenderup J, Velazquez AM, Cahill J, Rasmussen M, Wang, X, Min J, Zazula GD, Seguin-Orlando A, Mortensen A, Magnussen K, Thompson JF, Weinstock J, Gregersen K, Røed KH, Eisenmann V et al (2013) Recalibrating equus evolution using the genome sequence of an early Middle Pleistocene horse. Nature 49974–49978

Prado JL, Alberdi MT (2008) Restos de Hippidion y Equus (Amerhippus) procedentes de las barrancas de San Lorenzo, Pleistoceno tardío (provincia de Santa Fe, Argentina). Rev Esp Paleontol 2:225–236

Prado JL, Alberdi MT (2012) Équidos y Gonfoterios del Pleistoceno tardío de San Pedro, provincia de Buenos Aires, Argentina. Est Geol 68(2):261–276

Prado JL, Alberdi MT (2014) Global evolution of equidae and gomphotheriidae from South America. Integr Zool 9:434–443

Prado JL, Alberdi MT, Reguero MA (1998) El registro más antiguo de Hippidion Owen, 1869 (Mammalia, Perissodactyla) en América del Sur. Est Geol 54 (1–2):85–91

Prevosti FJ, Forasiepi A, Zimicz N (2013) The evolution of the cenozoic terrestrial mammalian predator guild in South America: competition or replacement? J Mamm Evol 20(1):3–21

Prieto A (1991) Cazadores Tempranos y Tardíos en Cueva del Lago Sofía 1. An Inst Patagonia 20:75–99

Prothero DR, Campbell KE, Beatty BL, Frailey CD (2014) New late Miocene dromomerycine artiodactyl from the Amazon basin: implications for interchange dynamics. J Paleontol 88:434–443

Recabarren O, Pino M, Cid I (2011) A new record of equus (Mammalia: Equidae) from the late pleistocene of central-south Chile. Rev Chil Hist Nat 84:535–542

Recabarren O, Pino M, Alberdi MT (2014) La Familia Gomphotheriidae en América del Sur: evidencia de molares al norte de la Patagonia chilena. Est Geol 70(1):1–11

Rincón AD, Alberdi MT, Prado JL (2006) Nuevo registro de Equus (Amerhippus) santaeelenae (Mammalia, Perissodactyla) del pozo de asfalto de Inciarte (Pleistoceno Superior), estado Zulia, Venezuela. Ameghiniana 43:529–538

Roth S (1899) Descripción de los restos encontrados en la caverna de Ultima Esperanza. Revista Mus La Plata 9:421–453

Salinas P, Naranjo JA, Marshall LG (1991) Nuevos restos del perezoso gigante (Megatheriidae, Megatherium medinae) de la Formación Chiu-Chiu, Cuenca del Río Loa, Calama, Norte de Chile. In: Abstracts of the VI Congreso Geológico Chileno 1(6):306–309

Sallaberry M, Soto-Acuña S, Yury-Yáñez R, Alarcón J, Rubilar-Rogers D (2015). Aves fósiles de chile. In: Ei: Rubilar-Rogers, D, Otero R, Vargas A, Salaberry M (eds) Vertebrados fósiles de Chile. Publ ocas - Mus Nac Hist Nat, pp 265–292

Scherer CS (2005) Estudo dos Camelidae (Mammalia, Artiodactyla) do Quaternário do Estado do Rio Grande do Sul. Dissertation, Universidade Federal do Rio Grande do Sul

Scherer CS (2009) Os Camelidae Lamini (Mammalia, Artiodactyla) do Pleistoceno da América do Sul: Aspectos taxonômicos e filogenéticos. Dissertation, Universidade Federal do Rio Grande do Sul

Scherer CS (2013) The Camelidae (Mammalia, Artiodactyla) from the quaternary of South America: cladistic and biogeographic hypotheses. J Mamm Evol 20:45–56

Scherer CS, Ferigolo J, Ribero AM, Cartelle C (2007) Contribution to the knowledge of Hemiauchenia paradoxa (Artiodactyla, Camelidae) from the Pleistocene of southern Brazil. Rev Bras Paleontolog 10:35–52

Spillmann F (1938) Die fossilen Pferde Ekuadors der Gattung Neohippus. Palaeobiologica 6:372–393
Woodburne MO, Cione AL, Tonni, EP (2006) Central American provincialism and the great American biotic interchange. In: Carranza-Castañeda Ó, Lindsay EH (eds) Advances in late tertiary vertebrate paleontology in Mexico and the great American biotic interchange, vol 4. Univ Nac Aut Mex. Inst Geol Centro Geoscien Publ Espec, pp 73–101
Villavicencio N, Lindsey EL, Martin FM, Borrero LA, Moreno PI, Barnosky AD (2015) Combination of humans, climate, and vegetation change triggered Late Quaternary megafauna extinction in the Última Esperanza region, southern Patagonia, Chile. Ecography 38:125–140
Webb SD (1991) Ecogeography and the great American interchange. Paleobiology 17:266–280
Weinstock J, Shapiro B, Prieto A, Marín JC, González BA, Gilbert MT, Willerslev E (2009) The late pleistocene distribution of vicuñas (Vicugna vicugna) and the 'Extinction' of the gracile Llama ('Lama gracilis'): new molecular data. Quat Sci Rev 28(15–16):1369–1373
Woodburne MO (2010) The great American biotic interchange: dispersals, tectonics, climate, sea level and holding pens. Mamm Evol 17:245–264

Chapter 6
Sporormiella Fungal Spores as a Proxy for Megaherbivore Abundance and Decline at Pilauco

Mario Pino, Nathalie Cossio-Montecinos and Benazzir Pinto

Abstract The study of coprophilous fungal spores associated with herbivorous faeces has been used to determine, among others, the presence, abundance and decrease of the megafauna populations on land at the end of the Pleistocene. *Sporormiella* sp. is the most abundant spore species in pollen samples, and it is exclusively restricted to faeces from domestic and wild herbivores. The present study encompasses the analysis and interpretation of *Sporormiella* sp. concentrations from two sediment columns with different resolution from the archaeo-paleontological Pilauco site. In both cases, the concentration of *Sporormiella* sp. reaches maximum values up to ~1.920 spores per cm^{-3} within the sedimentary layers PB-7 and PB-8, followed by a decline at the base of PB-9 layer corresponding to 12.800 cal. year BP. The disappearance of *Sporormiella* sp. across the PB-8/PB-9 erosional unconformity might be explained by a local decline of the megafauna producing this spore at Pilauco. Additional proxies and records are needed to further confirm the regional extinction of megafauna towards the end of the Pleistocene in north-western Chilean Patagonia.

Keywords Coprophilous fungal spores · Pilauco

6.1 Background

The end of the upper Pleistocene is marked by the last massive extinction of megafauna (see Chap. 2, this volume), which is concurrent with a cold time period referred to as the Younger Dryas. This event is characterized by the disappearance of at least 2/3 of all megafauna species (Rozas-Dávila et al. 2016; van der Kaars et al. 2017). Barnosky and Lindsay (2010) registered the extinction of at least 52

M. Pino (✉)
Instituto de Ciencias de la Tierra and Transdisciplinary Center for Quaternary Research in the South of Chile, Universidad Austral de Chile, Valdivia, Chile
e-mail: mariopinoquivira@gmail.com

N. Cossio-Montecinos · B. Pinto
LabNat Pilauco, Facultad de Ciencias, Universidad Austral de Chile, Valdivia, Chile

M. Pino and G. A. Astorga (eds.), *Pilauco: A Late Pleistocene Archaeo-paleontological Site*, The Latin American Studies Book Series,
https://doi.org/10.1007/978-3-030-23918-3_6

and 34 megafauna genera in South and North America, respectively. The presence of megafauna (weight > 1000 kg, sensu Haynes and Klimowicz 2015) towards the end of the Pleistocene has been studied with different methods including the identification of life traces from extinct species, datings of extinction periods, modelling of human and faunal population sizes, among others (Haynes 2009).

Currently, there are four different hypotheses aimed to explain the late Pleistocene massive extinction event. Frequently suggested and widely discussed scenarios include the notorious climate change at the end of the Pleistocene (i.e. Wroe et al. 2013), and the overhunting by humans associated with their expansion and surviving strategies (i.e. Fiedel and Haynes 2004). Many studies conclude that the combination of both, the arrival of humans and climate change, was the main trigger for the extinction of megafauna (Koch and Barnosky 2006; Lorenzen et al. 2011; Araujo et al. 2017). These factors caused significant changes in vegetation associations (Villavicencio et al. 2016) resulting in disequilibrium between different taxa. According to these authors, the extinction of mega-carnivores might be the result of their possible competition with humans. In turn, the coexistence of humans, horses, camelids and mylodons over thousands of years rather discards overhunting as a causal factor.

The remaining hypotheses focus on diseases transferred by humans (MacPhee and Marx 1997; Scott 2010), and more recently, on an asteroid impact that caused significant environmental changes triggering the Younger Dryas cold period (Firestone et al. 2007). This event occurred between 12,800 and 11,600 cal. year BP and lasted between 1,150 and 1,300 years according to results from different Greenland ice cores (Alley 2000). An alternative hypothesis suggests that cooling climate was related to the complete drainage of the palaeo-glacial Lake Agassiz (central North America) and associated suppression of thermo-haline circulation. Nonetheless, this idea still lacks sufficient geological evidence to support it (Broecker 2006). The Younger Dryas is mainly reflected by climate change in the North Atlantic realm, but there are also some evidences from the Southern Hemisphere. Some of them derive from the Central Andes and the Amazon Basin, but most of the glacial, palynological, and paleoentomological evidences originate from Chile and Argentina (Björck 2007).

There are a variety of environmental indicators commonly applied by recent studies, including i.e. the study of carbon and nitrogen isotopes, pollen and coal. The non-pollen palynomorphs in microscopic samples, including spores from coprophilous fungi associated with herbivorous faeces, have recently gained in importance for paleoecological and paleontological studies (Etienne and Jouffroy-Bapicot 2014). This method has been described and applied in the study of the presence, abundance and decline of megafauna populations on several continents (Davis 1987; Ficcarelli et al. 2003; Comandini and Rinaldi 2004; Davis and Shafer 2006; van Geel and Aptroot 2006; Ekblom and Gillson 2010; Feranec et al. 2011; Parker and Williams 2012).

Coprolites correspond to fossil faeces found in sediments, and potentially provide first-order information about dietary habits of their producers. Coprolites of herbivores conserve pollen, but also obligatory coprophilous fungi, which require large mammals or birds to complete their life cycle. The spores are ingested by the herbivorous animal while feeding on plants. They survive the digestive tract and subsequently germinate within the dung to become preserved over time. Many of these spores are being released, deposited and finally conserved in the sediments. The study of Baker et al. (2013) evaluated the three coprophilous spores considered as most reliable proxies for the activity of large herbivores: *Sporormiella* sp., *Sordaria* sp. type and *Podospora* sp type. *Sporormiella* sp. has been most frequently used to study the presence and decline/extinction of Pleistocene megafauna, as well as grazing activities from the Holocene. The widespread application of this spore is due to its frequent abundance in pollen samples, as well as the distinctive morphology of its cells (van Geel and Aptroot 2006).

Sporomiaceae family and the genus *Sporormiella* sp. belongs to the Ascomycota Phylum. This is the largest phylum of fungi with well over 33,000 identified/denominated species and many others still to be described. The ascus is the common feature among all Ascomycota species. These asci are contained in the ascomata/fruiting body and produce ascomycete sexual spores involved in sexual reproduction among some of the species. The shape and size of the asci and the spores they release vary between species. The most distinctive feature of these spores is their obligatory coprophilous character. They are typically found in faeces from domestic and wild herbivorous animals, both in cold and temperate climate environments. The spores are usually dark brown and have a pronounced sigmoid germination pore (Ahmed and Cain 1972; Mungai et al. 2012).

The correlation between the presence of fossil *Sporormiella* sp. in sediments and their association with faeces from extinct megaherbivores was established after the discovery of a Pleistocene dung layer in the Bechan cave, southern Utah (Davis 1987; Davis and Shafer 2006). Following the pioneer studies by Davis and co-authors, the variation in the concentration of these spores in Pleistocene and Holocene sedimentary sequences has been repeatedly used as a proxy for the presence, decline and extinction of megafauna on the continents (Koch and Barnosky 2006; Johnson et al. 2015).

In the Americas, several studies document high abundance of spores during the late Pleistocene, low abundance during most of the Holocene, and a renewed pronounced increase associated with the introduction of large grazing animals by the first European conquerors (Borrero et al. 1998; Barnosky et al. 2004; Koch and Barnosky 2006; Borrero 2008). Therefore, the disappearance of megafauna from the fossil record is temporally correlated with the decline in *Sporormiella sp.* spores in sediments (Davis 1987; Davis and Shafer 2006; Burney and Flannery 2005; Graf and Chmura 2006; Gill et al. 2013; Ekblom and Gillson 2010; Feranec et al. 2011; Raper and Bush 2009; Wood and Wilmshurst 2013). This type of fungi is generated during the live period of the producing animals and can be considered as evidence for the continuous presence of herbivores. Furthermore, the production of spores is proportional to the excrement biomass, i.e. large quantities and greater diversity of megaherbivores would be reflected by higher concentrations of fungi (Davis and Shafer 2006; Parker and Williams 2012; Etienne et al. 2013; Gill et al. 2013).

The spores are characterized by relatively low degree of dispersion. The regional transport occurs through the herbivore itself, when migrating and defecating directly above the ground. On the other hand, the spores can be transported directly from the faeces across relatively short distances and by different transport media (Feranec et al. 2011; Baker et al. 2013, 2016). The abundance of spores in the faeces and their correlation with contemporaneous herbivore concentrations need to be calibrated by considering a variety of other factors such as biological, taphonomic and physical variables. Nonetheless, the relationship proves to be unambiguous and characterized by a significant degree of correlation (Baker et al. 2016). *Sporormiella* sp. has been used in different studies worldwide to analyse the presence/abundance of megafauna and its decline. These include the studies of moa from New Zealand, giant turtles from the Galapagos Islands, giant lemurs, elephant birds and dwarf hippopotamus from Madagascar (Wood et al. 2008; Raczka et al. 2016).

6.2 The Paleontological and Chronostratigraphic Setting at Pilauco

The sedimentary record at Pilauco is marked by a clear erosional unconformity between the PB-8 and PB-9 layers. The unconformity has been interpreted to reflect significant environmental changes and is characterized by an abrupt decrease of pollen and increase of coal contents above it (Pino et al. 2019). Megafauna bones and coprolites can be found until a couple of centimetres below the unconformity, which has been dated as the base of the Younger Dryas chronozone (Chap. 15, this volume). The youngest fossil element at Pilauco corresponds to a coprolite sample dated at 11,004 ± 186 ^{14}C years BP (12,870 cal. year BP). At the adjacent archaeo-paleontological site Los Notros, an *Antifer* sp. antler sample dated at 10,860 ± 60 ^{14}C years BP (12,720 cal. year BP) has been registered immediately below the unconformity. The oldest fossil sample is a gomphothere tusk (Chap. 14, this volume) dated at 13,585 ± 81 ^{14}C years BP (16,310 cal. year BP) accompanied by at least 7 other fossil bone samples with an average age of 15,840 ± 44 cal. year BP. Consequently, the megafauna and human inhabitants at Pilauco and Los Notros sites were occupying the same space for at least 3,500 years.

The recovering of fossil megafauna bones from an area encompassing merely some dozens of square metres is a challenging task and depends on chance, as well as on a series of taphonomic factors. For example, the dispersion of *Sporormiella* sp. essentially depends on the transport media, such as wind and water, but its deposition in the sediments is not random. Since there are no records of megafauna bones in the PB-9 layer, the PB-8/PB-9 unconformity might be temporally associated with the local decline/extinction of megafauna. In this case, it should be also associated with a drastic decrease or total disappearance of *Sporormiella* sp. in the PB-9 sedimentary layer located immediately above the unconformity. Based on this hypothesis, the present study aims to evaluate the potential of *Sporormiella* sp. as a proxy to track the extinction of megafauna at Pilauco.

6.3 Materials and Methods

From a total of 300 available samples, each of 1 cm thickness (column 14AD), we prepared 76 sections (Silva 2014). For this purpose, a small drop of previously homogenized material was extracted from each vial of packed subsamples. Each drop was placed and uniformly distributed on a specimen slide and covered with a cover glass. Subsequently, the samples were sealed with transparent nail polish in order to avoid loss of material along the sides of the section margins.

In a second step, 17 samples (1 cm thickness each) were extracted from the column located in the grid 10AD. This sedimentary column encompasses only the top of layer PB-8 and the bottom of layer PB-9 (540–557 cm of local elevation). Each sample was processed for traditional counting of pollen and micro-coal elements by removing all organic matter, cellulose, minerals and clay (Faegri and Iversen 1989; see Chap. 9, this volume). For both types of samples, the counting included spores from *Sporormiella sp.* and *Lycopodium*. *Lycopodium* spores are always added to pollen and spore samples to calibrate the concentration (see Chap. 9, this volume).

Additionally, we estimated the concentration of *Sporormiella* sp. from aliquots previously obtained from four coprolite samples (MHMOP/PI/585, MHMOP/PI/586, MHMOP/PI/587 and MHMOP/PI/588; see Chap. 13, this volume) applying the first methodology described above. In general, all counting was realized on an optical microscope at ×400-magnification and after confirming the distinctive morphology of the counted spores, i.e. the brown/dark brown colour and the pronounced sigmoid germination pore (Ahmed and Cain 1972; Mungai et al. 2012). The *Sporormiella* sp. concentration (spore cm^{-3}) in each sample was calculated with the software Tilia 2.0.41.

6.4 Results

Figures 6.1 and 6.2 illustrate the general characteristics of columns 14AD and 10AD, respectively. Table 6.1 indicates the corresponding local elevation for all 76 samples (column 14AD) and the number of counted *Lycopodium* and *Sporormiella* sp. spores. In this column, the limit between layers PB-8 and PB-9 corresponds to 535 cm of local elevation, i.e. between samples 533 and 537. The minimum and maximum counts for *Lycopodium* was 136 and 2228, respectively.

The number of counted spores from *Sporormiella* sp. varies between 0 and 32. The samples between 355 and 388 cm of local elevation correspond to layer PB-7 and yield the highest numbers of spores coinciding with the highest concentration of gomphothere remnants (*Notiomastodon* aff. *N. platensis*, see Chaps. 4 and 8, this volume). Another maximum occurs in layer PB-8 between 517 and 529 cm, where the gomphotheres are replaced by horse fossils (*Equus* (*Amerhippus*) *andium*).

Fig. 6.1 Stratigraphic section of column 14AD with top-to-bottom layers PB-1, PB-6, PB-7, PB-8, and PB-9 (for details see Chap. 3, this volume). The irregularities on the outcrop face are due to the extraction of sediment columns for different analyses

Figure 6.3 shows the temporal variability of *Sporormiella* sp. concentrations. The highest concentrations of spores (>1500 spores cm^{-3}) are registered in the basal centimetres of layer PB-7 and at the top of layer PB-8, and the spores were present in all samples from these layers, except at 404 cm (Table 6.1).

The unconformity PB-8/PB-9 in column 14AD is located at 543 cm local elevation. Table 6.2 shows the number of *Sporormiella* sp. and *Lycopodium* spores between 540 and 557 cm. Within this interval, the *Sporormiella* sp. spores are abundant in the first few centimetres (540, 541, and 542) corresponding to layer PB-8, and subsequently decrease to zero at the unconformity (543 cm). Figure 6.4 shows the *Sporormiella* sp. concentration between 540 and 557 cm local elevation with similar distribution, i.e. higher concentrations in the first few centimetres (PB-8) and subse-

Fig. 6.2 Sedimentary column 10AD with layer PB-8 (brown) and the bottom and overlying layer PB-9 (dark-olive). The white arrow indicates the discordance between these layers

quent decrease to zero at 543 cm. Figure 6.5 shows some examples of *Soporomiella* sp. spores from samples from column 10AD.

Only three of the coprolite samples from Pilauco contain *Sporormiella sp.* spores in low concentrations; 35 spores cm^{-3} from sample MHMOP/PI/588) (PB-7 layer), and 65/72 spores cm^{-3} from samples MHMOP/PI/587 and/586 (PB-8 layer). Sample MHMOP/PI/585 did not contain *Sporormiella sp.* spores.

6.5 Discussion and Conclusions

The analysis of the 14AD column indicates the presence of *Sporormiella sp.* spores in all sedimentary layers (PB-7, PB-8, and PB-9). The fossil samples from *Notiomastodon* aff. *N. platensis* are mainly associated with layer PB-7, whereas *cf. Hemiauchenia paradoxa* (Labarca et al. 2013) and *Equus (Amerhippus) andium* (Recabarren et al. 2011; see this chapter, this volume) dominate in layer PB-8. This coincidence would explain the high concentrations of *Sporormiella* sp. spores in both layers as associated with the presence of megafauna.

Nonetheless, the content of fossil megafauna is much higher in layer PB-7 and generally variable throughout the sedimentary record (see Chaps. 4, 5 and 8, this volume). Consequently, it could be expected that the concentrations of associated spores would follow similar trends. Nonetheless, the highest *Sporormiella* sp. concentrations in layer PB-7 and PB-8 are similar (Figs. 6.3 and 6.4), despite the different fossil content. Notably, the record of fossil bone samples at the site depends on the taphonomic and cultural conditions, but also on some random factors, particularly considering that the site has been only partially excavated so far. Generally, the record

Table 6.1 Local elevation and number of *Lycopodium* y *Sporormiella* sp. spores in the 76 samples analysed in column 14AD. Layer PB-7 (brown), PB-8 (green), and PB-9 (grey)

elevation	*Lycopodium*	*Sporormiella* sp.
629	310	12
627	275	3
625	381	3
621	327	0
617	424	0
613	748	3
609	198	2
605	469	2
601	320	0
597	606	0
593	386	0
589	378	0
585	168	0
581	136	0
577	327	0
573	241	0
569	137	0
565	378	0
561	335	0
557	330	0
553	462	0
549	443	0
545	427	0
541	684	0
537	521	0
533	569	8
529	317	16
525	331	15
521	213	6
517	336	14
513	380	7
509	327	9
505	343	5
501	341	5
497	470	5
493	321	2
489	530	7
485	202	4
481	364	2
476	255	1
472	314	7
468	488	13
464	403	3
460	204	2
456	525	5
452	632	4
448	757	7
444	480	5
440	447	9
436	806	6
432	483	4
428	367	9
424	547	6
420	546	7
416	642	1
412	535	4
408	415	1
406	658	3
404	282	0
400	446	1
396	1752	6
392	1917	7
388	1445	10
384	1009	3
380	832	9
375	1239	9
371	854	5
367	754	12
363	1899	32
359	1292	10
355	1523	14
351	1712	6
347	756	6
343	843	6
339	675	14
335	2288	3

derived from spore concentrations is likely to be more representative and reliable, because of their potential for spatial dispersion.

The deposition rate in Pilauco averages ~13 year cm^{-1} for layer PB-7, and ~17 year cm^{-1} for layer PB-8. A significant increase up to ~0.5 year cm^{-1} occurs immediately above the PB-8/PB-9 unconformity, followed by a notorious decrease down to ~28 year cm^{-1} within the PB-9 layer (Pino et al. 2019). The significant difference in average deposition rates between PB-8 and PB-9 layers, could be reflected by the concentration of spores, but it can not explain the absolute absence of *Sporormiella* sp. over the unconformity.

In both analysed columns (14AD and 10AD), the maximum concentrations of *Sporormiella* sp. spores are close to 2000 spores cm^{-3} and occur close to the PB-8/PB-9 erosional unconformity. The recurrence of the spore in layer PB-9 in

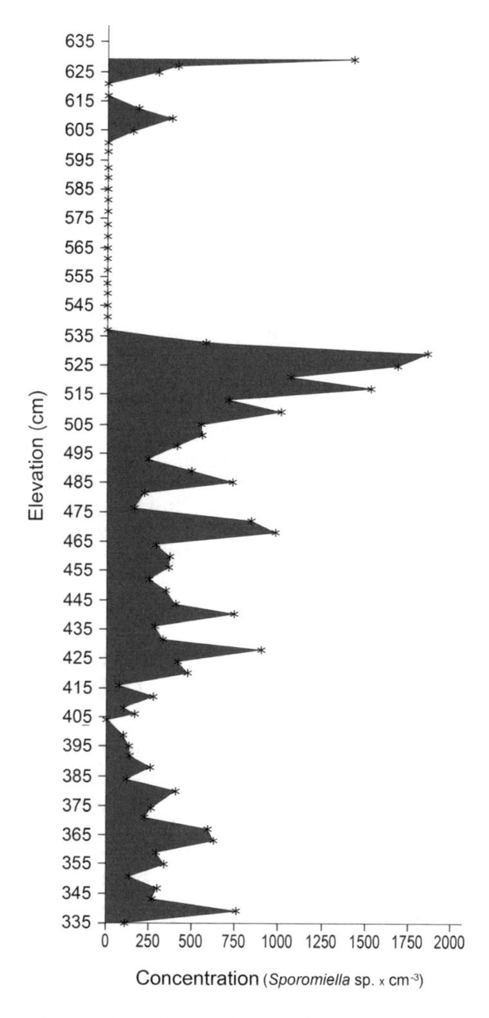

Fig. 6.3 Concentration of spores from *Sporormiella* sp. between layers PB-7 and PB-9. The spores disappear at 533 cm local elevation, which corresponds to the erosional unconformity between layers PB-8 and PB-9. Resembling the Lazarus effect (Wignall and Benton 1999), the spores appear again at the top of layer PB-9

Holocene times is also concurrent, although at different stratigraphic levels (Figs. 6.3 and 6.4).

Studies of modern settings commonly apply pollen and spores' traps at terrestrial sites, for example using Bison manadas (Gill et al. 2013). It has been confirmed that the trap records identify rather local sources for both microfossils, whereas the signal derived from lakes and wetland has a much more regional character (Levetin et al. 2000; Gill et al. 2013). According to Gill (2014), the strong potential of *Sporormiella* sp. as a proxy for bison populations rather discards the possibility that Sporormiella

Table 6.2 Local elevation (cm) and number of *Lycopodium* and *Sporormiella* sp. spores in the 17 samples from the 10AD column. Layer PB-8 (green) and PB-9 (grey)

elevation	*Lycopodium*	*Sporormiella* sp.
557	257	6
556	196	7
555	211	5
554	256	3
553	160	0
552	231	0
551	193	3
550	187	1
549	204	0
548	167	0
547	191	0
546	177	0
545	91	0
544	100	0
543	186	1
542	116	6
541	204	5
540	388	15

fluctuations were driven by the diversity of smaller mammalian herbivores, such as lagomorphs in North America (Feranec et al. 2011). Small herbivores in Pilauco (like pudu and coipo, González et al. 2014; Chap. 7, this volume) appear to rather set the *Sporormiella* sp. background level, against which a megafaunal signal can be detected. A threshold value of 2.8% (related to the total pollen + spores' concentration) has been proposed to distinguish between traps in ungrazed vs grazed megaherbivore sites (Gill 2014), which is consistent with the 2% value previously proposed by Davis (1987). The experiments with bison dung identify higher concentrations of *Sporormiella* sp. (16%) in the dung compared to those obtained from traps (Gill et al. 2013). Similar abundances have been recorded for *Sporormiella* sp in mammoth coprolites (Davis 1987). In Pilauco, the average percentage of spores between 540 and 543 cm is 2.3% (S.D. 0.95) (Fig. 6.4). In the Holocene section above the unconformity (550 and 551 cm) the values are higher (8.3 and 31.9%), probably due to the very low pollen content.

The concentration of spores has also been studied based on the variable abundance of beef cattle around a small lake serving as their watering place (Raczka et al. 2016). The concentrations of *Sporormiella* sp. in the sediments were high (between 1.200 and 6.740 spores cm^{-1}) on spots with increased densities of drinking animals (more than 40 animals per day), and much lower (between 0 and 340 spores cm^{-1}), where animal presence was scarce (less than 5 animals per day). The concentration of spores

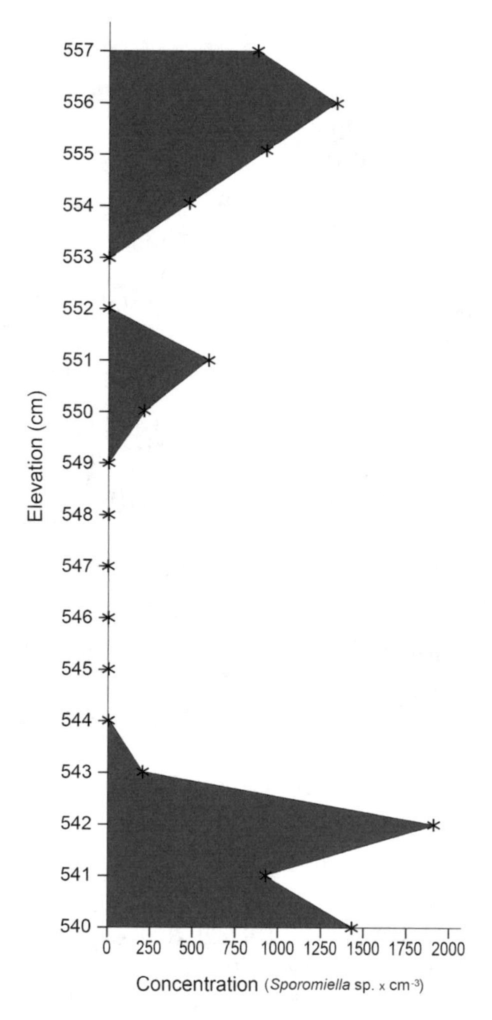

Fig. 6.4 Concentration of *Sporormiella* sp. spores in the column 10AD (spores cm^{-3})

was always decreasing towards the sediments in the centre of the lake. The highest spore concentrations at Pilauco are close to experimental values corresponding to 40 beef cattle, which are equivalent in weight to 10 gomphotheres.

The content of pollen from the analysed aliquots in samples from grid 14AD was not calculated, therefore it was not possible to obtain the percentage of *Sporormiella* sp. On the other hand, the low concentration or complete absence of *Sporormiella* sp. in coprolites from Pilauco remain unclear. Parasite eggs were recovered from all samples, which confirms their coprolitic character (Chap. 13, this volume); however, they might not be well-represented by the aliquot method.

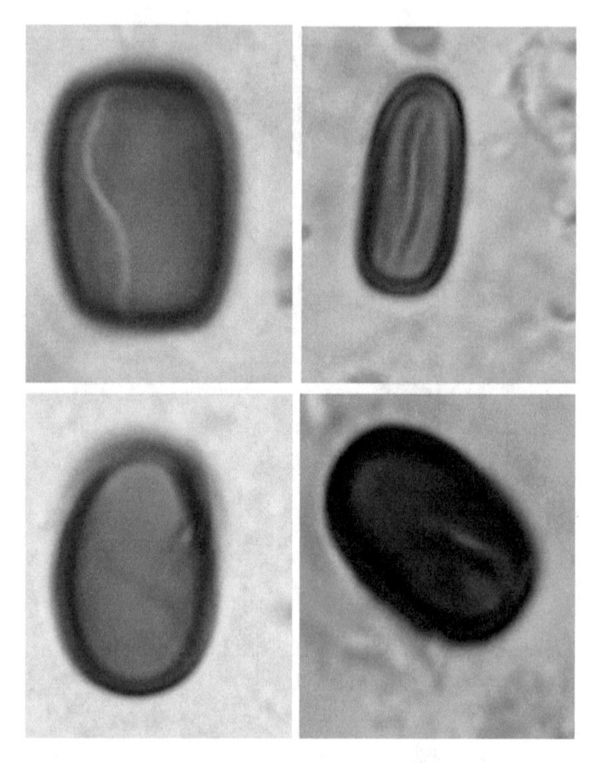

Fig. 6.5 Examples of *Sporormiella* sp. spores registered in column 10AD. The spore's length is 10 μm

Numerous studies suggest that the extinction of megafauna on both American continents occurred ~12.800 cal. year BP (e.g. Barnosky and Lindsey 2010; Brook and Barnosky 2012; Villavicencio et al. 2016; Pino et al. 2019). The contact between layers PB-8 and PB-9 at Pilauco has been dated with ~12.800 cal. year BP, and both analysed columns record an abrupt decrease/disappearance of *Sporormiella* sp. spores at this level (at 540 and 544 cm local elevation at 14AD and 10AD, respectively). The recurrence of these spores in the columns 14AD (at 610 cm) and 10AD (at 551 cm) at 12.200 and 12.460 cal. year BP, respectively, might be due to the presence of Holocene megaherbivores such as guanaco (*Lama guanicoe*) and south Andean deer (huemul, *Hippocamelus bisulcus*). The absence of *Sporormiella* sp. spores above the PB-8/PB-9 erosional unconformity is concurrent with the disappearance of pollen and an increase of coal contents (Chap. 3, this volume). Importantly, these trends are also coincident with an increase of extraterrestrial material, which might suggest that the associated wild fires (increased coal content) and the extinction of megafauna could both be associated with the proposed asteroid impact (Pino et al. 2019; see Chap. 15, this volume). Strictly speaking, the disappearance of *Sporormiella* sp. spores from the fossil record at Pilauco is an unambiguous indication for a local decline of megafauna. This proxy is more robust compared to the fossil bone register, given that the experimentally derived maximum dispersion distance of the spore is around 100 m (Gill et al. 2013; Raczka et al. 2016). More evidences from contemporaneous sites in north-western Patagonia are necessary to confirm the hypothesized process of extinction.

References

Ahmed SI, Cain RF (1972) Revision of the genera *Sporormia* and *Sporormiella*. Can J Bot 50(3):419–477

Alley RB (2000) The Younger Dryas cold interval as viewed from central Greenland. Quat Sci Rev 19:213–226

Araujo BB, Oliveira-Santos LGR, Lima-Ribeiro MS, Diniz-Filho JAF, Fernandez FA (2017) Bigger kill than chill: The uneven roles of humans and climate on late Quaternary megafaunal extinctions. Quat Int 431:216–222

Baker AG, Bhagwat SA, Willis KJ (2013) Do dung fungal spores make a good proxy for past distribution of large herbivores? Quat Sci Rev 62:21–31

Baker AG, Cornelissen P, Bhagwat SA, Vera FW, Willis KJ (2016) Quantification of population sizes of large herbivores and their long-term functional role in ecosystems using dung fungal spores. Methods Ecol Evol 7(11):1273–1281

Barnosky A, Lindsey E (2010) Timing of Quaternary megafaunal extinction in South America in relation to human arrival and climate change. Quat Int 217(1–2):10–29

Barnosky AD, Koch PL, Feranec R S, Wing SL, Shabel AB (2004) Assessing the causes of late Pleistocene extinctions on the continents. Sci 306(5693):70–75

Björck S (2007) Younger Dryas oscillation, global evidence. 1987–1994. In: Elias SA (ed) Encyclopedia of quaternary science, vol 3. Elsevier B.V. Oxford, pp 1985–1993

Borrero LA (2008) Extinction of Pleistocene megamammals in South America: The lost evidence. Quat Int 185(1):69–74

Borrero LA, Zárate M, Miotti L, Massone M (1998) The Pleistocene Holocene transition and human occupations in the southern cone of South America. Quat Int 49(97):191–199

Broecker WS (2006) Was the Younger Dryas triggered by a flood? Science 312(5777):1146–1148

Brook BW, Barnosky AD (2012) Quaternary extinctions and their link to climate change. In Hannah L (ed) Saving a million species island press/center for resource economics saving a million species: extinction risk from climate change. Island Press, pp 179–198

Burney D, Flannery T (2005) Fifty millennia of catastrophic extinctions after human contact. Trends Ecol Evol 20(7):395–401

Comandini O, Rinaldi A (2004) Tracing megafaunal extinctions with dung fungal spores. Mycologist 18:140–142

Davis OK (1987) Spores of the dung fungus *Sporormiella*: increased abundance in historic sediments and before Pleistocene megafaunal extinction. Quat Res (Orlando) 28(2):290–294

Davis OK, Shafer DS (2006) *Sporormiella* fungal spores, a palynological means of detecting herbivore density. Palaeogeogr Palaeoclimatol Palaeoecol 237(1):40–50

Ekblom A, Gillson L (2010) Dung fungi as indicators of past herbivore abundance, Kruger and Limpopo National Park. Palaeogeogr Palaeoclimatol Palaeoecol 296(1–2):14–27

Etienne D, Jouffroy-Bapicot I (2014) Optimal counting limit for fungal spore abundance estimation using *Sporormiella* as a case study. Veg Hist Archaeobot 23(6):743–749

Etienne D, Wilhelm B, Sabatier P, Reyss JL, Arnaud F (2013) Influence of sample location and livestock numbers on Sporormiella concentrations and accumulation rates in surface sediments of Lake Allos, French Alps. J Paleolimnol 49(2):117–127

Faegri K, Iversen J (1989) Textbook of pollen analysis, 4th edn. Wiley, London, UK

Feranec R, Miller N, Lothrop J, Graham R (2011) The *Sporormiella* proxy and end-Pleistocene megafaunal extinction: a perspective. Quat Int 245(2):333–338

Ficcarelli G, Coltorti M, Moreno-Espinosa M, Pieruccini P, Rook L, Torre D (2003) A model for the Holocene extinction of the mammal megafauna in Ecuador. J S Am Earth Sci 15(8):835–845

Fiedel S, Haynes G (2004) A premature burial: comments on Grayson and Meltzer's 'Requiem for overkill'. J Archaeol Sci 31:121–131

Firestone RB, West A, Kennett JP, Becker L, Bunch TE, Revay ZS, Schultz ZS, Belgya TD, Kennett J, Erlandson JM, Dickenson OJ, Goodyear AC, Harris RS, Howard GA, Kloosterman JB, Lechler P, Mayewski
PA, Montgomery J, Poreda R, Darrah T, Que Hee SS, Smith AR, Stich A, Topping W, Wittke JH, Wolbach WS (2007). Evidence for an extraterrestrial impact 12,900 years ago that contributed to the megafaunal extinctions and the Younger Dryas cooling. Proc Natl Acad Sci USA 104:16.016–16.021

Gill JL (2014) Ecological impacts of the late Quaternary megaherbivore extinctions. New Phytol 201(4):1163–1169

Gill JL, McLauchlan KK, Skibbe AM, Goring S, Zirbel CR, Williams JW (2013) Linking abundances of the dung fungus *Sporormiella* to the density of bison: implications for assessing grazing by megaherbivores in palaeorecords. J Ecol 101(5):1125–1136

González E, Labarca R, Chávez M, Pino M (2014) First fossil record of the smallest deer cf. Pudu Molina, 1792 (ARTIODACTYLA, CERVIDAE) in the Late Pleistocene of South America. J Vert Paleontol 34 (2):483–488

Graf M, Chmura G (2006) Development of modern analogues for natural, mowed and grazed grasslands using pollen assemblages and coprophilous fungi. Rev Palaeobot Palynol 141(1–2):139–149

Haynes G (2009) American Megafaunal Extinctions at the end of the Pleistocene Springer

Haynes G, Klimowicz J (2015) Recent elephant-carcass utilization as a basis for interpreting mammoth exploitation. Quat Int 35919–37

Johnson CN, Rule S, Haberle SG, Turney CS, Kershaw AP, Brook BW (2015) Using dung fungi to interpret decline and extinction of megaherbivores: problems and solutions. Quat Sci Rev 110:107–113

Koch PL, Barnosky AD (2006) Late Quaternary extinctions: state of the debate. Annu Rev Ecol Evol Syst 37, 215–250

Labarca R, Pino M, Recabarren O (2013) Los Lamini (Cetartiodactyla: Camelidae) extintos del yacimiento de Pilauco (Norpatagonia chilena): aspectos taxonómicos y tafonómicos preliminares. Estud Geol 69(2):255–269

Levetin E, Rogers CA, Hall SA (2000) Comparison of pollen sampling with a Burkard Spore Trap and a Tauber Trap in a warm temperate climate. Grana 39:294–302

Lorenzen ED, Nogués-Bravo D, Orlando L, Weinstock J, Binladen J, Marske Ugan A, Borregaard MKM, Gilbert TP, Nielsen R, Ho SYW, Goebel T, Graf KE, Byers D, Stenderup JT, Rasmussen M, Campos PF, Leonard JA, Koepfli KP, Froese D, Zazula G, Stafford TW, Aaris-Sørensen K, Batra P, Haywood AM, Singarayer JS, Valdes PJ, Boeskorov G, Burns JA, Davydov SP, Haile J, Jenkins DL, Kosintsev P, Kuznetsova T, Lai X, Martin LD, McDonald HG, Mol D, Meldgaard M, Munch K, Stephan E, Sablin M, Sommer RS, Sipko T, Scott E, Suchard MA, Tikhonov A, Willerslev R, Wayne RK, Cooper A, Hofreiter M, Sher A, Shapiro B, Rahbek C, Willerslev E (2011) Species-specific responses of Late Quaternary megafauna to climate and humans. Nature 479(7373):359–364

MacPhee RDE, Marx PA (1997) The 40,000-year plague: Humans, hyperdisease, and first-contact extinctions. In: Goodman SM, Patterson BD (eds) Natural change and human impact in Madagascar. Smithsonian Institution Press, Washington

Mungai P, Njoguc J, Chukeatirotea E, Hyde K (2012) Coprophilous ascomycetes in Kenya: *Sporormiella* from wildlife dung. Mycology 3(4):234–251

Parker NE, Williams JW (2012) Influences of climate, cattle density, and lake morphology on *Sporormiella* abundances in modern lake sediments in the U.S. Great Plains. The Holocene 22(4):475–483

Pino M, Martel-Cea A, Astorga G, Abarzúa AM, Cossio N, Navarro X, Lira MP, Labarca R, Lecompte MA, Adedeji V, Moore CR, Bunch TE, Mooney C, Wolbach W S, West A, Kennett, JP (2019). Sedimentary record from Patagonia, southern Chile supports cosmic- impact triggering of biomass burning, climate change, and megafaunal extinctions at 12.8 ka. Sci Rep 9(1) 4413

Raczka MF, Bush MB, Folcik AM, McMichael CH (2016) *Sporormiella* as a tool for detecting the presence of large herbivores in the Neotropics. Biota Neotrop 16(1)

Raper D, Bush MB (2009) A test of *Sporormiella* representation as a predictor of megaherbivore presence and abundance Quat Res (Orlando) 71:490–496.

Recabarren O (2007) Análisis de restos óseos de gonfoterios del área comprendida entre los 39° 39′ y 42° 49′ S, centro - sur de Chile. Undergraduate Dissertation. Universidad Austral de Chile., Escuela de Ciencias, Universidad Austral de Chile

Recabarren O, Pino M, Cid I (2011) A new record of Equus (Mammalia: Equidae) from the Late Pleistocene of central-south Chile. Rev Chil Hist Nat 84(4)

Rozas-Dávila A, Valencia V, Bush M (2016) The functional extinction of Andean megafauna. Ecology 97(10):2533–2539

Scott E (2010) Extinctions, scenarios, and assumptions: Changes in latest Pleistocene large herbivore abundance and distribution in western North America. Quat Int 217(1–2):225–239

Silva N (2014) Uso de esporas de *Sporormiella* sp. como proxy para determinar extinción de megafauna en el sitio Pilauco, Osorno. Undergraduate Dissertation, Universidad Austral de Chile

van der Kaars S, Miller GH, Turney CS, Cook EJ, Nürnberg D, Schönfeld J, Lehman SJ (2017) Humans rather than climate the primary cause of Pleistocene megafaunal extinction in Australia. Nat Commun 8:14142

van Geel B, Aptroot A (2006) Fossil ascomycetes in Quaternary deposits. Nova Hedwigia 82(3–4):313–329

Villavicencio NA, Lindsey EL, Martin FM, Borrero LA, Moreno PI, Marshall CR Barnosky AD (2016) Combination of humans, climate, and vegetation change triggered Late Quaternary megafauna extinction in the Última Esperanza region, southern Patagonia, Chile. Ecography 39(2):125–140

Wignall PB, Benton MJ (1999) Lazarus taxa and fossil abundance at times of biotic crisis. J Geol Soc (London)156 (3):453–456

Wood JR, Wilmshurst JM (2013) Accumulation rates or percentages? How to quantify *Sporormiella* and other coprophilous fungal spores to detect late Quaternary megafaunal extinction events. Quat Sci Rev 77:1–3

Wood JR, Rawlence NJ, Rogers GM, Austin JJ, Worthy TH, Cooper A (2008) Coprolite deposits reveal the diet and ecology of the extinct New Zealand megaherbivore moa (Aves Dinornithiformes). Quat Sci Rev 27(27):2593–2602

Wroe S, Field JH, Archer M, Grayson DK, Price GJ, Louys J, Faith JT, Webb GE, Davidson I, Mooney SD (2013) Climate change frames debate over the extinction of megafauna in Sahul (Pleistocene Australia–New Guinea). Proc Natl Acad Sci USA 110(22):8777–8781

Chapter 7
Micromammals (Mammal: Rodentia) from Pilauco: Identification and Environmental Considerations

Patricia Canales-Brellenthin

Abstract Rodents are small-sized animals that form one of the most extensive groups of mammals. They share highly specialized masticatory organs and very unique molar shapes allowing some species to be identified based on a single tooth. Additionally, rodents share some biological characteristics with other groups of small mammals (micromammals) making them particularly informative species in terms of their chronological occurrence and environmental conditions, which explains their extensive study in paleontological and archaeological sites. There are two groups of rodents in Chile—Caviomorpha and Myomorpha—with most of the related species represented in the fossil record. At Pilauco, representatives from both groups have been recorded and assigned to the species *Myocastor* cf. *coypus* and *Loxodontomys micropus*. Both species currently inhabit southern South America and their ecological demands suggest humid environments and temperate climate for the period of deposition of the fossil-bearing layer PB-7. The inferred habitat of *M. coipus* corresponds to a wetland landscape with slow water currents and abundant hydrophilic vegetation, whereas open lands with dense shrub vegetation cover or brushwood would provide shelter for *L. micropus*. A transition zone (e.g., an ecotone wetland-forest), covering the necessities of the two species, might also explain their common occurrence at Pilauco.

Keywords Rodents · Caviomorpha · Myomorpha · Pilauco

P. Canales-Brellenthin (✉)
Departamento de Geodinámica, Estratigrafía y Paleontología, Facultad de Ciencias Geológicas, Universidad Complutense de Madrid, Madrid, Spain
e-mail: paty.brellenthin@gmail.com

Transdisciplinary Center for Quaternary Research in the South of Chile, Universidad Austral de Chile, Valdivia, Chile

M. Pino and G. A. Astorga (eds.), *Pilauco: A Late Pleistocene Archaeo-paleontological Site*, The Latin American Studies Book Series,
https://doi.org/10.1007/978-3-030-23918-3_7

7.1 Background

Rodents are small-sized animals from the order Rodentia and the most extensive group among mammals with over 2,000 described species comprising an estimate of 42% of all mammal species (Musser and Carleton 2005). The most distinctive characteristic shared by all species of rodents is a dentition highly specialized to gnaw (Fig. 7.1). All rodents have only two pairs of incisors, which grow continuously over the course of entire lifetime. The front surface of the incisors is covered with enamel, whereas the rear surface is only dentine (Fig. 7.1). The rodent's incisors remain always sharp by wearing down the softer dentine, leaving a sharp, chisel-like edge. The canine teeth are missing, and the incisors are separated from the posterior teeth by a gap, or diastema (Fig. 7.1).

The rear portion of the dentition is reduced and, in most cases, composed of molars. Some rodents exhibit one, and very few species two premolars preceding the molar row. When present, the premolars are molarized and difficult to distinguish from molar teeth considering their shape, size, and function. Therefore, in rodents' teeth from the rear portion of the dentition are usually referred to as molariforms without specifying whether these are molars or premolars.

The height of the rear dental crowns in rodents is variable and usually related to diet. Generally, rodents are seeds-, fruits-, and fungi-eating species that exhibit low-crowned teeth (brachydont) (Fig. 7.2), whereas species feeding on abrasive plants are characterized by having tall teeth crowns (hypsodont) extending toward the gums (Fig. 7.2). Furthermore, the occlusal surfaces of the rear teeth in rodents are complex and specialized for each group. Therefore, it is common to identify particular families, genera, and species based on a single dental piece.

It is a common practice in paleontology to create artificial classification of animals outside the formal taxonomic norms based on factors such as work methodology or specific characteristics of the fossil remains. Such is the case for micromammals, which represent an informal group without taxonomic validity based on the size of fossil remains. Micromammals encompass small-sized ($\leq$5 kg) terrestrial

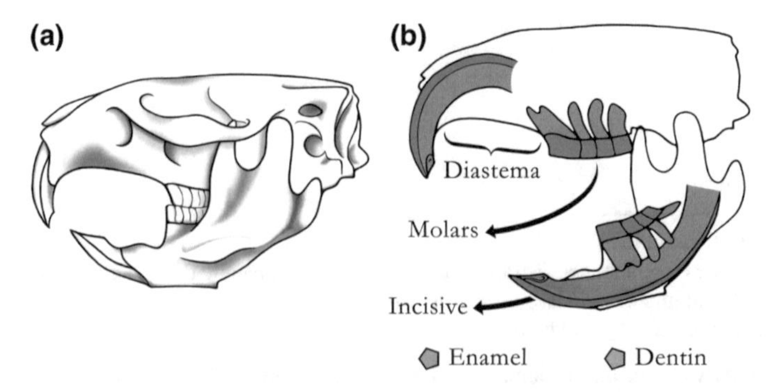

Fig. 7.1 Schematic representation of a rodent skull showing the external morphology (**a**), and detail of the dental structure (**b**)

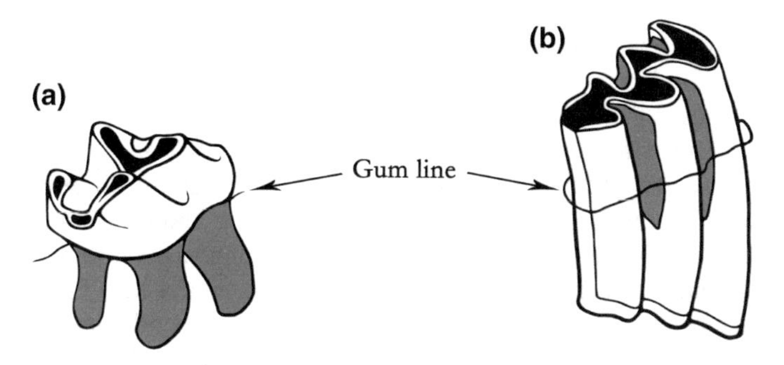

Fig. 7.2 Types of teeth depending on crown height. **a** Brachydont superior molar. **b** Hypsodont superior molar

animals including Rodentia, other groups of mammals such as Lagomorpha (hares and rabbits), Soricomorpha (moles and shrews), and Chiroptera (bats), among others.

Although an informal group, micromammals share certain biological characteristics such as elevated metabolic necessities, limited mobility due to their small size, and generally high rates of reproduction and evolution. Furthermore, a significant number of micromammals are particularly sensitive to environmental changes requiring specific climatic conditions for their survival. Therefore, many species are characterized by a rather limited range of geographic distributions and occupy a reduced number of biotopes. For the above, the study of this group of animals is widely extended in paleontological and archaeological sites, because them an excellent tool for paleoenvironmental and paleoclimatic interpretations (Flynn et al. 2003; Hernández-Fernández 2006; Reed and Denys 2011).

Isolated fossil findings of small mammals are usually not founded, mainly because their death is often associated with predation rather than being accidental in nature. Furthermore, their particularly small bones tend to easily decompose, except when special conditions favoring fossilization are given. Therefore, most of the fossil micromammals are found as concentrations of remains, which favors their fossilization (Fernández-López 2006). There are two non-excluding hypotheses to explain the origin of the micromammal fossil assemblages. The fluvial hypothesis (Wolff 1973; Korth 1979) states that micromammal remains are incorporated in water streams and are subjected to the processes of transport and hydrodynamics (e.g., selection, aggregation, fragmentation, abrasion). In turn, the eschatological hypothesis (Mellet 1974) proposes that most of the fossil of micromammals have passed through the digestive system of predators and are later deposited as excrements or pellets.

There are 61 species of rodents representing more than half of the described Chilean mammals and many of them have been recorded as fossils. According to the classification proposed by Wilson and Reeder (2005), two groups of rodents can be found in Chile: Caviomorpha (suborder Hystricomorpha) and Myomorpha. Additionally, the suborder Castorimorpha is represented by the recently introduced American beaver (*Castor canadensis*).

Caviomorphs have been linked to species of rodent from the Old World, especially Africa. During the 1990s, some authors emphasized the low probability that all rodent forms in the New World originated from a single radiation event and considered the term "Caviomorpha" not appropriate. Nevertheless, molecular studies have confirmed that all caviomorphs form a monophyletic group, differing from the Hystricomorpha of the Old World (Phiomorpha) (Huchon and Douzery 2001).

The oldest South American fossil record of Caviomoropha dates back to the early Eocene (Antoine et al. 2012), although is the Oligocene (34 Ma) the key period in the evolutionary history of the group given the great abundance and diversity of fossil remains assigned to this time period (Vucetich et al. 2014). Presumably, all Caviomorpha originated from African rodents. However, their arrival to South America is still under debate, because during the Oligocene, before the establishment of the Panama land bridge, the continent was still an isolated land mass. Thus, the "raft" hypothesis proposes that rodents arrived from Africa floating on trunks or agglomerations of plant remains traveling 1000–1500 km across the South Atlantic Ocean. Alternatively, it has been proposed that the Caviomorpha might have originated from a common Asian ancestor arriving to South America through North America or Australia and Antarctica. This novel hypothesis is based on the discovery of new living species in Asia that share similar molecular and morphological characteristics with caviomorphs. However, this scenario still lacks fundament due to the absence of pre-Oligocene Hystricognathi fossils along the proposed dispersion routes.

According to their morphology, the Caviomorpha are considered large rodents. They exhibit four tall-crowned rear teeth (1 premolar and 3 molars) characterized by complex occlusal surfaces exhibiting multiple enamel folding structures perpendicular to the longitudinal axes of the tooth maximizing its abrasive function. This type of dental structure is termed lophodont (Fig. 7.3), and most of the Caviomorpha are tetra-lophodont.

The skull of caviomorphs is wide and very robust, with extensive muscular insertions in order to accommodate the particular musculature of animals specialized for gnawing. The extremely enlarged infraorbital foramen is similar in size to the ocular orbit and is occupied by one of the 3 branches of the masseter muscle inserted in the mandible. This type of cranium is termed hystricomorphous and denominates the suborder Hystricomorpha.

The main rodent group in Chile is represented by the subfamily Sigmodontinae from the suborder Myomorpha. The sigmodontines are cricetids (Family Cricetidae) endemic to South America and form one of the most diverse mammal groups. Although their appearance is similar to common rats and mice, they have a wide morphological and ecological diversity and they occupy a wide variety of lifestyles (e.g., fossorial, arboreal, terrestrial and semi-aquatic). For this reason, the sigmodontines have successfully occupied different biomes (e.g., deserts, forests, wetlands, steppes). Moreover, these rodents are an essential element of the South American fauna, representing a principal prey for numerous carnivorous mammals and raptors.

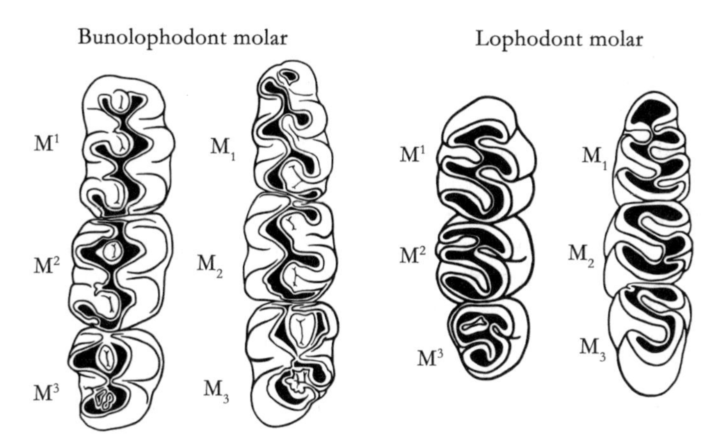

Fig. 7.3 Types molars in rodents according to characteristics of the occlusal surface. M^1–M^3: Upper row from the first to the third molar. M_1–M_3: Lower row from the first to the third molar

The phylogenetic relationships between groups or tribes of sigmodontines are generally not entirely clear. Nevertheless, it is commonly accepted that most of the sigmodontines are endemic to South America, but their direct ancestors originate from North America with the fossil record indicating a South American invasion in the late Miocene. The oldest records of these rodents date back to the end of the Miocene in Argentina between 5.8 and 7.1 Ma (Verzi and Montalvo 2008; Nasif et al. 2009). Such very old findings are particularly scarce, whereas the fossil record of the genera and species that form the current communities of sigmodontines dated back to the early-to-middle Pleistocene (781 ka). The oldest record in Chile is from the upper Pleistocene, 126 ka ago (Simonetti and Rau 1989; Borrero 2003).

In contrast to the Caviomorpha, the sigmodontines are rather small with the biggest species weighing less than 500 g and rarely exceeding 300 mm in length. Their cranial characteristics are variable, but all sigmodontines lack premolars, summing up a total of 16 teeth. The only exception is the species *Neusticomys oyapocki*, which exhibits reduced number of teeth with only 2 molars on each jawbone and a total of 12 teeth. The molars can vary from brachydont (low-crowned teeth) to hypsodont (high-crowned teeth). The occlusal zone exhibits a biserial disposition of the cuspids with complicated patterns presenting a gradual transition from a bunodont to a lophodont tooth (Fig. 7.3).

7.2 Rodents from Pilauco

In archaeo- and paleontological sites, rodent remains are usually found in the sediment sieving, where the small-sized bone fragments not detected during the excavation process are recovered. However, in Pilauco the entire fossil record of rodents was found in-situ, thus with the exact positions of each fossil remain. In addition, all rodent fossil remains were found within the layer PB-7 (Chap. 3, this volume).

The fossil rodent specimens recorded at Pilauco correspond to five remains: 3 large-sized isolated incisors, two of which were found in neighboring grids and sharing similar poor state of preservation. They are highly fractured making difficult to identify their upper or lower jaw position. The remaining incisor is better preserved and corresponds to the upper left mandible (Fig. 7.4). The fourth specimen is a large-sized and well-preserved lower molar (MHMOP/PI/38, Fig. 7.4). Finally, the last specimen corresponds to a skull fragment of small size preserving the complete upper right dentition. Two cm away, another incisor was found, which in term of size and morphology corresponds to the right incisor lacking from the skull and was therefore considered part of the same finding.

The identification of these remains was based on specialized bibliography providing the diagnostic characteristics of each group. Also, the specimens were compared to modern material from the mammal collections held at Universidad Austral de Chile. These analyses reveal the presence of both groups, Caviomorpha and Sigmodontinae at the Pilauco site.

The morphological characteristics of the isolated molar identify this specimen as a third right side lower molar corresponding to the species *Myocastor coypus* (Fig. 7.4). The 3 isolated incisors were also assigned to the same species:

Order Rodentia Bowdich (1821)
Suborder Hystricomorpha Brandt (1855)
Parvorder Caviomorpha Wood (1955)
Family Myocastoridae Ameghino (1904)
Genus *Myocastor* Kerr (1792)
Myocastor cf. *coypus* (Molina 1782)
Material for reference: MHMOP/PI/38 (Fig. 7.4).

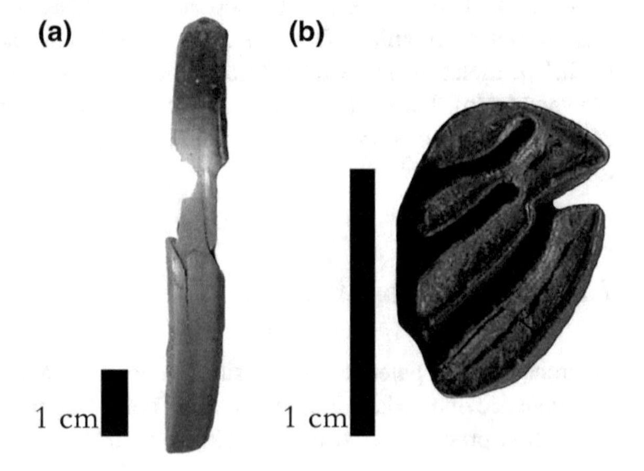

Fig. 7.4 Isolated fossils from *Myocastor cf. coypus* from Pilauco. **a** Upper left incisor. **b** Third right-side lower molar (MHMOP/PI/38)

Myocastor coypus, commonly known as coipo, coypu, or nutria, resembles a large-sized and robust rat (Fig. 7.5). The coipo often weighs between 2 and 4 kg, but adult males can reach 8 kg. The average length of wild adults is 50 cm, with the male generally larger than the female. It has a curved body and a long, uncoated tail. The head is large with triangular shape and small eyes placed in its upper part, which appear above the water when swimming. The rear feet are much longer than the front ones and the fingers exhibit interdigital webbing membranes (Woods et al. 1992).

Myocastor coypus is a semi-aquatic mammal endemic to South America with occurrence in Argentina, Bolivia, Brazil, Paraguay, and Chile. In Chile, the species inhabit all types of wetlands from Coquimbo to Tierra del Fuego (30–55°S) spanning a wide geographical range. It occupies aquatic biomes such as rivers or quiet creeks, lakes, estuaries, and swamps, generally preferring vegetated coastal areas. These gregarious animals form groups of 2–13 individuals and construct burrows for shelter with material from the local vegetation. The structures may be a simple tunnel or a complex system containing passages that extend from 6 to 15 m. The animals feed mainly on (aquatic) plants, which are consumed with roots, leaves, stems, and bark. They also consume terrestrial plants and occasionally molluscs and small arthropods.

The small fractured skull was identified as *Loxodontomys micropus* based on direct comparison of the dentition to living species. As part of the skull was well preserved, a series of standard measurements could be performed and compared to those from modern individuals.

Order Rodentia Bowdich (1821)
Suborder Myomorpha Brandt (1855)
Family Cricetidae Fischer (1817)
Subfamily Sigmodontinae Wagner (1843)
Genus *Loxodontomys* Osgood (1947)
Loxodontomys micropus (Waterhouse 1837)
Reference material: MHMOP/PI/39 (Fig. 7.6).

Fig. 7.5 **a** Modern specimen of *Myocastor coypus*. **b** Schematic view of the isolated molar (MHMOP/PI/38) found at Pilauco in its mandibular position

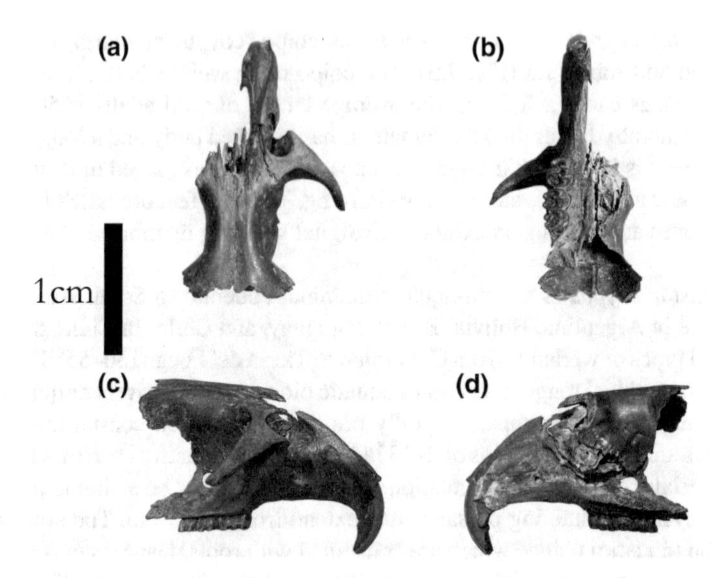

Fig. 7.6 fossil remains of *Loxodontomys micropus* (MHMOP/PI/39) from Pilauco. **a** Dorsal view, **b** ventral view, **c** lateral right-side view, **d** lateral left-side view

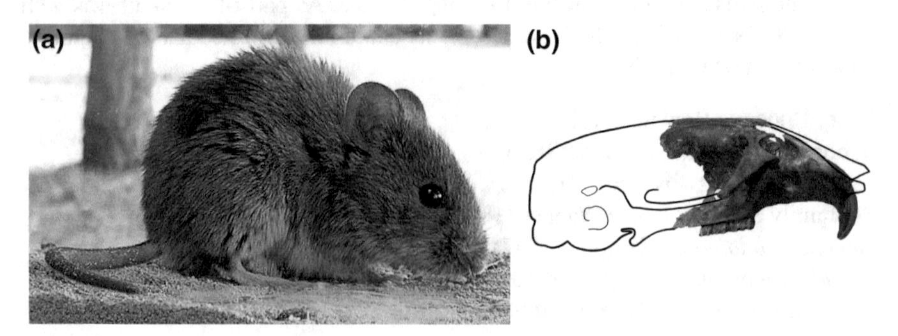

Fig. 7.7 **a** Modern specimen of *Loxodontomys micropus* (Photograph by Daniel Udrizar Sauthier). **b** Schematic view of the cranium fragment found at Pilauco

Loxodontomys micropus, commonly known as "pericote patagónico" or southern big-eared mouse, is a moderately large cricetid. It has a plump body and can barely exceed 100 g weight and measures 230–240 mm in total length (Fig. 7.7), including its tail that corresponds to 75% of the entire body length. The cranium (Fig. 7.7) is robust and exhibits relatively large ears (Teta et al. 2008).

Loxodontomys micropus inhabits southern Chile and south-western Argentina. In Chile it can be found from the Andean and sub-Andean region to the Cordillera Nahuelbuta. In Chile, its latitudinal habitat extends from Ñuble (36°S) to the Strait of Magellan (54°S), including the Island of Chiloé, and from 35°S to the southern extreme of the Santa Cruz province (52°S) in Argentina. The species occupies a wide variety of habitats but is mainly found in the temperate rainforest around Valdivia, in *Nothofagus* forests, areas with shrubs, and in the Patagonian steppe covered with dense humid pasture. This animal is mainly nocturnal and herbivorous, consuming mosses, mushrooms, fruits, and seeds.

7.3 Discussion and Conclusions

Numerous fossil remains of large-sized mammals have been found at Pilauco site, while the record of small-mammals is scarce and restricted to one single layer. This might be due to the relatively easy gravitational, wind and water transport of such small remains (Dodson and Wexler 1979; Korth 1979; Montalvo et al. 2012). Nevertheless, the micromammal fossil record from Pilauco provides fossil evidence of the two large groups of modern rodents present in Chile. Caviomorpha is represented by *Myocastor* cf. *coypus,* which is the only living species from the family Myocastoridae, and Sigmodontinae is represented by *Loxodontomys micropus.* Both species are currently found in southern South America.

Based on actualism principles that modern species have undergone little change through time (i.e., niche conservation) we can extrapolate the current habitat preferences of species to the past. Hence, the Pilauco specimens hint towards the presence of wetlands or landscapes with humid pastures and water ponds with dense hydrophilic vegetation, providing habitat and food for *Myocastor* cf. *coypus.* The surroundings were likely characterized by grasslands covered with dense vegetation or an ecotone between wetland and forests (see Chap. 9, this volume) matching the ecological needs of *Loxodontomys micropus.*

Both species have been described in younger than Pilauco archaeological sites, but *Myocastor coypus* has been particularly associated with human activities because of the predation for its meat (Sartori and Colasurdo 2012). Additionally, *Loxodontomys micropus* has been frequently found in caves and shelters alternately occupied by other animals and humans (Andrade and Teta 2003; Simonetti and Rau 1989), although the consumption of its meat by humans could not be clearly established. Overall, there is no evidence of human consumption of *Myocastor coypus* at Pilauco, as this kind of fossil remains never show cutting marks. *Myocastor coypus* and *Loxodontomys micropus* were probably part of the fauna inhabiting the wetland and the northern hills, and after its death the remains were accumulated in the sediments that later formed the site.

References

Andrade A, Teta P (2003) Micromamíferos (Rodentia y Didelphimorphia) del Holoceno tardío del sitio arqueológico Alero Santo Rosario (Provincia de Río Negro, Argentina). Atek Na 1:273–287
Antoine PO, Marivaux L, Croft DA, Billet G, Ganerod M, Jaramillo C, Martin T, Orliac MJ, Tejada J, Altamirano AJ, Duranthon F, Fanjat G, Rousse S, Salas Gismondi R (2012) Middle Eocene rodents from Peruvian Amazonia reveal the pattern and timing of caviomorph origins and biogeography. Proc R Soc Biol Sci Ser B 279:1319–1326
Borrero LA (2003) Taphonomy of the Tres Arroyos 1 Rockshelter, Tierra del Fuego, Chile. Quat Int 109–110:87–93
Dodson P, Wexler D (1979) Taphonomic investigation of owl pellets. Paleobiology 5:275–284
Fernández-López SR (2006) Taphonomic alteration and evolutionary taphonomy. J Taphon 4(3):111–142
Flynn JJ, Wyss AR, Croft DA, Charrier R (2003) The Tinguiririca Fauna, Chile: biochronology, paleoecology, biogeography, and a new earliest Oligocene South American Land Mammal 'Age'. Palaeogeogr Palaeoclimatol Palaeoecol 195(3–4):229–259
Hernández-Fernández M (2006) Rodent paleofaunas as indicators of climatic change in Europe during the last 125,000 years. Quat Res (Orlando) 65(2):308–323
Huchon D, Douzery E (2001) From the Old World to the New World: a molecular chronicle of the phylogeny and biogeography of hystricognath rodents. Mol Phylogenet Evol 20(2):238–251
Korth WW (1979) Taphonomy of microvertebrate fossil assemblages. Ann Carnegie Mus 48(15):235–285
Mellet JS (1974) Scatological origin of microvertebrate fossil accumulations. Science 185(4148):349–350
Molina GI (1782) Saggio sulla storia naturale del Chile. Stamperia di S. Tomnaso d' Aquino, Bologna (Italy)
Montalvo C, Cheme Arriaga L, Tallade P, Sosa R (2012) Owl pellet dispersal by wind: observations and experimentations. Quat Int 278(8):63–70
Musser GG, Carleton MD (2005) Order Rodentia. In: Wilson DE, Reeder DM (ed) Mammal species of the world, 3rd edn. The Johns Hopkins University Press
Nasif NL, Esteban GI, Ortiz PE (2009) Novedoso hallazgo de egagrópilas en el Mioceno tardío, Formación Andalhuala, Provincia de Caramarca, Argentina. Ser correl geol 25:105–114
Reed DN, Denys C (2011) The taphonomy and paleoenvironmental implications of the Laetoli micromammals. In: Harrison T (ed) Paleontology and geology of Laetoli: human evolution in context. Vertebrate paleobiology and paleoanthropology series. Springer, Dordrecht
Sartori J, Colasurdo MB (2012) La recurrencia del coipo (Myocastor coypus) en los registros arqueofaunísticos de la cuenca inferior del Río Paraná (Argentina). Arqueol Iberoam 13:23–36
Simonetti JA, Rau J (1989) Roedores del Holoceno temprano de la Cueva del Milodón, Magallanes, Chile. Not Mens Mus Nac Hist Nat 315:3–5
Teta P, Pardiñas UFJ, Udrizar DE, D'Elía G (2008) Loxodontomys micropus (Rodentia: Cricetidae). Mamm Species 8371–11
Verzi DH, Montalvo CI (2008) The oldest South American Cricetidae (Rodentia) and Mustelidae (Carnivora): Late Miocene faunal turnover in central Argentina and the Great American Biotic Interchange. Palaeogeogr Palaeoclimatol Palaeoecol 267:284–291
Vucetich MG, Dozo MT, Arnal M, Pérez ME (2014) New rodent (Mammalia) from the late Oligocene of Cabeza Blanca (Chubut) and the first rodent radiation in Patagonia. Hist Biol 27(2):236–257
Waterhouse GR (1837) Characters of new species of the genus Mus, from the collection of Mr. Darwin. Proc Zool Soc London 1837:15–21, 27–29

Wilson DE, Reeder DM (2005) Mammal species of the world: a taxonomic and geographic reference, vol 1, 3rd edn. Johns Hopkins University Press
Woods CA, Contreras L, Willner-Chapman G, Whidden HP (1992) Myocastor coypus. Mamm Species 398:1–8
Wolff RG (1973) Hydrodynamic sorting and ecology of a Pleistocene mammalian assemblage from California (USA). Palaeogeogr Palaeoclimatol Palaeoecol 13:91–101

Chapter 8
Taphonomy of the Pilauco Site, Northwestern Chilean Patagonia

Rafael Labarca

Abstract Pilauco shows two distinct layers containing remains of Pleistocene mammals (PB-7 and PB-8). The site is spatially divided into two sectors, East (45 m^2) and West (27 m^2). The current study is centered in the Western sector, where the majority of the materials come from layer PB-7 (%NISP = 92.5). Overall, this layer does not show signs of weathering, exhibiting instead trampling marks, and in a lesser quantity, large carnivore tooth marks. The fragmentation level is low, particularly for the fossils of Gomphotheriidae, for which most of the fractures occurred when the fossils were not fresh. No human marks of any kind were identified. The impact of these distinct factors in the formation of the record of PB-7 was evaluated using the available data and concluded in an in situ death of a gomphothere, to which would have been added anatomical elements of other taxa, redeposited coluvially and/or through vertical migration as a result of trampling. Carnivores would have been primarily responsible for the alteration and possibly subtraction of skeletal remains; as of now there is no evidence of human impact in this process. The materials recovered in PB-8 layer could also have been deposited by colluvial processes, although the sample is very small to discuss the taphonomic processes that have occurred in this layer.

Keywords Taphonomy · Carnivore · Trampling · Tooth · Marks · Colluvial

8.1 Background

Taphonomy is the study of all of the processes through which living organisms pass during their transition from death to their incorporation into the fossil record (Lyman 1994). Particularly in late Pleistocene archeological sites, where the cultural imprint is sometimes weak and/or debatable, and where distinct formational agents

---

R. Labarca (✉)
Escuela de Arqueología, TAQUACH, Transdisciplinary Center for Quaternary Research in the South of Chile, Universidad Austral de Chile, Puerto Montt, Chile
e-mail: r.labarca.e@gmail.com

M. Pino and G. A. Astorga (eds.), *Pilauco: A Late Pleistocene Archaeo-paleontological Site*, The Latin American Studies Book Series,
https://doi.org/10.1007/978-3-030-23918-3_8

are stratigraphically combined, taphonomical interpretations play a fundamental role (Borrero 2015; Martin 2013; López et al. 2016; Suárez et al. 2014, when referring to South American cases). Following the work of Gifford-González (1991), one of the objectives of this discipline is to precisely determine the post-depositional influences on a certain bone assemblage in order to ponder the agents that affect it and thus describe its depositional history.

The Pilauco site is not an exception to this. Since scientific work began at the site, it has been suggested that it has a complicated formational history, not only from a stratigraphic point of view, but also paleontological and archeological, due to the spatial association between fossil remains and human artifacts (Navarro-Harris et al. 2019). Additionally, the absolute dates obtained for the layer containing these associations, which although are more or less contemporary with the occupation of Monte Verde II (Dillehay 1997), are located in a chronological range that exceeds the Pleistocene-Holocene boundary by several millennia (Pino et al. 2013; see Chap. 3, this volume).

Despite the fact that a taphonomic study would allow for the identification of all the processes involved in the formation of a fossil association, the paleontological studies developed in Pilauco centered almost exclusively in taxonomically identification of the bone fossil assemblage (Recabarren et al. 2011; González et al. 2010, 2014; Labarca et al. 2013, 2014). The limited work that addressed taphonomic issues (e.g., Labarca et al. 2013, 2014) was carried out from the perspective of a single taxon, that is, only the description of marks on certain bones were considered. The relationship between the described assemblage and the remaining taxon, and how these are associated and spatially arranged in the sedimentary matrix, were not considered. However, these studies allowed the generation of a primary characterization of the taphonomic processes occurring on the site, where processes such as trampling, abrasion, and large carnivores would have apparently had an important role in the preservation of the site's fossil record.

8.2 Materials and Methods

The fossil materials comes from layers PB-7 and PB-8 in the western sector of the site (N = 81), recovered during the campaigns of 2010 through 2017. The fossil assemblage from the Eastern sector of the site (excavated during 2007 and 2008) were not considered in the present research due to differences in registry methodologies. Nevertheless, some taphonomic characteristics observed in those assemblages will be used as comparisons. The fossil materials of rodents, camelids, equids, deers, and gomphotheres have already been taxonomically studied elsewhere (e.g., Labarca et al. 2013, 2014; Recabarren et al. 2014; see Chaps. 4, 5, and 7, this volume), so the assignments made previously were respected (Table 8.1). As such, only one coracoid of a bird from layer PB-7 and two specimens of artiodactyl from PB-8 were determined in this work, for which reference skeletons in the author's possession and general manuals of osteology were consulted (e.g., Baumel and Witmer 1993;

Table 8.1 Taxonomic richness of the Pleistocene layers of the Pilauco site (sectors E and W). In accordance with the taxonomic assignments of González et al. (2010, 2014), Pino et al. (2013), Labarca et al. (2013), and Recabarren et al. (2014), see also Chaps. 5 and 7, this volume

Taxa	PB-7	PB-8
Conepatus sp.	–	X
Pilosa indet.	X	X
Notiomastodon aff. *N. platensis*	X	–
Equus (*A.*) *andium*/*Equus* sp.	X	X
cf. *Hemiauchenia paradoxa*	X	X
cf. *Pudu* sp.	–	X
Artiodactyla indet.[a]	–	X
Myocastor cf. *M. coypus*	–	X
Loxodontomys micropus	–	X
Anatidae indet.[a]	X	–
Birds indet.	–	X
N TAXA	5	8

[a]Assigned in this work

Smuts and Bezuidenhout 1987). Age of death of the specimens, was estimated considering epiphyseal fusion, eruption, and dental wear, according to Silver (1963), Kaufmann (2009), and Mothé et al. (2010). Sex estimation of gomphothere remains was estimated considering a metric approximation. This assumes the existence of sexual dimorphism between males and females in the Gomphotheriidae family, similar to the dimorphism documented for living members of the order Proboscidea (Haynes 1991). In this regard, several authors (e.g., Haynes 1991; Mothé et al. 2010; Ferretti 2010; Tassy 1996) have made ethological, ecological, and biological connections between extinct and living proboscideans, including members of the family Gomphotheriidae (e.g., Mothé et al. 2010). Measurements of long bones of gomphotheres were taken following Ferretti (2010).

The quantification of the remains was expressed in terms of NISP (Number of Identified Specimens), MNE (Minimum Number of Elements), MNI (Minimum number of Individuals), and MAU (Minimum standardized Anatomical units) (Binford 1978; Grayson 1984; Lyman 2008, among others). The fragmentation of the assemblage was estimated according to the quotient between the NISP and the MNE (Lyman 1994).

Several taphonomic modifications were considered, observed using a stereoscopic magnifying glass with magnification up to 40x. These included (1) marks generated due to consumption by carnivores, following the nomenclature of Binford (1981) and Muñoz et al. (2008); (2) rodent teeth marks; (3) trampling marks, for which the characteristics proposed by Domínguez-Rodrigo et al. (2009, 2012) were followed; (4) marks left by imprints of roots (Lyman 1994); signs of weathering (Bherensmeyer 1978) and (5) anthropic modifications (basically cut marks, scraping marks and percussion marks, see Mengoni-Goñalons (1999) for general definitions).

The fractures were classified as fresh or not fresh based on the morphology of their edge and surface (Outram 2001), independent of their causal agent. With regard to fragmentation, a specimen was considered complete if at least 90% of the original specimen was represented. Modern marks and fractures, which have resulted from the reexposure of the materials and their subsequent handling, were not considered in the present study.

With respect to trampling, five specimens of *Notiomastodon* aff. *N. platensis* (MHMOP/PI/611/614/615/629y/630) were selected. A quantification of the total number of marks was carried out in an area of 20 × 10 cm, recording the section, depth, shape of the groove's sides, and presence/absence of certain morphological characteristics such as striation and micro-abrasion of each mark, in accordance with the proposal of Domínguez-Rodrigo et al. (2009, 2012).

8.3 Taxonomic Richness and Abundance

Table 8.2 summarizes the absolute frequencies of the fossil remains per taxa in layers PB-7 and PB-8, sector W. The entire sample was taxonomically assigned at least at the order level (Labarca et al. 2013; Recabarren et al. 2014; see Chap. 5, this volume). As will be seen later, this is due to a practically complete absence of fragmentation. The vast majority of the remains identified were found in PB-7. Additionally, this layer showed the greatest taxonomic richness in this sector with a total of five taxa as compared to the three identified in layer PB-8 (Table 8.2). On this subject, Artiodactyla undet. was considered a different taxon other than cf. *H. paradoxa*, because the anatomical elements identified (portion of cervical vertebra and rib fragment) were smaller than the camelid bones recognized in the site. This assures the presence of a different taxon, not ruling out cf. *Pudu* sp., identified in the eastern sector, layer PB-8 (González et al. 2014). The values of the taxonomic richness of sector W contrast those reported for sector E, in which layer PB-8 shows the greatest number of identified taxa (PB-7 taxa number = 4; PB-8 taxa number = 8; Pino et al. 2013). Noteworthy here is the presence of four extant taxa (*Conepatus* sp., cf. *Pudu* sp., *Loxodontomys micropus y Myocastor* cf. *M. coypus*, Table 8.1). Evidently, despite being contiguous and belonging in the same layer, different depositional processes would have affected sectors E and W. Some differences in taphonomic alterations between the two sectors seem to confirm this impression (see below). In contrast, layer PB-7 does not vary greatly in terms of its taxonomic richness between the two sectors, only highlighting the presence of Anatinae undet. (portion of coracoid) in Sector W.

In PB-8, the few remains identified are distributed equally between equids, camelids, and indetermined artiodactyls, but the remains of gomphotheres are absent. In PB-7 however, the findings seem largely dominated by remains of proboscideans (%NISP = 79.5%), with many fewer representatives of specimens assigned to equids, camelids, and indetermined pilosans (Table 8.2; Fig. 8.1).

Table 8.2 Absolute frequencies (NISP) of taxa identified in sector W of the Pilauco site

Taxa	PB-7 (lower)	PB-7 (upper)	PB-7 (total)	PB-8
Notiomastodon aff. *N. platensis*	60	0	60	0
Equus sp.	1	2	3	2
Pilosa undet.	2	2	4	0
cf. *Hemiauchenia paradoxa*	3	4	7	2
Artiodactyla undet.	0	0	0	2
Anatinae undet.	1	0	1	0
Total	67	8	75	6

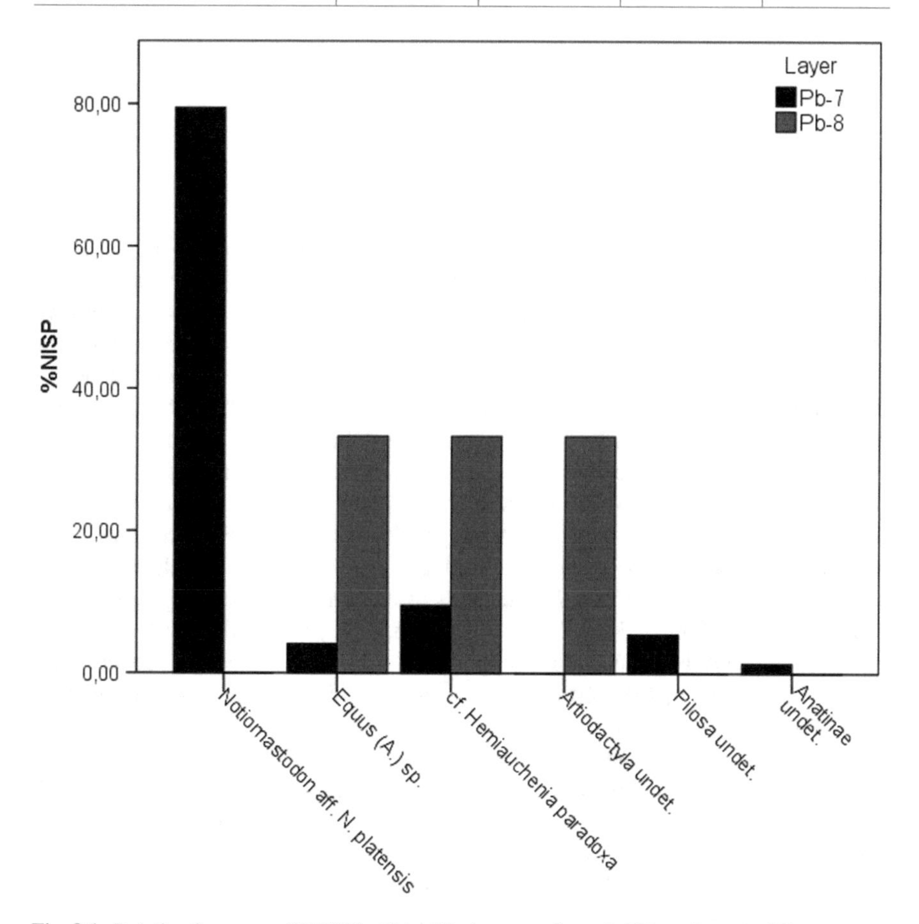

Fig. 8.1 Relative frequency (%NISP) of identified taxa per layer in Pilauco's sector W

Through the study of the vertical deposition of the fossil remains within layer PB-7, two depositional events were identified. They were designated as upper and lower PB-7, each separated by a few centimeters. The ages calibrated to the 2 σ range indicate that both events are statistically indistinguishable, although the lower event is stratigraphically older (see Chaps. 3 and 16, this volume). Based on the NISP, it is evident that the main depositional event corresponds to lower PB-7, which is dominated largely by remains of gomphotheres. In contrast, in upper PB-7 the register of gomphotheres is nonexistent, containing only scarcely represented remains of equids, camelids, and osteoderms of undetermined pilosans (Table 8.2). The upper segment of PB-7 corresponds taxonomically to PB-8, indicating a transition to this last layer. Considering the scarce sample of upper PB-7 and its apparent concurrence with lower PB-7, both assemblages were treated as a single unit.

8.4 Frequency of Skeletal Parts

In layer PB-8, the equid remains correspond to a proximal portion of fused first phalanx and a pm 3–4. These give an account of a single adult of more than two and a half years of age, judging by the presence of a permanent third or fourth premolar (Silver 1963). The identified remains of camelids correspond to a diaphyseal portion of a tibia and a practically complete radio-ulna. It is not possible to infer an age of death when the epiphysis is not preserved. However, the tibia segment could belong to a young individual, judging by a porous area toward the proximal section (Table 8.3).

In PB-7, three horse bone remains (fragment of mandibular ramus, portion of a pelvis, and astragalus) were recovered. These give an account of a single individual of older than 1.5–2 years, considering the presence of a pelvis with its bones completely fused (Silver 1963). On the other hand, the remains of camelids are more abundant (NISP = 7, MNE = 7). Elements of the axial skeleton and appendicular were registered (Table 8.3), and the presence of a single individual was estimated. Among the materials, a practically complete mandible with a very worn dental series stands out. Judging by the proposal of Kaufmann (2009), this individual would have been between 7 and 9 years at its time of death.

With regard to the fossils of Gomphotheriidae, in the PB-7 layers exposed until the year 2017 in sector W were recovered 60 specimens, which give an account of 49 bone elements, corresponding to 24% of a complete proboscidean skeleton. The rib cage is undoubtedly the most represented anatomical segment on the site. It constitutes more than half of the identified elements of gomphotheres (% MNE = 54.3). The %MAU indicates that more than 73% of total ribs found are registered (Table 8.4). Based on the morphology and curvature of the ribs, they were categorized into anterior, intermediate, and posterior (Ferretti 2010). From this categorization it was clear that there were a similar number of anterior (MNE = 9) and intermediate ribs (MNE = 9), while the posterior ribs appear less represented (MNE = 6). Six of the ribs could not be determined. Thoracic and lumbar vertebrae appeared to have values close to or equal to 40% of the MAU, while cervical and caudal vertebrae are poorly

Table 8.3 Absolute frequencies (NISP) of anatomic units of artiodactyls, equids, and pilosans from sector W of the Pilauco site

	Equus sp.		cf. *H. paradoxa*		Artiodactyla undet.		Pilosa undet.	
	PB-7	PB-8	PB-7	PB-8	PB-7	PB-8	PB-7	PB-8
Mandible	1	–	1	–	–	–	–	–
Cervical V.	–	–	–	–	–	1	–	–
Lumbar V.	–	–	1	–	–	–	–	–
Rib	–	–	–	–	–	1	–	–
Scapula	–	–	1	–	–	–	–	–
Humerus	–	–	1	–	–	–	–	–
Radius-ulna	–	–	–	1	–	–	–	–
Pelvis	1	–	1	–	–	–	–	–
Tibia	–	–	–	1	–	–	–	–
Astragalus	1	–	1	–	–	–	–	–
Phalanx 1	–	1	–	–	–	–	–	–
Dermal ossicles	–	–	–	–	–	–	4	–
Premolar 3–4	–	1	–	–	–	–	–	–
Incisor	–	–	1	–	–	–	–	–
Total	3	2	7	2		2	4	

represented. The sacral vertebrae are not recorded in the sample studied. The front legs are entirely absent, even though a virtually complete left scapula was identified. Its right pair is located in the North profile of the excavation and has not yet been removed. The pelvis is absent, as are the femurs and patella. In contrast, the distal segment of the hind leg is represented through several right-side anatomical units, some of which articulate with each other: tibia, fibula, astragalus, calcaneus, and cuboid (Fig. 8.2). A fragment of left calcaneus is added to this record. All anatomical units are fused, including the skull sutures, which are completely closed. The only exception is a thoracic vertebral disk segment. As such, the anatomical elements identified give an account of a single adult individual. The calibrated radiocarbon dates made on remains of gomphotheres in PB-7 (UCIAMS101670, UCIAMS101830, PSUAMS2416, PSUAMS2421, PSUAMS2418, UCIAMS101831, PSUAMS2420, and PSUAMS2419) varies between 15,110 and 15,890 yr BP (Chap. 3, this volume). Thus, the dates are not consistent with this impression. These differences could be due to the variability in the processing techniques used by the two laboratories responsible for dating.

It is possible to estimate the age of death of the Pilauco gomphothere by comparing it with the epiphyseal fusion tables available for modern elephants (*Loxodonta africana* Linnaeus, Haynes 1991). Within these taxa, females reach sexual maturity before males. This signifies, among other things, that males continue to grow for a longer period of time (Krumrey and Buss 1968; Haynes 1991). The above implies

Table 8.4 Frequency of skeletal parts of *Notiomastodon* aff. *N. platensis* in layer PB-7 of Pilauco

Anatomical unit	Number of elements	NISP	MNE	MNI	MAU	%MAU
Skull	1	4	1	1	1	100
Mandible	2	0	0	0	0	0
Atlas	1	1	1	1	1	100
Axis	1	0	0	0	0	0
Cervical V.	5	1	1	1	0.2	20
Toracic V.	17	8	7	1	0.411	41
Lumbar V.	5	2	2	1	0.4	40
Sacral V.	5	0	0	0	0	0
Caudal V.	22	1	1	1	0.042	5
Rib	34	30	25	1	0.735	74
Scapula	2	3	1	1	0.5	50
Humerus	2	0	0	0	0	0
Radius-ulna	2	0	0	0	0	0
Carpals	8	0	0	0	0	0
Metacarpals	10	0	0	0	0	0
Pelvis	2	0	0	0	0	0
Femur	2	0	0	0	0	0
Patella	2	0	0	0	0	0
Tibia	2	1	1	1	0.5	50
Fibula	2	1	1	1	0.5	50
Astragalus	2	1	1	1	0.5	50
Calcaneus	2	2	2	1	1	100
Tarsals	16	1	1	1	0.25	25
Metatarsals	10	0	0	0	0	0
Phalanx 1	20	1	1	1	0.05	5
Phalanx 2	12	0	0	0	0	0
Phalanx 3	20	0	0	0	0	0
Epiphysis undet.		1	1	1		
M2/m2		2	2	1		
Total	204	60	49	1		

that the epiphyses of the long bones are fused earlier in females than males (Haynes 1991). Thus, the sex of the gomphotheres must first be determined in order to more accurately apply the epiphyseal fusion tables of modern elephants.

Assuming that within the Gomphotheriidae family there would have been sexual dimorphism between males and females (Ferretti 2010; Mothé et al. 2010), the measurements of the tibiae MHMOP/PI/614 was compared with four specimens combined from three fossiliferous locations in Central Chile: Taguatagua (SGO.PV.687, 242), Tierras Blancas (SGO.PV.7) and Quereo, Los Vilos (SGO.PV.267) (Table 8.5).

As shown in Fig. 8.3, the measurements are clearly divided into two size groups. This division is interpreted as a product of sexual dimorphism. Thus, the specimen MHMO/PI/614, and therefore the individual deposited in layer PB-7, would correspond to a male. Haynes (1991) suggests that the tibia of male *Loxodonta africana* is completely fused around 32 years for the proximal epiphysis and between 28 and 32 years for the distal epiphysis. Thus, an age greater than 32 years is estimated for the time of death of the gomphothere in Pilauco sector W.

The presence of an unfused vertebral plate does not contradict this result. According to Haynes (1991), the vertebral plates are one of the last skeletal segments to fuse. A second way to estimate the age of death is from dental wear. Mothé et al. (2010) combined the stages of dental wear of Simpson and Paula-Couto (1957) with current data of elephants (Moss 1996), in order to calculate the age of death of a bone assemblage of gomphotheres from Brazil. Among the remains of Pilauco were identified two M2/m2 (MHMOP/PI/627 and/628) with similar states of wear (Category 3, Simpson and Paula-Couto 1957). Using this data could be proposed an age of death between 30 and 35 years, which corresponds to an individual adult, according to the categories proposed by Haynes (1991). Both estimates are therefore coherent with each other.

8.5 Horizontal and Vertical Distribution

The spatial distribution of the remains found in the PB-7 layer indicates a main sector of at least 16 m^2, composed by gomphotheres bones (Fig. 8.4). However, the entire deposition area has not been completely excavated, as evidenced by the presence of fossils in the North profile of excavation (a scapula in the 14AD unit and a gomphothere anterior rib in the 14AB unit).

It may be assumed that the skull is relatively in situ (15AD unit), because its size and weight make its displacement unlikely (ca. 120 kg for a male African elephant, Haynes and Klimowicz 2015). Therefore, it is possible to suggest that an important part of the ensemble, composed primarily of ribs and thoracic vertebrae, would show relative anatomical coherence in a general SW-NE orientation (Fig. 8.4). However, the arrangement of the ribs according to their location on the rib cage is not in line with this impression, since ribs of the three segments of the thorax are spatially associated. On the other hand, in a scenario of high anatomical coherence, it is expected to find

Table 8.5 Measurements (mm) of tibiae from gomphotheres of central Chile and Pilauco

	MHMOP/PI/614	SGO.PV.687 (TT2-C4H69)-Left	SGO.PV.687 (TT2-C2H19)-Right	SGO.PV. 7-Left	SGO.PV. 267-Left	SGO.PV. 242-Left
Maximum length	567	470	475	590	590	ND
Lateral length	490	375	380	480	483	ND
Medial length	550	457	450[a]	573	570	ND
Breath at mid diaphysis	115	73	78	115	105	83
Depth at mid diaphysis	123	88	95	126	115	92
Breadth of proximal articular surface	225	180	185[a]	283	225	ND
Depth of proximal articular surface	ND	135	125	197	155	ND
Breath of distal articular surface	179	143	142	203	180	175
Depth of distal articular surface	129	108	115	148	133	115

[a]approximated
ND: No Data

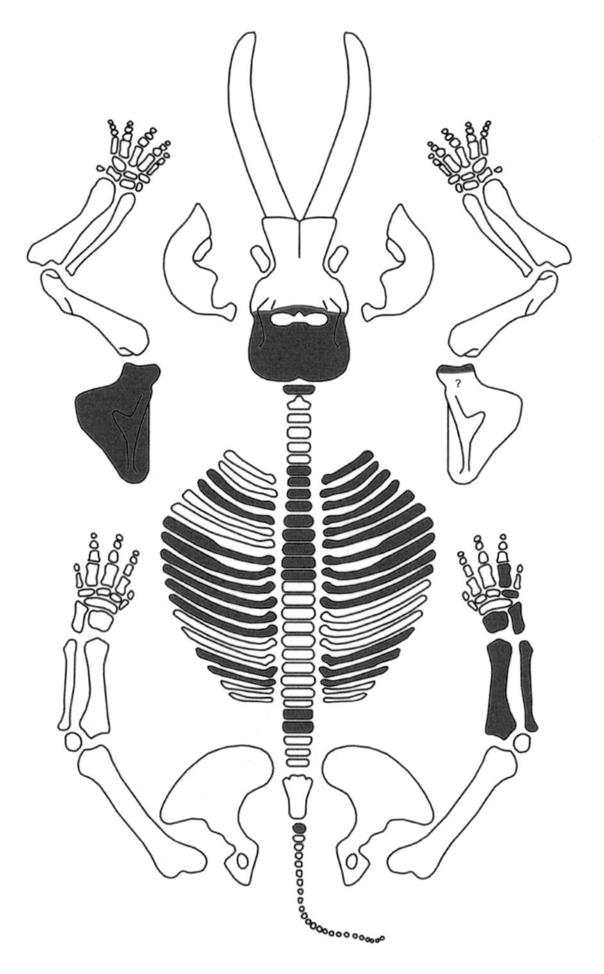

Fig. 8.2 Schematic sketch of a proboscidean skeleton showing the anatomical units (MNE) of the gomphotheres of Pilauco. The locations of vertebrae, ribs and phalanges are approximate. Modified from Santucci et al. (2016)

bones of the front limb and the scapular girdle close to the head. However, these are absent, except for the right scapula found in the profile of the 14AD unit. Conversely, the left scapula is located almost four meters from its right pair (15A unit), while less than a meter away from the skull is located five bones of the hind limb that articulate with each other (tibia, fibula, astragalus, calcaneus, and cuboid, mainly located in 14AC unit). The bones of the hind limbs and pelvis should hypothetically be found at the opposite end from the skull area, at the eastern boundary of the excavation. However, pelvis, femurs, or patellas have not been recovered. Moreover, beneath the skull, one of the two lumbar vertebrae were identified (Fig. 8.4). It is important to clarify that the posterior appendicular segment is not found in the east excavation of

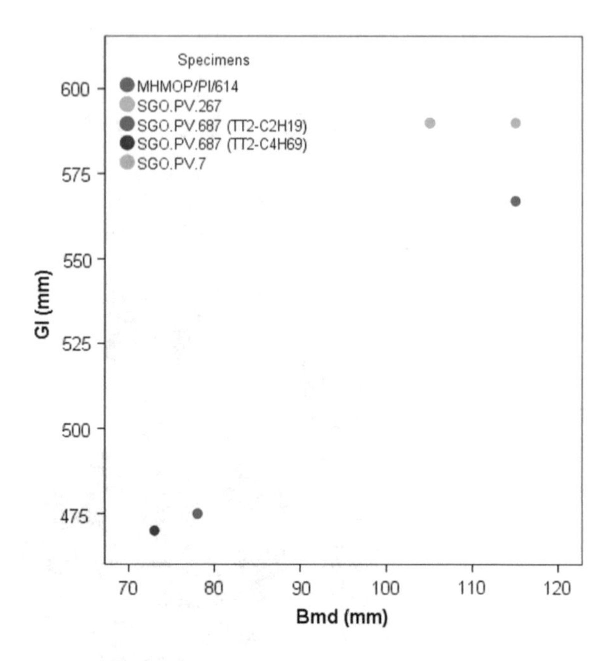

Fig. 8.3 Diagram of the dispersion of measurements of gomphothere tibia. Gl: Maximum length; Bmd: Breadth at mid diaphysis

the site (campaigns 2007–2008), as bones of a second gomphothere were exhumed from there, located more than six meters from the Western concentration (Pino 2008). Thus, it may be postulated that the bones would have undergone different degrees of displacement from their original location, assuming an in situ death of the animal, although in a relatively limited area of dispersion. The bones of the front limbs may be located in the unexcavated areas north, perhaps in conjunction with the remaining cervical and thoracic vertebrae. However, the posterior appendicular bones, pelvis, and perhaps part of the lumbar vertebrae have apparently been removed from the site.

In contrast, artiodactyls and perissodactyls are scattered throughout the excavated area of PB-7, without forming defined concentrations. Some are even located outside the main concentration area, as is the case of the camelid diaphysis recovered in the grid 18A (Fig. 8.4). As noted above, some of the remains of both orders are located above the bones of gomphotheres. That could indicate at least a second depositional event (for example, in the 15AD and 14AC units, Fig. 8.4).

Based on the orientation of the fossils registered in layer PB-7, a rose diagram was plotted to examine the orientation of the materials (Fig. 8.5). The diagram did not show a clear trend. The fossils have a general NE-SW orientation, and to a lesser extent SE-NW. Those with an E-W orientation were less represented.

It is possible that the vertical arrangement of the parts indicates that not all parts are placed horizontally in the sedimentary matrix. In PB-7, 31 anatomical units (43% of the total NISP) present some degree of verticality, with elevation variations that range between 2 and 25 cm between opposite sides. This variation is more evident in the ribs due to their marked length. Part of the layout of the fossils may be

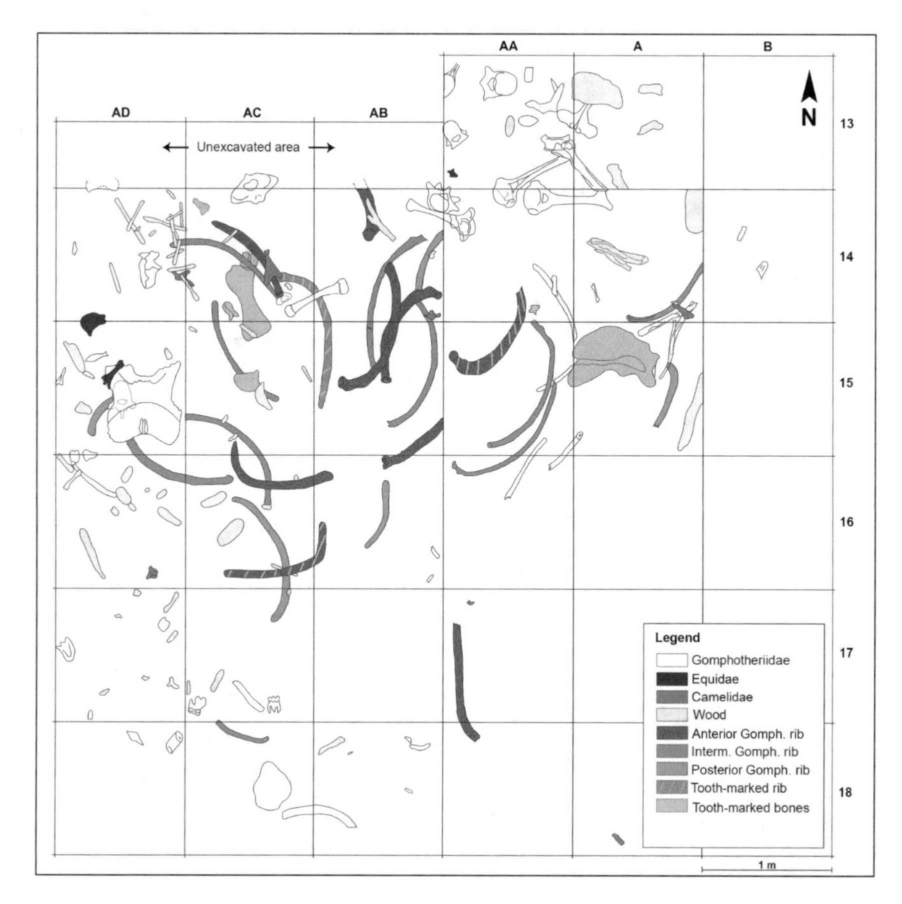

Fig. 8.4 Excavation layout for layer PB-7, sector W of Pilauco

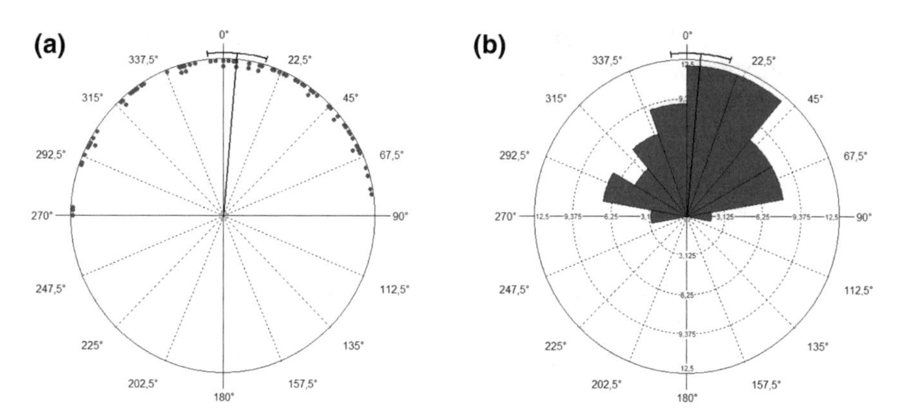

Fig. 8.5 Rose diagram (intervals of 22.5°) with the orientation of skeletal remains of layer PB-7, sector W of Pilauco

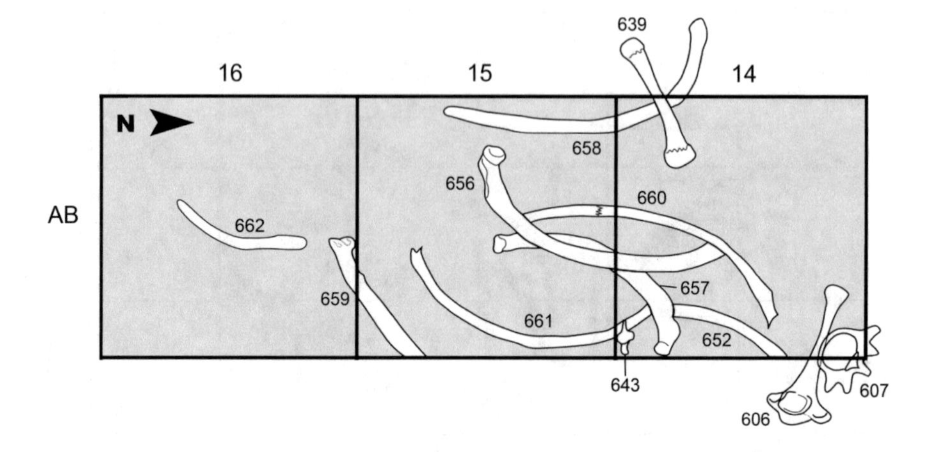

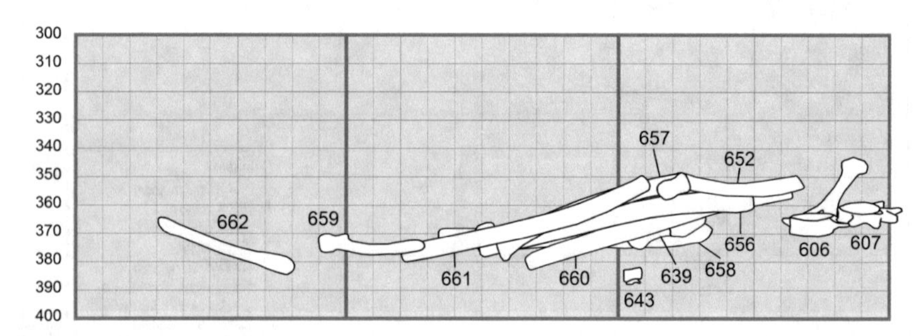

Fig. 8.6 Horizontal (up) and vertical (down) layout of some of the bones from layer PB-7, according to the depth defined locally for the site. The numbers correspond to the last three numbers of the Museum codes assigned to each piece

seen in Fig. 8.6. In PB-8 a relatively similar situation is observed, where two pieces (a rib of indeterminate Artiodactyl and a radio-ulna of cf. *Hemiauchenia paradoxa* were recovered in a practically vertical arrangement.

8.6 Fragmentation

The bone fragmentation is minimal, since the ratio MNE/NISP delivered a value of 0.81 for the remains of gomphotheres in layer PB-7. This is largely due to the considerable quantity of complete elements (%NISP = 49.3; NISP = 37, Table 8.4), but also due to each fragmented specimen is itself an element. Within this subsample are 14 ribs of *Notiomoastodon* aff. *N. platensis*, which are extremely fragile largely due to their notable length and low thickness. As shown in Table 8.4, in general the MNE is constructed from a very low NISP, which is in part due to the aforementioned complete pieces, because the incomplete bones represent a significant proportion of this (over 70% approximately).

In Equidae and Artiodactyla a similar phenomenon occurs, although only one piece is complete (Equid tooth). In all other cases the NISP is equal to the MNE (Table 8.3). Later in this article, this point will be addressed when the causes of deposition of bone remains are discussed. However, neither splinters, fragments of diaphysis, portions of spongy tissue nor flat bones were recorded within the sample, except for a portion of gomphothere epiphysis. Therefore, strictly speaking there is no "fragmentary" bone. This resulted in 100% of the sample being identified, to at least the order level.

8.7 Surface Modifications

The inspection of the surface of the bones could be performed on a total of 68 remains (%NISP = 89.3) registered in layer PB-7. The remaining pieces showed adhered sediments or were dental pieces, which due to their structure generally are not affected by certain taphonomic processes (e.g., root action, tooth marks). In PB-8, the study was carried out on five specimens (%NISP = 83.3). The taphonomic modifications are not the same for all the sample (Fig. 8.7). Surface, shape, and angle of the fractures were considered (Outram 1994). It was determined that in PB-7 40% showed a fresh fracture (NISP = 35), while 54% showed non-fresh fractures and 6% showed both attributes (Table 8.6). Although the number of Equidae and Camelidae bones are low compared to Gomphotheriidae, it may be noted that in this last taxonomic group the non-fresh fractures predominate. However, in camelids, the situation is reversed.

The non-fresh fractures were located mainly in the dorsal and ventral areas of gomphothere ribs (NISP = 15). This was to be expected because they are the anatomical unit most represented in the sample. The fresh fractures do not present a clear trend, registering in gomphotheres on relatively small bones (vertebral disc, indeterminate epiphysis, cuboid, and astragalus) and fragments of skull (NISP = 3). From this last sample, a diaphysis of cf. *H. paradoxa* humerus stands out. It presents a sinuous profile fracture, smooth edges, and sharp angles (Fig. 8.8). However, it does not have evidence of fracture by direct percussion (negative impact, flakes, etc.) (Fig. 8.8). This piece is located at about 2.5 m from the main concentration area in the 18A unit (Fig. 8.4). The agents producing the fractures will be discussed later.

There are no bones in Pilauco with signs of weathering (Fig. 8.7), which implies that the fossils were at the surface for a short time, insufficient to allow the production of macroscopic modifications. Local conditions significantly affect the speed with which the bones are weathered (Behrensmeyer 1978). In the case of Pilauco, a humid and anoxic deposit environment undoubtedly led to a very good preservation of bone remains. It has been suggested that in temperate latitudes proboscideans bones need about 20 years or more to exhibit cracks and splinters. (Haynes 1991). The low rate of weathering is probably due to the anatomical peculiarities of the proboscideans. The thickness of the skin functions like a barrier, avoiding the disassembling and the weathering of the bones (Haynes and Klimowitz 2015).

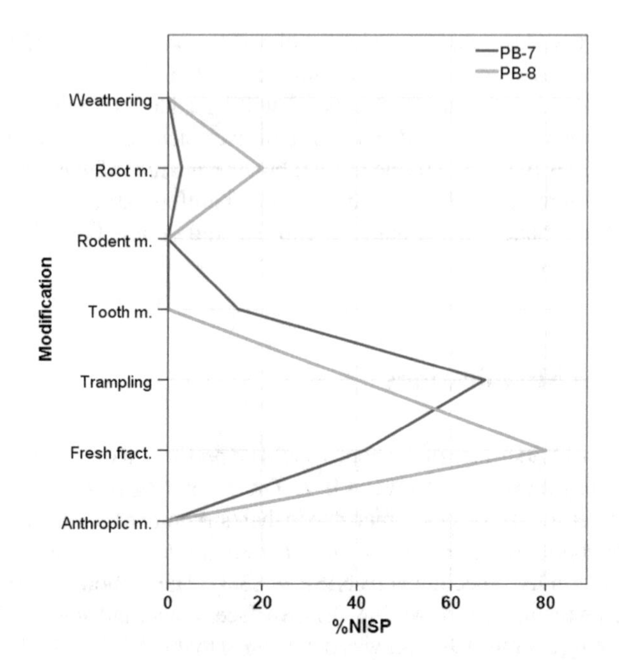

Fig. 8.7 Relative frequency of surface modifications in the sample studied. m: Marks

Table 8.6 Absolute and relative frequency (in parentheses) of fracture types per taxa in layer PB-7

Taxa	Fresh	Not fresh	Combination	Total
Notiomastodon aff. *N. platensis*	9 (32)	18 (64)	1 (4)	28
cf. *H. paradoxa*	3 (75)	1 (25)	0	4
Equus sp.	1 (50)	0	1 (50)	2
Anatinae undet.	1 (100)	0	0	1
Total	14 (40)	19 (54)	2 (6)	35

Due to most of the gomphotheres bones in layer PB-7showed non-fresh fractures, the absence of weathering could indicate that the fractures would have occurred once the bones were already in the sedimentary matrix, or the weathering was insignificant in this time.

Roots marks are scarce in the PB-7 sample. They are proportionately more abundant in PB-8, although here it is a very small sample (Fig. 8.7). In general, the marks did not affect the units completely. In two cases (*Notiomastodon* aff. *N. platensis* ribs), they were seen on only one side, suggesting some kind of post-depositional stability.

There was a significant proportion of evidence of trampling in the bone assemblage of PB-7 (%NISP = 67.1%). This modification is also recorded in the PB-8 layer, although it appears to be less significant (Fig. 8.7). The details of these modifications will be described below.

Fig. 8.8 MHMOP/PI/642. Diaphysis of the humerus of cf. *Hemiauchenia paradoxa* showing fresh fracture, from layer PB-7 (for an archaeological interpretation see Chap. 16, this volume)

Finally, it is noted that no human marks of any kind (combustion marks, cut marks, scraping marks, percussion marks, etc.) were recorded in the sample analyzed, despite the study of the surface of the bones with magnification up to 40x. Special emphasis was placed on this topic due to the archeological character of the site. However, the marks that could be mistaken for cut marks correspond in fact to marks of trampling, according to the characteristics described in the literature (Olsen and Shipman 1988; Domínguez-Rodrigo 2009, 2012, among others).

8.8 Teeth Marks

The bones have no evidence of rodent marks, but there are traces of carnivore teeth exclusively in bones recovered from layer PB-7. The latter were observed in about 14% of the sample studied, affecting only remains of *Notiomastodon* aff. *N. platensis* (Fig. 8.7). Labarca et al. (2014) described four gomphothere anatomical units with teeth marks in layer PB-7, produced during consumption of the bones by a large carnivore, possibly an extinct felid. These pieces corresponded to a tibia, astragalus, calcaneus, and an indeterminate epiphysis. After that, in the detailed study of the whole sample, six more bones were identified with tooth marks, increasing the number of pieces with this component to ten.

The scapula MHMOP/PI/640 has clear, very intensive, evidence of consumption by carnivores in its proximal segment, mainly in its lateral border. Consumption of scapular cartilage and of part of the proximal edge is observed, especially near to the spine. Large amounts of scoring are recorded on the spongy tissue in different directions. The edges of the blade toward its medial border are irregular and serrated. The bone was consumed from lateral border to medial border, because the medial border has less evidence of consumption than the lateral (Fig. 8.9). The specimen MHMOP/PI/664 (anterior rib) exhibits an evident furrowing in the neck area, with removal of bone tissue concentrated toward the caudal edge. Scoring and irregular edges are observed at the dorsal part of the head, toward the caudal segment (Fig. 8.9). Also, the specimen MHMOP/PI/611 (anterior rib) shows an irregular border fracture at the dorsal part of the head (currently detached), with clear furrowing in a "U" shape. This sector, however, was modified in an undetermined time after the feline consumption, so it is possible that there were more marks that could not be registered. In the body of the specimen, three short, shallow, and relatively flat scorings are observed by the caudal edge. A longer scoring completely and diagonally traverses the piece. It has a U-shaped section and irregular edges, with a maximum length of ca. 38 mm by a maximum width of 6 mm.

The intermediate rib MHMOP/PI/658 presents a possible furrowing in the articulation of the head with the neck. This piece, however, has recently lost bone tissue, so its modifications are uncertain. The lateral and posterior portions of the right cuboid MHMOP/PI/679 are significantly altered by carnivores. Portion of the articular surface for the calcaneus and for the fourth metatarsal are missing. There is a deep lateral groove (13.6 mm of maximum length) that crosses the piece in its middle section, reaching to its posterior face. It is associated with other minor furrows. The fifth metatarsal articular surface and the remnant of the fourth metatarsal articular surface present on their posterior and lateral edges, respectively, at least three scorings, and some furrowing, which gives the edges a sinuous appearance (Fig. 8.10). The first has a maximum width of 10.1 mm, the second, 11.7 mm. They are parallel and separated from each other by about 14.3 mm.

Finally, the specimen MHMOP/PI/650, a first phalanx, shows abundant tissue removal by carnivores in its distal right lateral side. There is clear evidence of furrowing, which produced a deep U-shaped cavity of ca. 7 mm. In addition, in the distal side toward the left there is a very shallow scoring, arranged dorso-plantar/palmar, of 18.7 mm in length and 6.5 mm in width.

Labarca et al. (2014) noted that although the intensive consumption of bone remains is not characteristic of current felids, the dimensions of the pits and scorings observed in the Pilauco sample are not compatible with any reference canid. Rather they approximate the measurements of modern lions, according to the actualistic data of Domínguez-Rodrigo and Piqueras (2003) and Delaney-Rivera et al. (2009). Additionally, based on the morphology of the marks, they could not have been generated by bears (Haynes 1983; Saladié et al. 2011). Borrero et al. (1997) and Martin (2008, 2013) described that at least the Pleistocene jaguar of southern Patagonia (*Panthera onca mesembrina*) regularly consumed bone fragments of camelids, equids, and mylodontids. Marean and Ehrhardt (1995) documented teeth marks on unfused proboscidean bones that were produced by a species of felid in the family Machairodontinae family (*Homotherium* sp.) in a North American Pleistocene context. In agreement with these findings, Van Valkenburgh and Hertel (1993)

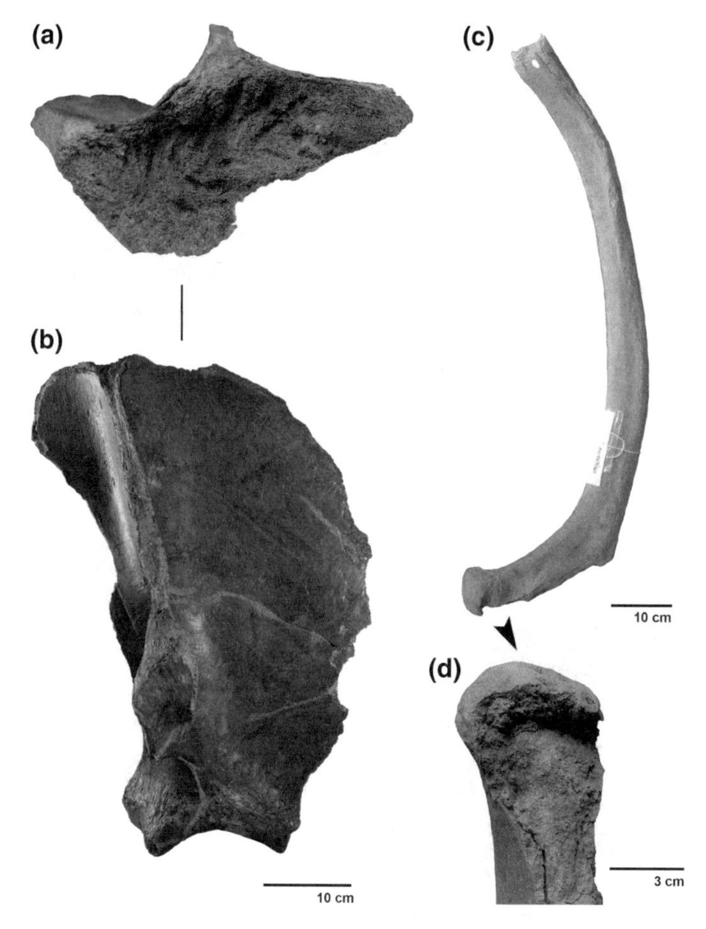

Fig. 8.9 Teeth marks of carnivores in bones of *Notiomastodon* aff. *N. platensis*. **a** and **b** Scapula MHMOP/PI/640 in proximal and lateral view, respectively; **b** and **c** Rib MHMOP/PI/640 in caudal and dorsal view, respectively

found a high rate of fractures in teeth of large carnivores at the Rancho La Brea site (USA). This was interpreted as an indication of greater consumption of bones than the current species. Considering all of the above, Labarca et al. (2014) suggest that a similar situation may have occurred in the Pleistocene of northern Patagonia. A big felid, then most likely extinct, may have regularly incorporated bones into its diet, and could have been responsible for the teeth marks observed in Pilauco.

The new pieces found with tooth marks keep anatomical and/or spatial coherence with previously documented specimens (see below). In addition, they show similar metric attributes with the already documented marks, such as length of the scoring in the epiphysis. The newly observed marks were recorded primarily in the epiphyses or anatomical segments with spongy tissue (e.g., neck of the ribs), a situation similar

Fig. 8.10 Carnivore teeth marks in bones of *Notiomastodon* aff. *N. platensis*. **a** and **b** Cuboids MHMOP/PI/679 in plantar and posterior view respectively; **c** and **d** Phalange 1 MHMOP/PI/650 in dorsal and latero-dorsal view, respectively

to the one already described. Thus, it is possible to point out that the producer of the new marks would have been the same that consumed the bones described by Labarca et al. (2014).

The horizontal distribution of bones with carnivorous tooth marks clearly show two spatially segregated loci where the carnivore would have consumed the skeletal remains. The first is located near the skull, around the grids 14AC and 15AC. Located here are a tibia, astragalus, calcaneus, cuboid, phalanx, and an intermediate rib (Fig. 8.4). The first four pieces have the same laterality (left), while the tibia articulates with the fibula and astragalus. Therefore, it is possible to suggest that this segment would have been displaced by the carnivore(s) from its original position, to be later consumed. Additionally, in this first locus, the individual(s) would have partially consumed the dorsal portion of a rib. The second locus is situated on the

15A and 15AA grids. It is associated with a left scapula, a left astragalus, and an anterior rib (Fig. 8.4). Assuming that the right scapula is relatively in situ (13AD grid), it is feasible to assume that the left scapula would also have been displaced from its original location to be consumed. The same can be postulated for the anterior rib, if it is assumed that the skull has not undergone major resettlement from its original location. The right calcaneus with teeth marks is noteworthy, as it corresponds to the only piece of the right posterior limbs recovered in the site, possibly suggesting that both hind legs are distal. They were consumed and displaced, but only one of them remained in the original bone concentration area.

8.9 Trampling Marks

Trampling marks considerably affected ~67% the bone ensemble excavated from layer PB-7. They were also detected in the PB-8 layer, affecting ~40% of the sample (Table 8.7). In the PB-7 layer, marks affect all taxa, even though they are proportionately more common in bones of gomphotheres (Table 8.8). With regard to the position over the bone, the majority were observed covering the complete specimen (%NISP = 75.5). This situation is expressed more clearly in the remains of proboscideans, where 82% of the sample shows traces of trampling on their entire surface. In camelids and equids, although the sample is small, the situation is inverted (Table 8.7). Various studies have pointed out the possibility that trampling marks could be confused with cut marks (Olsen and Shipman 1988; Behrensmeyer et al. 1986; Domínguez-Rodrigo 2009, 2012), as they may share some morphological attributes.

Considering the antiquity of the PB-7 layer and the presence of lithic artifacts spatially associated with the remains of extinct mammals, attempting to discern the cause of the striations was necessary. This was done in accordance with the protocol proposed by Domínguez-Rodrigo et al. (2009, 2012), who defined a series of visible morphological attributes with low magnification (40x). These could be analyzed together in order to reliably differentiate between types of marks.

Table 8.8 shows the quantification result of a five-bone sample from the PB-7 layer (Fig. 8.11) in accordance with the protocol proposed by Domínguez-Rodrigo et al.

Table 8.7 Absolute frequency and relative frequency (in parentheses) of trampling marks per taxa, with consideration of their position, for the specimens of layer PB-7

	NISP/%NISP	Location	
		Complete	Partial
Notiomastodon aff. *N. platensis*	38 (71.7)	31 (81.5)	7 (18.4)
cf. *H. paradoxa*	4 (66.7)	1 (25)	3 (75)
Equus sp.	1 (33.3)	0	1 (100)
Pilosa undet.	2 (50)	2 (100)	0
Total	45	34 (75.5)	11 (24.5)

Table 8.8 Summary of the absolute and relative quantification (in parentheses) of the strations attributes in bones of *Notiomastodon* aff. *N. platensis* (layer PB-7)

	MHMOP/PI/611	MHMOP/PI/614	MHMOP/PI/615	MHMOP/PI/630	MHMOP/PI/629
	Rib	Tibia	Rib	Skull	Atlas
Analyzed marks	116	21	64	81	12
Micro-abrasion	p	p	p	P	p
Overlapping estriae	p	p	p	P	a
Groove trajectory straight	12 (10)	4 (19)	2 (3)	11 (14)	1 (8)
Groove trajectory sinuous	104 (90)	17 (81)	62 (97)	70 (86)	11 (92)
Barb	0	0	0	0	0
Sides of the groove symmetrical	6 (6)	1 (5)	0	12 (15)	0
Sides of the groove asymmetrical	110 (95)	20 (95)	64 (100)	69 (85)	12 (100)
_/shape Groove	116 (100)	21 (100)	64 (100)	81 (100)	12 (100)
V shape Groove	0	0	0	0	0
Shallow Groove	106 (91)	18 (86)	64 (100)	81 (100)	12 (100)
Deep Groove	10 (86)	3 (14)	0	0	0
Internal microestriation	8 (7)	1 (5)	0	1 (1)	0
Microestriation on bottom	8 (7)	1 (5)	0	1 (1)	0
Microestriation on walls	0	1 (5)	0	0	0
Microestriation straight	0	1 (5)	0	1 (1)	0
Microestriation irregular	8 (7)	0	0	0	0
Microestriation continuous	1 (1)	0	0	0	0
Microestriacion discontinuous	7 (6)	1 (5)	0	1 (1)	0

(continued)

Table 8.8 (continued)

	MHMOP/PI/611	MHMOP/PI/614	MHMOP/PI/615	MHMOP/PI/630	MHMOP/PI/629
	Rib	Tibia	Rib	Skull	Atlas
Shoulder effect	0	1 (5)	0	0	0
Flaking on shoulder	0	0	0	0	0
Max length (mm)	22.7	14.7	28.2	33.8	20.9
Max width (mm)	0.9	0.3	1.1	0.8	0.4

(2009, 2012). First, it highlights the importance of the number of marks presented by each sample, considering that only a surface of 20 × 10 cm was studied in each bone. Olsen and Shipman (1988) have pointed out that the cut marks are comparatively less numerous than those generated by trampling, because they are produced by specific butchery actions. For this reason, they have also suggested that cut marks are located in specific parts of the bones, usually associated with muscle and/or ligament insertions, which contrasts with the Pilauco samples. In Table 8.8 and Fig. 8.12, a great homogeneity is observed in the proportions of the various attributes shown for each bone, independent of the number of striations. The marks have mainly a sinuous orientation, do not present barbs, exhibit asymmetrical sides, have sections in a "_/" shape, are mostly superficial, have very low internal micro-striation and they are irregular and discontinuous. To these characteristics should be added the presence of micro-abrasion, overlap of striations inside the incision, and the recurrent presence of marks that overlap each other. The proportions of the attributes recorded in the bones were compared with the experimental results obtained by Domínguez-Rodrigo et al. (2009) (Fig. 8.12). Through this comparison it was possible to establish similarities between the fossil bones and the trampling marks, dispelling the indicated proportions for the cut marks generated by unifacial stone flakes. Considering these results, we propose that the marks present on the surface of the bones studied correspond to trampling marks. Thus, the presence of cut marks in the sample analyzed should be discarded. The remaining bone units with incisions were studied in a less intensive manner, but generally present the same pattern (Fig. 8.13).

8.10 Discussion and Conclusions

Borrero (2015) noted that all archeological evidence of the initial settlement of the continent should be studied as diligently as possible, without discarding unusual a priori evidence, and avoiding blind acceptance without discussion of information. The taphonomic study presented here is oriented precisely in that direction.

The results attained in this work confirm the general impressions indicated in previous publications. The Pilauco site presents an intricate formational history, where diverse taphonomic agents have aided in the configuration of the fossil record.

The spatial distribution in sector W of layer PB-7 showed a locus mainly consisting of a single gomphothere, associated with some isolated remains of equids, camelids, and dermal ossicles of an undetermined pilosa. The relative anatomical coherence presented by the remains of proboscideans indicated a relatively in situ death of the animal, even though some elements of the skeleton have not been recovered. Although the absence of a part of the gomphothere assemblage may be due to excavation decisions, it is possible that a portion of it may have been removed from its original place of deposit to other areas outside the site.

The PB-7 and PB-8 layers interfinger to the south with fluvial deposits of the ancient Damas. River. Therefore, could be possible that after the postmortem process of natural disarticulation (Hill 1979), the bones could have been perturbed by

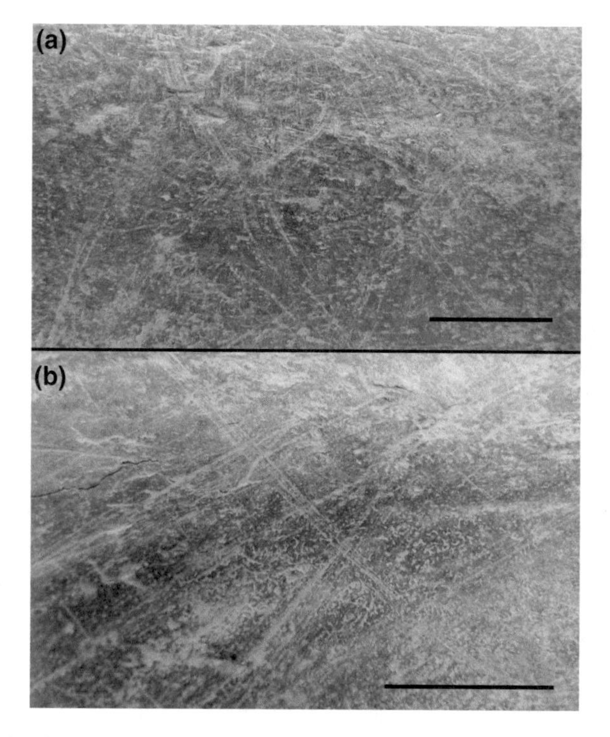

Fig. 8.11 Marks of trampling on bones of gomphotheres from layer PB-7. **a** Rib MHMOP/PI/611; **b** Skull MHMOP/PI/630. The bar measures 1 cm

some type of rearrangement/displacement by hydric causes. However, there are no stratigraphic or sedimentological evidence of fluvial deposits in PB-7 and PB-8 layers (see Chap. 3, this volume).

Since the initial works of Voorhies (1969), several experimental approaches have been developed to define which pieces are most susceptible to fluvial displacement (i.e., Boaz and Behrensmeyer 1976; Kaufmann 2009; Evans 2015; Domínguez-Rodrigo et al. 2017; Frison and Todd 1986). The results of these works have been somewhat dissonant due to the different variables incorporated in the studies (type of taxa, state and articulation of the bone remains, velocity of flow, etc.). According to the main taxa recovered at Pilauco and considering the low fragmentation of the sample, indices that were created from remains of African elephants (*Loxodonta africana*) were used (fluvial transport, FTI and saturated weight, SWI). The correlation between the % MAU of *Notiomastodon* aff. *N. platensis* and the FTI yielded positive but no statistically significant results ($\rho = 0.376$; $p = 0.151$; $n = 16$). However, the correlation between the same % MAU and the SWI delivered slightly negative but statistical no significant results ($\rho = -0.069$, $p = 0.779$, $n = 19$). These values suggest that the fluvial agents no modeled the Pilauco gomphotheres bone assemblage.

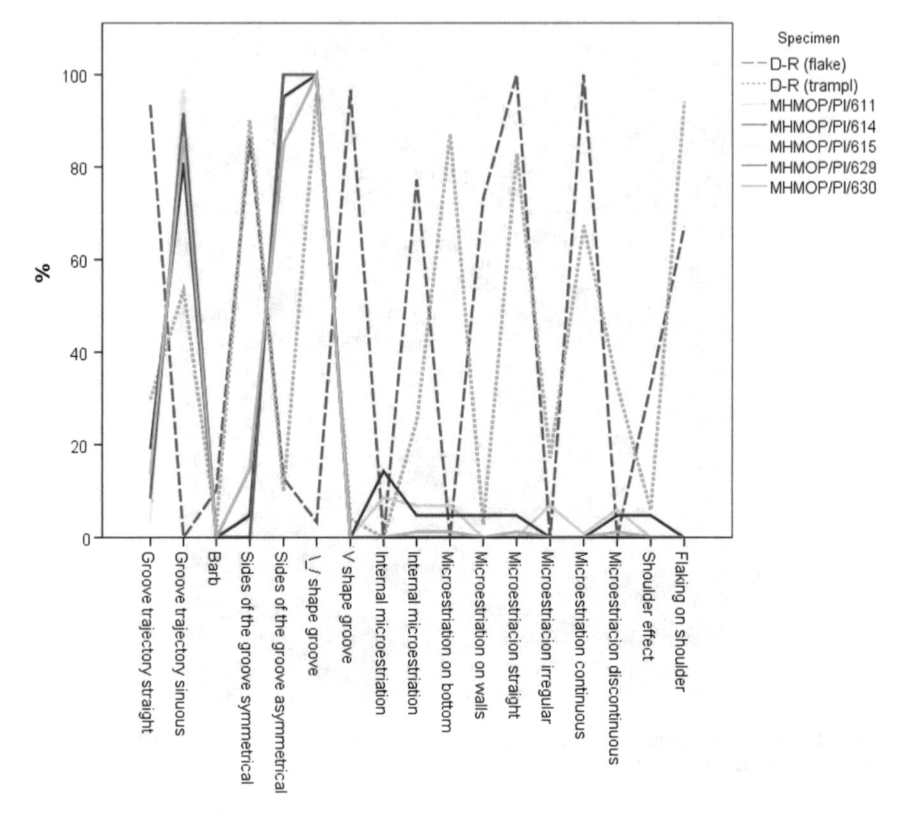

Fig. 8.12 Relative frequency of distinct morphological attributes of the marks registered in the selected sample from Pilauco, and the comparative data of Domínguez-Rodrigo et al. (2009)

Within the sample are bones representing high (e.g., calcaneus and astragalus), medium (e.g., ribs), and low (e.g., skull) displacement chances. The absence of a predominant direction in the orientation of the bone remains (Fig. 8.5), would indicate the same. This also takes into consideration that the current channel of the river has a general NW-SE orientation. Moreover, the bone remains of gomphotheres of layer PB-7 sector W do not exhibit abrasion or polish marks associated with the action of water (Fernández-Jalvo and Andrews 2003). The latter also applies to the remains of artiodactyls and perissodactyls, so the deposition causes for these bones should be related to other agents. All of the above is concordant with the sedimentological analyses, which indicate a primarily colluvial origin for the sediments of layers PB-7 and PB-8 (see Chap. 3, this volume).

It is important to note that some skeletal remains assigned to cf. *Hemiauchenia paradoxa* registered in the East site sector show abrasion marks that were interpreted as generated by fluvial transport (Labarca et al. 2013). Some of these pieces correspond to anatomical elements regularly associated with a high probability of transport (e.g., astragalus, Voorhies 1969; Kaufmann et al. 2011). These differences

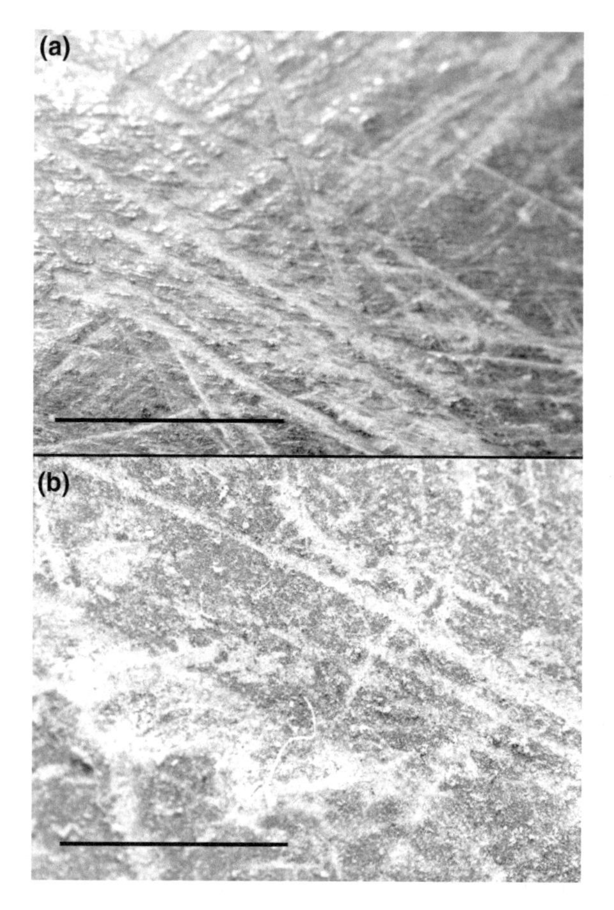

Fig. 8.13 Trampling marks on bones of gomphotheres from layer PB-7 of the Pilauco site. **a** Rib MHMOP/PI/611; **b** Rib MHMOP/PI/620. The bar measures 5 mm

between sectors could be related to different taphonomic histories within the site. Another possibility is that other non-hydric processes, such as wind action, could generate similar abrasion marks. This issue cannot yet be resolved with the information available for sector E.

Another agent that could have contributed to the relocation and displacement of anatomical elements in the PB-7 layer are the humans. This includes the presence of a lithic industry consisting of small unifacial stone tools and their waste. They are manufactured mainly in dacitic volcanic glass and basalt, which are spatially associated with the remains of Pleistocene fauna (Navarro-Harris et al. 2019; Chap. 16, this volume). Moreover, Navarro-Harris et al. (2019) suggested that Pilauco was a working place to handle meat, at least during two intervals, since part of the lithic ensemble shows characteristics consistent with cutting and scraping activities. As noted previously, the detailed surface analysis of the skeletal remains in Pilauco ruled out the presence of cultural cut marks. However, these could also be absent within a scenario of low-intensive butchering, which for now cannot be confirmed. Indeed, the actualistic studies developed by Haynes in Africa suggest that the possibility of leaving cut marks in bones of proboscideans is generally low. This is due to the large amount of muscle, fat, and thick periosteum that surround the skeletal remains (Haynes 1987, 1991; Haynes and Klimowicz 2015). This same author has suggested that skilled butchers can process an elephant without leaving any mark, although this depends on the intensity of the processing of the carcasses. If the animal is not intensively processed, it is unlikely for the tools to be able to reach the surface of the bones (Haynes and Klimowicz 2015). Thus, Haynes expects that archeological sites with remains of proboscideans should generally present a low frequency of cut marks on bones, with values around 1% of the total sample (Crader 1983; Haynes and Klimowicz 2015). Taphonomic studies of fossil materials in different parts of the world and with different chronologies seem to confirm this impression. In general, there is a limited presence of cut marks on bones of proboscideans in archeological sites (e.g., Saunders and Daeschler 1994; Yravedra et al. 2010; Saccà 2012).

Crader (1983) and Haynes and Krasinski (2010) recorded some anatomical units with percussion/chop marks in modern sites of elephant processing. In consideration of their experimental studies, De Juana et al. (2010) suggested that this is an effective technique to disarticulate anatomical units of large animals. This type of modifications leaves marks clearly visible, since it necessarily implies a contact between the bone and the butchering instrument. Despite the degree of anatomical disorder observed in the studied sample of layer PB-7 (Fig. 8.4), no marks of disarticulation by percussion have been observed in any bone in the collection. Therefore, the loss of the anatomical connection between the bones of the gomphotheres of Pilauco seems to be better explained by another process.

On the other hand, evidence of fracture by percussion (negative flakes, pits) in the diaphysis of long bones have been recorded in bone assemblages of archeological sites with remains of proboscideans (Santucci et al. 2016; Saccà 2012; Villa et al. 2005; Gaudzinski et al. 2005). These have been associated with obtaining raw materials rather than for bone marrow extraction (Saccà 2012). In Pilauco, no fractures or percussion marks that could be related to this type of activity have been identified in long bones of gomphotheres, as most of the fractures were made in a non-fresh state and on ribs. Therefore, the absence of cut and percussion marks (on diaphysis and epiphysis) in gomphothere bones enabled the ruling out of an intensive human processing of the carcass deposited in layer PB-7. It also led to the assumption that the anatomical disorder and rearrangement of the bone units would not have been due to human causes.

The study showed an absence of certain anatomical units, those that could have been moved to the residential camps for their high economic yield, such as femurs, coxal bones, and vertebrae (Binford 1978). However, the excessive weight of these bone units in the proboscideans makes this possibility unfeasible, except that the settlements are located near primary butchering area, as suggested by the ethnoarcheological evidence assessed with the Hadza people (e.g., Lupo 2006; Monahan 1998). Just the femur of an African elephant (without flesh) weighs between 14 and 23 kg, while the hip can weigh between 21 and 91 kg without flesh (Haynes and Klimowicz 2015). Fisher (1992) has documented ethnographically that the hip, along with the skull, jaw, and scapula, are bones that remain in the sites of elephant slaughter of the Ituri Forest People, while long bones were moved for the extraction of bone marrow. In these cases, temporary camps were set up 10 m from the processing site. With regard to Pilauco, it is not logical that high yield units such as ribs had been quickly butchered, while units of a similar yield, but heavier, such as femurs and coxal bones were moved toward the camps. In this way, the absence of anthropic marks (cut and percussion), the inconsistency in the eventual transport strategies and a lithic instrument not very suitable for intensive processing of large prey, makes it necessary to think of other alternatives that better explain the absence of the anatomical units of the pelvic girdle and rear extremity.

A third alternative is that the dispersal and selective extraction of the skeletal units have been carried out mainly by carnivores during their feeding process. This scenario has stronger empirical evidence, as teeth marks of large carnivores in the bones of proboscideans have been consistently documented in PB-7. In this regard, Labarca et al. (2014) were inclined to an extinct felid as the responsible agent, which until now is not represented in the fossil record of northern Patagonia. The consumption of bones by extinct Pleistocene felids has been well documented in southern Patagonia. Here the extinct jaguar (*Panthera onca mesembrina*) regularly incorporated bones of different species into their diet (Martin 2008, 2013). For the machairodonts (also identified in southern Patagonia through the genus *Smilodon populator* Lund, Prieto et al. 2010), the information available is comparatively limited. However, the paleontological research developed in North America suggests a certain tendency to avoid bone consumption due to the dental structure of Machairodontinae (Marean and Ehrhardt 1995). Labarca et al. (2014) suggest that the gomphothere carcass could have been scavenged by the carnivores, judging by the marks on the distal tibia and the fact that a felid could have hardly hunted an adult proboscid with an estimated mass of around three tons (Ferretti 2010). This impression is maintained here, postulating a scavenging scenario of the units that could have been found partially articulated, such as the distal part of the right hind leg. The anatomical disarray observed in the ribs could also be due to the action of carnivores, which is confirmed by the presence of teeth marks in two of these anatomical units. It is not possible to support the idea that the scavenging events had necessarily happened immediately after the death of the animal. Haynes (1991) suggests that the skeletal remains of proboscideans can be maintained for several months with some degree of articulation. This is due to their ligaments and the thickness of their skin, which helps to delay the natural processes of disarticulation. It is interesting to note that this same author suggests that from actualistic observation, the scavengers can accelerate the processes of disarticulation, especially in those cases that have been intensely scavenged. Indeed, Haynes (Haynes and Klimowicz 2015) suggests that the animals "*begin to feed (…), by working their way through the thinnest skin of the abdomen, moving upwards on rear limb elements to the innominates, toothmarking and breaking back the edges of the ilium, then continuing to feed forward to the ribs, which are broken either in midshaft or near the vertebral articulation, and eventually forward to clean tissue from scapulae where vertebral borders are toothmarked and broken*". The disarticulation of the appendicular elements would occur in a later stage of scavenging, ending with the separation of the skull and the jaw from the spine, and the segmentation of the latter in several parts. The presence of ribs, scapula, and distal leg segment marked with teeth is consistent with that described by Haynes and seems to account for a carcass that was intensely scavenged before its final burial. It is possible that this process occurred with limited temporal intervals. Within this scenario, it may be suggested that part of the bone units could have been moved outside the central deposition area by the carnivores themselves. Haynes and Klimowicz (2015) mention scavenged bones separated up to 30 m from their original place of deposit. We know that at least the Patagonian jaguar moved bones for later consumption (Borrero et al. 1997; Martin 2008, 2013), who analyzed several rock-shelters with faunal assemblages which were interpreted as dens of the Pleistocene jaguar. Among the transported bones have been identified fragments of skull of mylodon (*Mylodon darwini* Owen), so it is possible that the hemipelvis or femurs of the gomphotheres of Pilauco could have been transported by carnivores.

Another group of marks very conspicuous in the sample studied are those generated by trampling. These would have been caused by the contact of the bones with abrasive particles of the sedimentary matrix containing the bones (Domínguez-Rodrigo et al. 2009). Therefore, this process necessarily occurred once the bones were already completely skeletonized. Several fossils are completely covered by marks of this type, suggesting that it was a systematic and continuous process, even after the burial of the bones. The presence of bones in oblique and subvertical positions suggests that trampling may have also contributed to the displacement of some parts, both horizontally and vertically. It is possible that some of the taxonomic similarity between PB-8 and upper PB-7 is due to this process, as materials could have migrated from PB-8 to PB-7. Similarly, trampling may have contributed to the fragmentation of some ribs. This assumes that the fractures would have occurred in situ, perhaps once the materials were already buried, judging by the considerable presence of non-fresh fractures. Limited experimental studies of trampling developed with humans on bone remains of cattle (*Bos taurus* Linaneus) and sediments extracted from the Pilauco site have generated marks similar to those observed in the fossilized bones. However, these were recorded in lower intensity and appear to be more superficial (Guajardo 2013).

Finally, it is important to briefly discuss the depositional origin of the remains that do not correspond to proboscideans in the PB-7 layer. It was noted previously that the bones are scattered without defined concentrations, and some were even located around the remains of gomphotheres. In terms of modifications, they do not show teeth marks nor cultural marks, but rather marks of trampling. Ruling out fluvial transport as a potential factor, it is possible to suggest a gravity-driven colluvial transfer of the elements from the high sectors located north of the site (see Chap. 3, this volume). Once inserted in the sedimentary matrix, the pieces would be subjected to trampling. Considering that pieces from both PB-7 and PB-8 show this type of modification, it is possible to assume that some bone elements of PB-8 could have vertically migrated to PB-7 due to trampling. This scenario could explain the presence of similar taxa between upper PB-7 and PB-8. The pieces in PB-7 don't show particularly significant variations that would denote a different taphonomic history than the bones of gomphotheres (e.g., higher level of weathering, significant root marks). Despite this, it is notable that the bone assemblages of camelids and equids present a greater proportion of fresh fractures, including a diaphysis of a humerus with a very clear, fresh fracture.

Finally, the origin of the few materials recovered in PB-8 could be related to a colluvial redeposition, even if the sample is too small to develop an extensive discussion. It is interesting to note that in east sector the fauna ensemble is more diverse and apparently more abundant (Recabarren et al. 2011; Pino et al. 2013; González et al. 2014). Again, this may suggest differences in the depositional processes between the two sectors, which for the moment cannot be evaluated comparatively.

Through the analysis of the surface modifications of the skeletal remains and their horizontal and vertical distribution, this research was able to define a series of taphonomic processes occurring within the Pilauco site, particularly in sector W of layer PB-7. All of the skeletal evidence recovered so far suggests that the Pilauco site corresponded mainly to a place of feeding for large extinct felids. These animals intensely modified a gomphothere carcass, probably dead in situ, accelerating a process of natural disarticulation that had apparently already begun. The carnivores that fed on the carcass (perhaps on more than one occasion) would have been principally responsible for the relative anatomical disorder of the units recovered on the site, and perhaps the absence of some anatomical elements. The evidence rules out human incidence in the formation of the fossil assemblage. In addition, the fossil set was enriched by the contribution of bones of equids, camelids, and pilosas transferred colluvially to the site, from high sectors. The trampling would have contributed to the horizontal and vertical displacement of anatomical units, while significantly marking the surface of the bones. Some of the fractures recorded in the sample could be explained by this process, which seems to have continued even once the remains were completely inserted into the sedimentary matrix.

References

Baumel JJ, Witmer LM (1993) Osteologia. In: Baumel JJ, King AS, Breazile JE, Evans HE, Vanden Berge JC (eds) Handbook of avian anatomy: nomina anatomica avium, 2nd edn. Nuttall Ornithological Club, Cambridge, MA, pp 45–132

Behrensmeyer AK (1978) Taphonomic and ecologic information from bone weathering. Paleobiol 4(2):150–162

Behrensmeyer AK, Gordon KD, Yanagi GT (1986) Trampling as a cause of bone surface damage and pseudotools. Nature 319:768–771

Binford LR (1978) Nunamiut ethnoarchaeology. Academic Press, New York

Binford LR (1981) Bones: ancient men and modern myths. Academic Press, New York

Boaz NT, Behrensmeyer AK (1976) Hominid taphonomy: transport of human skeletal parts in an artificial fluviatile environment. Am J of Phys Anth 45:53–60

Borrero LA (2015) Con lo mínimo: los debates sobre el poblamiento de América del Sur. Intersecc 16(1):5–14

Borrero LA, Martin FM, Prieto A (1997) La Cueva Lago Sofía 4, Última Esperanza: una madriguera de felino del Pleistoceno tardío. An del Inst Patagon 25:103–122

Crader DC (1983) Recent single-carcass bone scatters and the problem of "butchery" sites in the archaeological record. In: Clutton-Brock J, Grigson C (eds) Animals and archaeology, hunters and their prey. BAR Int Ser 163(1):107–141

De Juana S, Galán AB, Domínguez-Rodrigo M (2010) Taphonomic identification of cut marks made with lithic handaxes: an experimental study. J Archaeol Sci 37:1841–1850

Delaney-Rivera C, Plummer TW, Hodgson JA, Forrest F, Hertel F, Oliver JS (2009) Pits and pitfalls: taxonomic variability and patterning in tooth mark dimensions. J Archaeol Sci 36(11):2597–2608

Dillehay TD (1997) Monte Verde: a late pleistocene settlement in Chile. The archaeological context and interpretationk, vol 2. Smithsonian Institution Press, Washington, DC

Domínguez-Rodrigo M, Piqueras A (2003) The use of tooth pits to identify carnivore taxa in tooth-marked archaeofaunas and their relevance to reconstruct hominid carcass processing behaviours. J Archaeol Sci 30(11):1385–1391

Domínguez-Rodrigo M, de Juana S, Galán AB (2009) A new protocol to differentiate trampling marks from butchery cut marks. J Archaeol Sci 36:2643–2654

Domínguez-Rodrigo M, Pickering TR, Bunn HT (2012) Experimental study of cut marks made with rocks unmodified by human flaking and its bearing on claims of 3.4-million-year-old butchery evidence from Dikika, Ethiopia. J Archaeol Sci 39:205–214

Domínguez-Rodrigo M, Cobo-Sánchez L, Yravedra J, Uribelarrea D, Arriaza C, Organista E, Baquedano E (2017) Fluvial spatial taphonomy: a new method for the study of post-depositional processes. Archaeol Anthropol 10(7):1769–1789

Evans TE (2015) Critical evaluation of our understanding of bone transport and deposition in fluvial channels. Dissertation, Montana State University

Fernández-Jalvo T, Andrews P (2003) Experimental effects of water abrasion on bone fragments. J Taphon 1(3):147–163

Ferretti MP (2010) Anatomy of Haplomastodon chimborazi (Mammalia, Proboscidea) from the late Pleistocene of Ecuador and its bearing on the phylogeny and systematics of South American gomphotheres. Geodiv 32(4):663–721

Fisher JW Jr (1992) Observations on the late pleistocene bone assemblage from the Lamb Spring site, Colorado. In: Stanford DJ, Day JS (eds) Ice age hunters of the rockies. Denver Museum of Natural History and University Press of Colorado, Denver, pp 51–82

Frison GC, Todd LC (1986) The Colby Mammoth Site: taphonomy and archaeology of a clovis kill in northern wyoming. University of New Mexico Press, Albuquerque, New Mexico

Gaudzinski S, Turner E, Anzidei AP, Álvarez-Fernández E, Arroyo Cabrales J, Cinq-Mars J, Dobosi VT, Hannus A, Johnson E, Münzel SC, Sher A, Villa P (2005) The use of proboscidean remains in everyday palaeolithic life. Quat Res 126–128:179–194

Gifford-González D (1991) Bones are not enough: analogues, knowledge, and interpretive strategies in zooarchaeology. J Anthropol Archaeol 10:215–254

González E, Prevosti FJ, Pino M (2010) Primer registro de Mephitidae (Carnivora: Mammalia) para el Pleistoceno de Chile. Magallania 38(2):239–248

González E, Labarca R, Chávez-Hoffmeister M, Pino M (2014) First fossil record of the smallest deer cf. Pudu Molina, 1782 (Artiodactyla, Cervidae), in the late pleistocene of South America. J Vert Paleont 34(2):483–488

Grayson D (1984) Quantitative zooarchaeology. Academic Press, Orlando

Guajardo A (2013) Análisis experimental de la presencia de marcas de pisoteo, en fósiles animales del sitio Pilauco, Osorno, Centro-Sur de Chile. Undergraduate Dissertation, Universidad Austral de Chile
Haynes G (1983) A guide for differentiating mammalian carnivores taxa responsible for gnaw damage to herbivore limb bones. Paleobiol 9(2):164–172
Haynes G (1987) Elephant-butchering at modern mass-kill sites in Africa. Curr Res Pleist 4:75–77
Haynes G (1991) Mammoths, mastodonts, and elephants: biology, behavior, and the fossil record. Cambridge University Press, Cambridge
Haynes G (2017) Taphonomy of the Inglewood mammoth (Mammuthus columbi) (Maryland, USA): green-bone fracturing of fossil bones. Quat Int 445:171–183
Haynes G, Klimowicz J (2015) Recent elephant-carcass utilization as a basis for interpreting mammoth exploitation. Quat Int 359–360:19–37
Haynes G, Krasinski KE (2010) Taphonomic fieldwork in southern Africa and its application in studies of the earliest peopling of North America. J Taphon 8(2–3):181–202
Hill A (1979) Disarticulation and scattering of mammalian skeletons. Paleobiol 5(3):261–274
Kaufmann C (2009) Estructura de edad y sexo en guanaco. Estudios actualísticos y arqueológicos en Pampa y Patagonia. Patagonia. Sociedad Argentina de Antropología, Buenos Aires
Kaufmann C, Gutiérrez MA, Álvarez MC, González E, Massigoge A (2011) Fluvial dispersal potential of guanaco bones (Lama guanicoe) under controlled experimental conditions: the influence of age classes to the hydrodynamic behavior. J Archaeol Sci 38:334–344
Krumrey WA, Buss IO (1968) Age estimation, growth, and relationships between body dimensions of the female African elephant. J Mammal 49(1):22–31
Labarca R, Pino M, Recabarren O (2013) Los Lamini (Cetartiodactyla: Camelidae) extintos del yacimiento de Pilauco (Norpatagonia chilena): aspectos taxonómicos y tafonómicos preliminares. Est Geol 69(2):255–269
Labarca R, Recabarren O, Canales-Brellenthin P, Pino M (2014) The gomphotheres (Proboscidea: Gomphotheriidae) from Pilauco site: scavenging evidence in the late pleistocene of the chilean patagonia. Quat Int 352:75–84
López P, Cartajena MI, Carabias D, Morales C, Letelier D, Valentina F (2016) Terrestrial and maritime taphonomy: differential effects on spatial distribution of a late pleistocene continental drowned faunal bone assemblage from the Pacific coast of Chile. Archaeol and Anthropol Sci 8(2):277–290
Lupo K (2006) What explains the carcass field processing and transport decisions of contemporary hunter-gatherers? Measures of economic anatomy and zooarchaeological skeletal part representation. J Archaeol Meth Theory 13:19–66
Lyman RL (1994) Vertebrate taphonomy. Cambridge Manuals in Archaeology, Cambridge
Lyman RL (2008) Quantitative paleozoology. Cambridge Manuals in Archaeology, Cambridge
Marean CW, Ehrhardt CL (1995) Paleoanthropological and paleoecological implications of the taphonomy of a sabertooth's den. J Hum Evol 29(6):515–547
Martin FM (2008) Bone-crunching felids at the end of the Pleistocene in Fuego-Patagonia, Chile. J Taphon 6:337–372
Martin FM (2013) Tafonomía y paleocología de la transición Pleistoceno-Holoceno en Fuego Patagonia. Interacción entre humanos y carnívorosy su importancia como agentes en la formación del registro fósil. Ediciones de la Universidad de Magallanes, Punta Arenas
Mengoni Goñalons G (1999) Cazadores de Guanacos de la estepa Patagónica. Colección tesis doctorales, Sociedad Argentina de Antropología
Monahan CP (1998) The Hadza carcass transport debate revisited and its archaeological implications. J Archaeol Sci 25:405–424
Moss C (1996) Getting to know a population. In: Kangwana K (ed) Studying elephants. Kenya, African Wildlife Foundation, pp 58–74
Mothé D, Avilla LS, Winck GR (2010) Population structure of the gomphothere Stegomastodon waringi (Mammalia: Proboscidea: Gomphotheriidae) from the pleistocene of Brazil. Ann da Academ Bras de Cien 82(4):983–996

Muñoz AS, Mondini NM, Durán V, Gasco A (2008) Los pumas (Puma concolor) como agentes tafonómicos. Análisis actualístico de un sitio de matanza en los Andes. Geobios 41:123–131

Navarro-Harris X, Pino M, Guzmán-Marín P, Lira MP, Labarca R, Corgne A (2019) The procurement and use of knappable glassy volcanic raw material from the late Pleistocene Pilauco site, Chilean Northwestern Patagonia. Geoarchaeology 1–21. https://doi.org/10.1002/gea.21736

Olsen SL, Shipman P (1988) Surface modification on bone: trampling versus butchery. J Archaeol Sci 15:535–553

Outram AK (2001) A new approach to identifying bone marrow and grease exploitation: why the "indeterminate" fragments should not be ignored. J Arch Sci 28(4):201–210

Pino M (2008) Pilauco, un sitio complejo del Pleistoceno tardío. Osorno, norpatagonia chilena. Universidad Austral de Chile, Imprenta América, Valdivia

Pino M, Chávez-Hoffmeister M, Navarro-Harris X, Labarca R (2013) The late Pleistocene Pilauco site, Osorno, south-central Chile. Quat Int 299:3–12

Prieto A, Labarca R, Sierpe V (2010) Presence of Smilodon populator lund (Carnivora, Felidae, Machairodontinae) in the late pleistocene of southern chilean patagonia. Rev Chi de Hist Nat 83(2):299–307

Recabarren O, Pino M, Cid I (2011) A new record of Equus (Mammalia: Equidae) from the late pleistocene of central-south Chile. Rev Chi de Hist Nat 84:535–542

Recabarren O, Pino M, Alberdi MT (2014) La Familia Gomphotheriidae en América del Sur: evidencia de molares al norte de la Patagonia chilena. Est Geol 70(1):e001

Saccà D (2012) Taphonomy of Palaeloxodon antiquus at Castel di Guido (Rome, Italy): proboscidean carcass exploitation in the lower palaeolithic. Quat Int 276–277:27–41

Saladié P, Huguet R, Díez C, Rodríguez-Hidalgo A, Carbonell E (2011) Taphonomic modifications produced by modern brown bears (Ursus arctos). Int J Osteoarch 23:13–33

Santucci E, Marano F, Cerilli E, Fiore I, Lemorini C, Palombo MR, Anzidei AP, Bulgarelli GM (2016) Palaeoloxodon exploitation at the Middle Pleistocene site of La Polledrara di Cecanibbio (Rome, Italy). Quat Int 406:169–182

Saunders JJ, Daeschler EB (1994) Descriptive analyses and taphonomical observations of culturally-modified mammoths excavated at "The Gravel Pit", near Clovis, New Mexico in 1936. Proc Acad Nat Sci Phil 145:1–28

Silver IA (1963) The ageing of domestic animals. In: Brothwell D, Higgs E (eds) Science in archaeology: a survey of progress and research, Revised edn. Thomas and Hudson, Great Britain, pp 283–302

Simpson GG, Paula-Couto C (1957) The mastodonts of Brazil. Am Mus of Nat Hist 112(2):131–145

Smuts M, Bezuidenhout AJ (1987) Anatomy of the dromedary. Clarendon Press, Oxford

Suárez R, Borrero LA, Borrazzo K, Ubilla M, Martínez S Perea D (2014) Archaeological evidences are still missing: a comment on Fariña et al. Arroyo del Vizcaíno site, Uruguay. Proc Roy Soc B 281:20140449

Tassy P (1996) Growth and sexual dimorphism among Miocene elephantoids: the example of Gomphotherium angustidens. In: Shoshani J, Tassy P (eds) The Proboscidea. Evolution and palaeoecology of elephants and their relatives. Oxford University Press, Oxford, pp 92–100

Van Valkenburgh B, Hertel F (1993) Tough times at La Brea: tooth breakage in large carnivores of the late pleistocene. Science 261:456–459

Villa P, Soto E, Santonja M, Pérez-González A, Mora R, Parcerisas J, Sesé C (2005) New data from Ambrona: closing the hunting versus scavenging debate. Quat Int 126–128:223–250

Voorhies MR (1969) Taphonomy and population dynamics of an early Pliocene vertebrate fauna, Knox County, Nebraska. Rocky Mount Geol 8:1–69

Yravedra J, Domínguez-Rodrigo M, Santonja M, Pérez-González A, Panera J, Rubio-Jara S, Baquedano E (2010) Cut marks on the Middle Pleistocene elephant carcass of Aridos 2 (Madrid, Spain). J Archaeol Sci 37(10):2469–2476

Chapter 9
Vegetation–Climate–Megafauna Interactions During the Late Glacial in Pilauco Site, Northwestern Patagonia

Ana M. Abarzúa, Alejandra Martel-Cea and Viviana Lobos

Abstract We present a vegetation reconstruction based on high-resolution analysis of two pollen, seeds, and charcoal records between ~16,000 and 12,800 cal yr BP. Pollen records show the dominance of non-arboreal and aquatic pollen types, mainly Poaceae, Asteraceae, Cyperaceae, and *Sagittaria*. At ~16,000 cal yr, BP recorded high percentages of the aquatic fern *Isoetes* sp., characteristic of shallow and oligotrophic water. The arboreal component of the pollen assemblages is represented by north Patagonian taxa. Pollen and seeds of *Prumnopitys andina* are recorded between ~14,000 and 12,800 cal yr BP, representing mild summers and very cold winters. Among the non-arboreal assemblage, taxa like *Gunnera*, *Acaena*, and Amaranthaceae could indicate some seasonality in water availability. A second detailed record was made only below and above of an erosional unconformity dated on ~12,800 cal yr BP. The sediments on top of the unconformity show the decline of pollen concentration, extremely high fire activity, and elevated sedimentation rates, probably related to intense colluvial erosion/transport. The high abundance of pioneer species and open-forest taxa during the complete record is not comparable to other pollen records from the region and can be interpreted as a high disturbance regime induced by megafauna and human activities.

Keywords Pollen · Charcoal · Reconstruction

A. M. Abarzúa (✉) · A. Martel-Cea · V. Lobos
Instituto de Ciencias de la Tierra and Transdisciplinary Center for Quaternary Research in the South of Chile, Universidad Austral de Chile, Valdivia, Chile
e-mail: anaabarzua@uach.cl

A. Martel-Cea
Escuela de Graduados, Facultad de Ciencias Forestales y Recursos Naturales, Universidad Austral de Chile, Valdivia, Chile

M. Pino and G. A. Astorga (eds.), *Pilauco: A Late Pleistocene Archaeo-paleontological Site*, The Latin American Studies Book Series,
https://doi.org/10.1007/978-3-030-23918-3_9

9.1 Background

Paleoecologic research focused on pollen fossil had evidenced how plants have responded to dramatic and rapid environmental millennial-scale changes, such as those occurring during the Last Glacial–Holocene Transition ca. 16,000–11,500 cal yr BP (Hoek 2008). This climatic transition produced a major reorganization of the vegetation that has been intensely investigated in the northern hemisphere, in comparison to the number of studies in southern latitudes. The vegetation of the Chilean Northwestern Patagonia has experienced significant changes over the course of the last glacial–interglacial cycle during the Pleistocene–Holocene (Moreno et al. 2015; Villagrán 2001; Denton et al. 1999; Heusser 1966). However, at local and regional scales, plant communities can be also altered by disturbance regimes, produced by fires, humans, and animals (Johnson 2009). Indeed, besides the close relationship between vegetation and climate, megafauna can act as gardeners affecting the late-glacial vegetation assemblages and therefore, the ecosystem structure (Malhi et al. 2016; Johnson 2009). For this reason, it is interesting to explore the synergy between climate change and extinct fauna on temperate rainforest in northern Patagonia.

The dramatic global change at the end of the last ice age was triggered by shifts in insolation, Arctic and Antarctic ice masses, concentrations of greenhouse gases, and other amplifying feedbacks at regional and global levels (Clark et al. 2012). An example of such changes is the close correlation between CO_2 concentration and Antarctic temperature during the Last Glacial Termination (Monnin et al. 2001). In recent decades, notable advances have been made on the basis of paleoclimatic records of Antarctic ice cores (Blunier and Brook 2001), marine records in high and midlatitudes (Pahnke et al. 2003; Charles et al. 1996), and land records, in southern Chile (Moreno et al. 2015, 2018; Denton et al. 1999). These records have shown the detailed occurrence of climatic events suggesting a variable degree of correlation between Northern and Southern Hemisphere records. During the Last Termination, the Antarctic ice cores show a reversal in the warming trend, known as Antarctic Cold Reversal (ACR; 14,500–12,600 cal yr BP) (Blunier et al. 1997; Jouzel et al. 1995). On the contrary, during this time a warming trend has been recorded in Greenland, and in a series of North Atlantic terrestrial and sea records (e.g., GISP2). Interestingly, the end of ACR in Antarctica is synchronous with the beginning of a rapid cooling and glacial advances in the Northern Hemisphere known as the Younger Dryas Chronozone (YDC; 12,800–11,500 cal yr BP, Rutter et al. 2000).

In southern South America, the evidence for the YDC is still controversial on current paleoclimatic research. An earlier cold phase, the Huelmo/Mascardi Cold Reversal, is partially contemporaneous with the Younger Dryas. It is not completely synchronous in the records of lake sediment in Chile and Argentina (HMCR; Massaferro et al. 2014; Hajdas et al. 2003; Moreno 2000; Ariztegui et al. 1997). On the other hand, this cooling trend is not evident in pollen records from Isla Grande de Chiloé in Chile (42° S; Abarzúa and Moreno 2008; Abarzúa et al. 2004; Villagrán 1985). On the contrary, increased fire activity and abrupt vegetation changes are inferred in several lacustrine records in northern Patagonia around ca. 12,000 cal yr BP, suggesting high variability in precipitation (Moreno et al. 2015; Jara and Moreno 2014) triggered by a southward position of SWW (Moreno et al. 2012) and increased atmospheric temperatures (Montade et al. 2019).

There are few continental records representing the transition between the Late Pleistocene and the Holocene in southern South America that may include the HMCR, the YDC, megafaunal record, and early human activity. In this context, the Pilauco archaeo-palaeontological site (40° S) presents an opportunity to provide new insights related to the climatic oscillations of the Pleistocene–Holocene transition and the effects of megafauna on vegetation assemblages.

9.2 Regional Setting

The modern climate of south-central Chile corresponds to Temperate Oceanic Climate (Cfb) based on the Köppen-Geiger climate classification (Peel et al. 2007). In the area of Pilauco, this climate is characterized by humid and temperate conditions with mean annual precipitation of ca. 1330 mm $year^{-1}$ occurring mainly during the austral fall and winter months. Mean annual temperature for the region corresponds to 11.9 °C, ranging from 9.8 °C in winter to 15.9 °C during summer (Cañal Bajo meteorological station; 40.6° S; 70.1° W, 61 m a.s.l.).

The site of Pilauco is located at the eastern lowland of the Cordillera de la Costa just near the Pacific shore (Fig. 9.1) within temperate evergreen and deciduous rainforests. The most common species belonging to the Nothofagaceae family, including *Nothofagus obliqua* and *N. dombeyi* together with *Eucryphia cordifolia*, *Weinmannia trichosperma*, *Persea lingue*, *Aextoxicon punctatum*, *Laurelia sempervirens*, *Gevuina avellana*, and *Luma apiculata*. The most important understory species are *Chusquea quila*, *Lapageria rosea*, *Boquila trifoliolata*, and *Cissus striata* (San Martin et al. 1991). Today, urban development, intense livestock farming, and crops practices have caused severe forest fragmentation, and forest remains are located mainly in national and private parks in the Andean and Coastal Ranges.

9.3 Materials and Methods

Several stratigraphic columns have been analyzed from 2012 to 2016 at Pilauco. Overall, the site excavation process involved the subdivision into 1 m^2 grids to identify the position of the stratigraphic columns, and paleontological findings (Chap. 1, this volume). The sediments for pollen analyses were collected in 2012 from a 280 cm-long and 5 cm width column (grid 14AD), including layers PB-7 to PB-9. Additionally, a 34 cm-long column (grid 10AD) was collected in 2014 in order to obtain a detailed sequence of the PB-8/PB-9 transition. Grain size and loss on ignition analyses were performed at 1 cm interval along both stratigraphic columns.

Fig. 9.1 **a** Location of the Pilauco site and other records mentioned in the text. Red dots correspond to the modern distribution of *Prumnopitys andina*. **b** Location of the Pilauco site in the city of Osorno. **c** A view southward of the site

The chronology of the sedimentary layers was constrained by 14 AMS radiocarbon dates, derived from the two columns. Radiocarbon ages were calibrated with CALIB 7 software using the Southern Hemisphere curve (SHCal13, Hogg et al. 2013). The age-depth model was developed using BACON in R (Blaauw and Christen 2013).

Pollen was extracted using standard palynological procedures following Faegri and Iversen (1989). The samples were successively treated with HCl, KOH, HF, acetolysis, and sieved through a 150-μm mesh to remove large debris after KOH treatment. Pollen counts were made on a Zeiss microscope at a magnification of 400–1000x. Taxonomic determinations of pollen grains were made by comparison to modern pollen/spores reference collections held at the Palynology and Environmental Reconstructions Laboratory, Universidad Austral de Chile, and the pollen taxonomic descriptions published by Heusser (1971). A minimum of 300 pollen grains from terrestrial taxa was counted from each slide with the exception of nine samples from column 14AD that presented very low pollen concentrations. For these samples, we reduce the pollen count to 200 pollen grains. Pollen concentrations were calculated based on counts of *Lycopodium clavatum* spores added to the preparations (Stockmarr 1971). The pollen diagrams were produced using the software Tilia (Grimm 2011). Percentages values shown in the pollen diagrams are based on the total sum of terrestrial pollen. Cluster analysis CONISS (Grimm 1987) was performed to support the zonation of pollen diagrams.

The macroscopic charcoal analysis was performed in column 10AD using 2 cm^3 of sediment taken at contiguous 1 cm intervals along the complete section. The samples were sieved using the methods outlined by Whitlock and Larsen (2001). The charcoal fractions (>0.125 mm) were counted in gridded Petri dishes under a stereomicroscope and expressed as charcoal concentration (particles cm^{-3}).

Seeds were collected by the naked eye during the excavation; other eight sediment samples were processed by flotation. To determine the taxonomic identity of the carpological structures, a seed reference collection from modern vegetation was performed in an altitudinal transect at 40° S.

9.4 Results

The base of layer PB-7 in column 14AD dated 16,290 cal yr BP, the contact between layer PB-7 and PB-8 correspond to 14,000 cal yr BP, whereas the erosional unconformity between the transition of layer PB-8 and PB-9 have an age of 12,800 cal yr BP. In this study, the youngest date in PB-9 is 10,239 cal yr BP (Table 9.1). The date of 10,416 ± 38 ^{14}C yr BP (12,222 cal yr BP) obtained from charcoal in column 10AD, is too young in relation to its stratigraphic position, is not supported by other radiocarbon ages and was therefore discarded from the age model.

The textural analysis from the top of layer PB-6 to the top of PB-9 (grid 14AD, 255 samples) indicates gravel content between 12 and 22% (Fig. 9.2). The main component in the three layers is a >60% sandy matrix. The average low content of organic matter (~10%) does not represent its real abundance since it has a low density in relation to the density of the mineral fraction (Chap. 3, this volume). The erosional unconformity between layer PB-8 and PB-9 is well developed in column 10AD showing significant differences in the textural composition of the two layers.

The mean percentage of sand and mud in layer PB-8 are 36% and 64%, respectively, whereas the same parameters in layer PB-9 correspond to 52% and 48%. The total organic carbon in both layers is very similar, near 12%, with a large S.D. near 6%.

The most distinctive textural characteristic of layer PB-9 is the presence of a large amount of pale-yellow volcanoclastic sediment transported by gravity from the northern hills, and large pieces of carbonized wood (Chap. 3, this volume).

Three pollen zones were outlined by CONISS in column 14AD:

PB-7 (350–420 cm elevation): The age of this section ranged between 15,900 and 14,000 cal yr BP. It is characterized by high pollen concentration (until 100,000 grains cm^{-3}), and the absence of charcoal particles. Non-arboreal taxa such as Poaceae, Asteraceae, Solanaceae, and aquatic flora dominate the pollen spectra (ca. 60%). Spores of aquatic ferns are mainly represented by *Isoetes* sp. (ca. 15%).

Table 9.1 AMS dates of the layers PB-7, PB-8, and PB-9 derived from the stratigraphic columns PB-AD14 and PB-AD10 in Pilauco site

Layer	Grid	Elevation (cm)	Dated material	Lab code	Radiocarbon age (^{14}C yr BP) ± SD	Median age probability (cal yr BP)
PB-9	14AD	639	Charcoal	UCIAMS110206	9135 ± 20	10,239
PB-9	AD10	542	Charcoal	AA108167	10,416 ± 38	12,224
PB-9	AD14	576	Charcoal	UCIAMS101684	10,710 ± 30	12,658
PB-8	AD14	560	Charcoal	UCIAMS101669	10,950 ± 30	12,758
PB-8	AD14	505	Seed	AA108168	11,079 ± 40	12,900
PB-8	AD14	420	Seed	AA108170	12,173 ± 42	14,021
PB-7	AD14	358	Wood	UCIAMS101672	12,860 ± 35	15,268
PB-7	AD14	353	Plant tissue	UCIAMS101770	13,045 ± 30	15,560
PB-7	AD14	352	Plant remain	UCIAMS101673	13,145 ± 35	15,727
PB-7	AD14	352	Plant remain	UCIAMS101769	13,175 ± 40	16,059
PB-7	AD14	351	Wood	UCIAMS101674	13,195 ± 35	16,110
PB-7	AD14	339	Seed	UCIAMS110203	13,570 ± 70	16,291

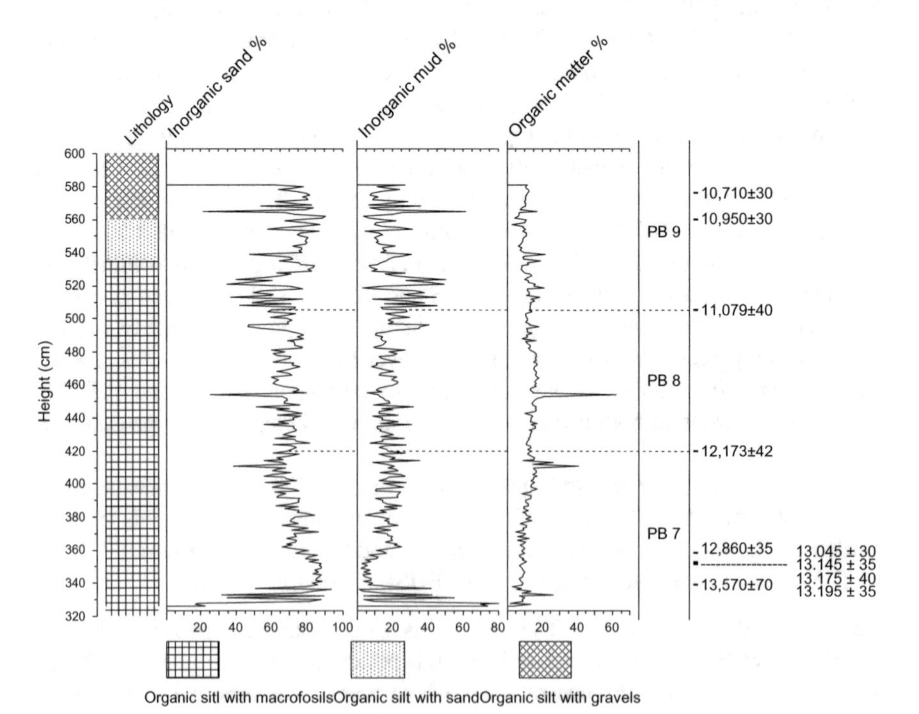

Fig. 9.2 Lithology and radiocarbon dates in column 14AD showing percentages of inorganic sand and mud and total organic matter

Forest taxa are mainly represented by evergreen tree species such as *Nothofagus dombeyi*-type, *Saxegothaea conspicua*, *Weinmannia trichosperma*, and Myrtaceae, reaching ca. 10% of the pollen assemblage. Pollen grains of *Fitzroya/Pilgerodendron*, *Maytenus* sp., *Drimys winteri*, and *Hydrangea* sp. are also present, but display low representation (<5%; Fig. 9.3).

PB-8 (420–535 cm elevation): This layer spans an age range between 14,000 and 12,800 cal yr BP. It is characterized by a slight decrease in pollen concentration at the transition from the previous zone, whereas microscopic charcoal appears for the first time in the record (up to 13,800 particles cm^{-3}). The pollen record shows a brief decrease of Poaceae, while arboreal taxa show the addition of *Prumnopitys andina*, *Nothofagus obliqua*-type, *Aristotelia chilensis*, and *Aextoxicon*/*Escallonia* with abundance <15% (Fig. 9.3).

PB-9 (535–600 cm elevation): This layer present age range between 12,800 and 12,500 cal yr BP, and is characterized by the decrease of Poaceae, aquatic taxa, ferns and the increment of Myrtaceae species. Total pollen concentration drops until the near absence of palynomorphs (from 70,000 to 1000 grains per cm^{-3}), whereas microscopic charcoal concentration reaches maxima values of the record (300,000 particles cm^{-3}; Fig. 9.3).

Three pollen zones were outlined by CONISS in column 10AD:

PAD1 (548–550 cm elevation): This first zone corresponds to the top part of the layer PB-8, which is characterized by high pollen concentration up to 90,000 grains per cm^{-3}. Dominant components in the pollen spectra are non-arboreal pollen taxa such as Poaceae (~35%), followed by forest tree taxa such as *Maytenus* (15%) and Myrtaceae (10–20%). Low charcoal concentration is recorded in this zone (Fig. 9.4).

PAD2 (550–559 cm elevation): This section corresponds to the top centimeter of layer PB-8, and the base of PB-9. The section displays a higher representation of Poaceae (50–70%), and an increase in the representation of the subfamily Asteroideae (up to 13%). The arboreal representation of components such as Myrtaceae and *Maytenus* declined, while *Aristotelia chilensis, Prumnopitys andina*, *Saxegothaea conspicua*, and *Weinmannia trichosperma* increment their representation. Aquatic taxa represented by *Triglochin*-type (Juncaginaceae), and the fern *Blechnum* show an important increment. However, the overall pollen concentration declines drastically to ca. 5000 grains cm^{-3}, while charcoal concentration increases rapidly from 22 to 405 particles cm^{-3} (Fig. 9.4).

PAD3 (559–661 cm altitude): Non-arboreal taxa display a mild decline with Poaceae dropping to 25%. Forest taxa such as Myrtaceae, *Nothofagus dombeyi*-type, *Hydrangea*, and *Prumnopitys andina* display a relative increment, while *Maytenus* and *Weinmannia trichosperma* are absent in this section. Pollen concentrations display the lower values of the record, ca. 1200 grains cm^{-3}, while charcoal concentration remains very high (Fig. 9.4).

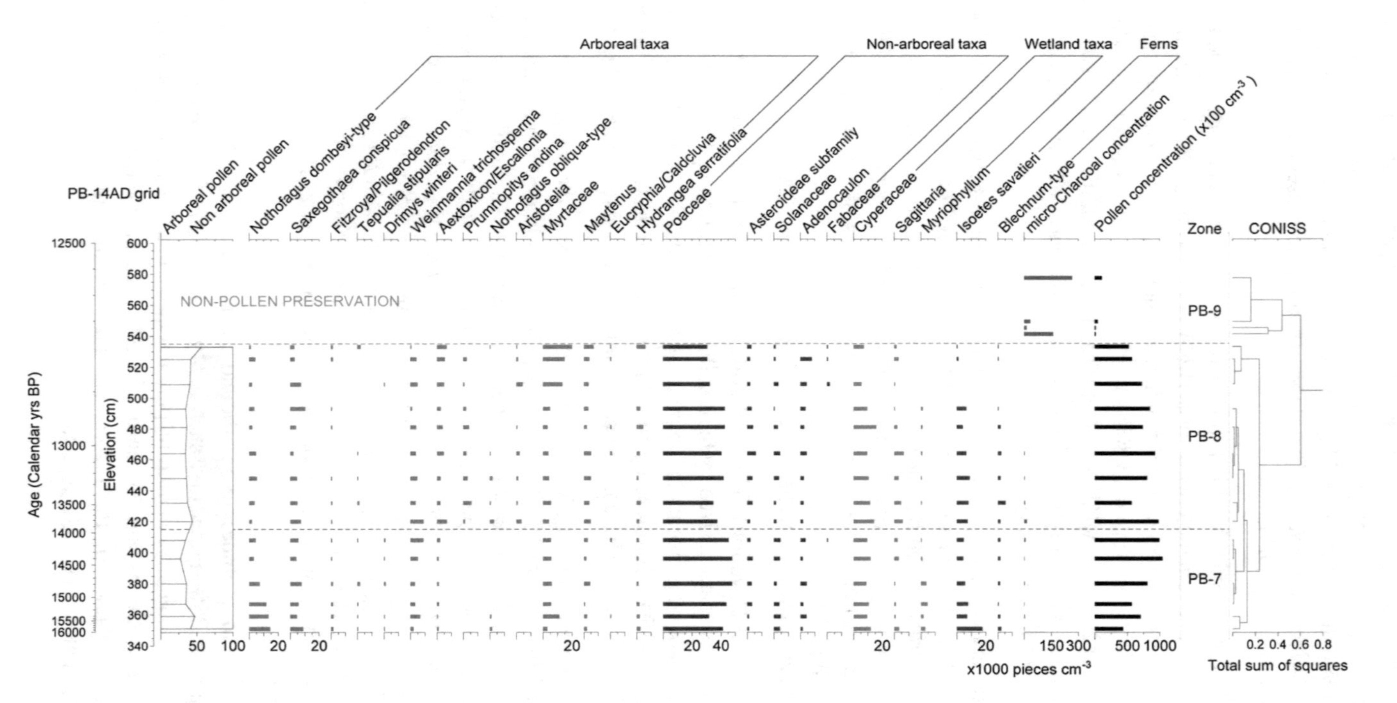

Fig. 9.3 Pollen diagram showing main plant taxa, concentration of charcoal microparticles, and pollen zones by CONISS from the column 14AD. Notice the high sedimentation rates between 13,000 and 12,500 cal yr BP

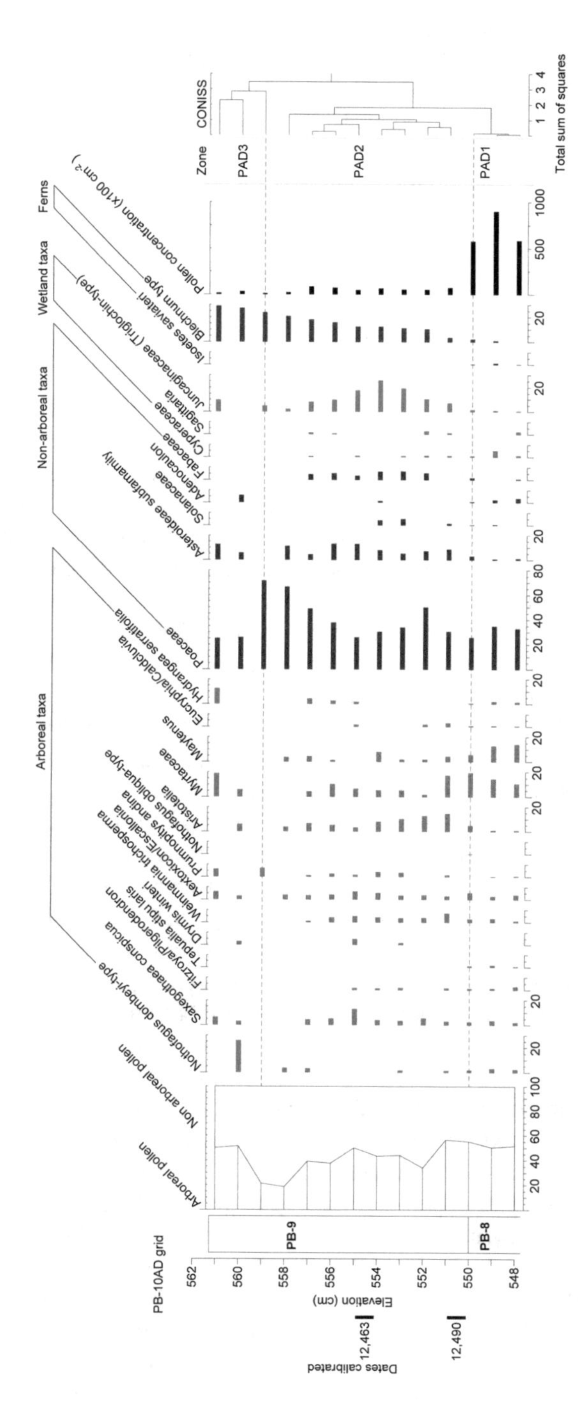

Fig. 9.4 Pollen diagram showing relative abundance of main plant taxa and zonation 10AD showing the transition between layers PB-8 and PB-9

The analysis of fossil seeds identified arboreal taxa such as *Aristotelia chilensis*, *Maytenus boaria*, and a wider variety of non-arboreal species such as *Gunnera tinctoria*, members of the families Asteraceae, Poaceae, Polygonaceae, Apiaceae, Cyperaceae, among others. The layer PB-8 is characterized by a particularly high abundance of the conifer *Prumnopitys andina*. A cultural perforated *P. andina* seed is described in Chap. 16, this volume). Figure 9.5 summarizes pollen and charcoal records.

9.5 Discussion and Conclusions

The palynological and carpological results obtained from Pilauco evidence the dominance of non-arboreal taxa indicating an open landscape associated with palustrine vegetation during most of the record (Abarzúa et al. 2016). The presence of micro and megaspores of the aquatic fern *Isoetes* sp. (cf. *Isoetes savatieri*, Macluf et al. 2003; Hickey et al. 2003) and *Myriophyllum* sp. suggest shallow, oligotrophic and temperate water conditions between 16,000 and 12,800 cal yr BP. The slight arboreal component represented by north Patagonian temperate species such as *Saxegothaea conspicua*, *Nothofagus dombeyi*-type, *Weinmannia trichosperma*, alongside with Myrtaceae, and traces of *Fitzroya/Pilgerodendron* between 16,000 and 14,000 cal yr BP suggests very humid and cold climate conditions.

By 14,000 cal yr BP, the Pilauco site records a reshaping of the arboreal component of the vegetation with the presence of *Prumnopitys andina, Nothogafus obliqua*-type, Myrtaceae species, and *Aristotelia chilensis* (Fig. 9.5). Based on their modern geographical distribution, we infer the onset of seasonality on precipitation and temperatures during this period. *Prumnopitys andina* spans an extremely restricted modern distribution from 36 to 39° S, at altitudes between 500 and 1200 m a.s.l. (Hechenleitner et al. 2005) far from the study area (Fig. 9.1). This conifer species is today distributed in a sub-Mediterranean climate with mild water deficit conditions, and non-extreme low temperatures. Furthermore, the forest patches in which this conifer tree grows show an edaphotopographical peculiarity, tending to be located on colluvial deposits of coarse rock blocks, often at the foot of cliffs or on steep slopes (Amigo et al. 2011). Thus, the northern hills surrounding Pilauco site may have been the perfect scenario for *Prumnopitys andina* in association with *Aristotelia chilensis*, *Nothofagus obliqua*, and *Maytenus boaria* between 14,000 and 12,800 cal yr BP. *P. andina* is also recorded in several pollen records from lowlands areas, from 34 to 40° S, during the late-glacial period (Fig. 9.1). In the site of Laguna Tagua-Tagua (34° S, 71° W; 200 m a.s.l.), its expansion is recorded between 18,000 and 11,000 cal yr BP (Heusser 1983). In the Purén Valley (38° S, 73° W; 70 m a.s.l.), it occurred between 15,000 and 11,500 cal yr BP (Abarzúa et al. 2014). Further south, in Laguna Las Ranas (39° S, 72° W; 400 m a.s.l.), the expansion is recorded between 19,000 and 12,000 cal yr BP (Abarzúa unpublished data), and in Rucañancu site (39° S, 72° W, 280 m a.s.l.) is recorded from 12,400 to 11,300 cal yr BP (Heusser 1984). Although the differences in the time of expansion and decline of *P. andina* can be

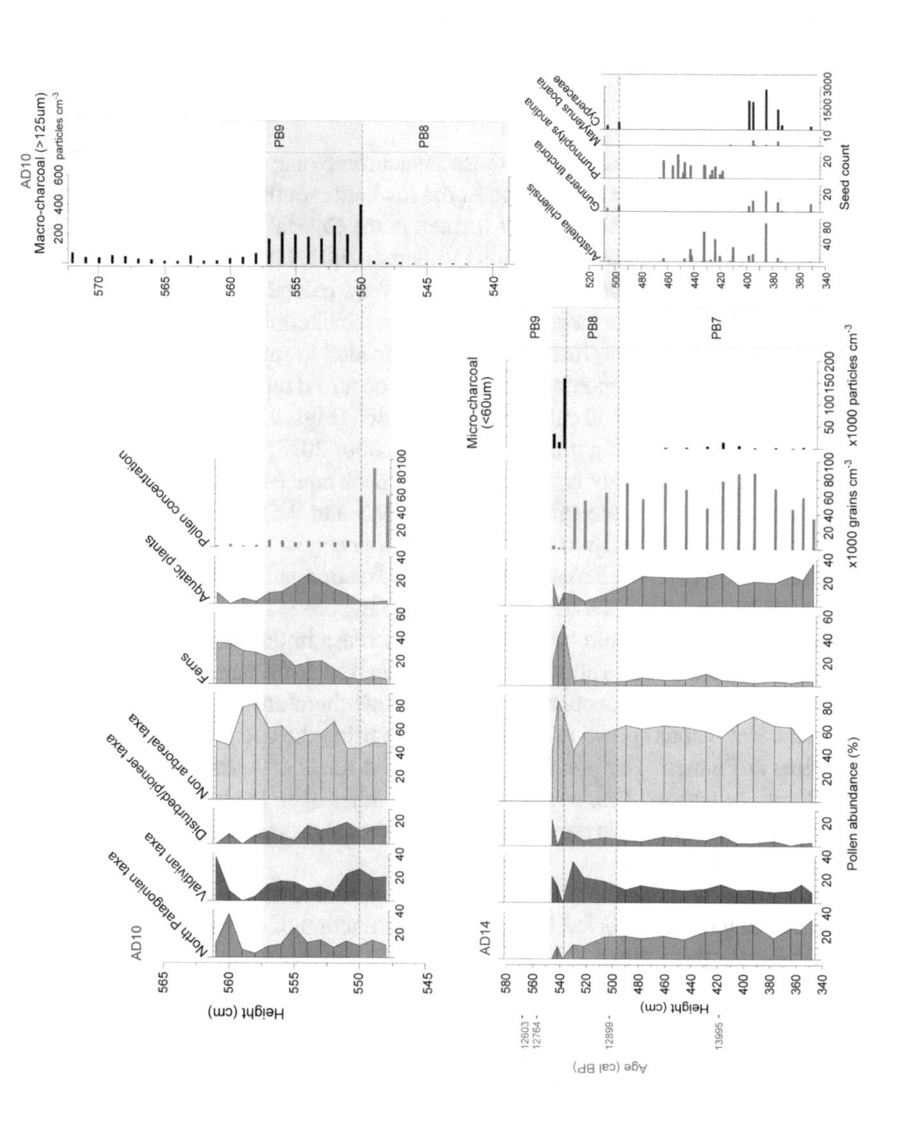
AD10
North Patagonian taxa
Valdivian taxa
Disturbed/pioneer taxa
Non arboreal taxa
Ferns
Aquatic plants
Pollen concentration
Height (cm)
AD10
Macro-charcoal (>125um)
particles cm-3
PB9
PB8
AD14
Micro-charcoal (<60um)
PB7
Pollen abundance (%)
x1000 grains cm-3
x1000 particles cm-3
Age (cal BP)
12603
12764
12899
13995
Aristotelia chilensis
Gunnera tinctoria
Prumnopitys andina
Maytenus boaria
Cyperaceae
Seed count

◂**Fig. 9.5** Summary of pollen and charcoal records. Above: pollen from grid 10AD, below: pollen and charcoal from grid 14AD, right: charcoal record from grid 10AD. Green bands represent pre-YDB sediments (PB-8) and light red bands are post-YDB (PB-9) sediments. Non-arboreal taxa made up most of the local vegetation (40–85%). Below the YDB layer, north Patagonian taxa were dominant reflecting cooler, wetter climatic conditions. Coincident with the YDB, the North Patagonian elements were replaced by Valdivian forest marking a major shift to a somewhat drier, warmer climate. Disturbed/pioneer taxa and ferns also show a dramatic increase at the beginning of the YDB layer associated with major fire regime

related to the accuracy of chronologies; these asynchronies can also be related to the reduction of a greater conifer distribution in the lowlands south-central Chile (from 34 to 40° S) toward higher altitudes sites located in the Coastal and Andean Ranges during the Pleistocene/Holocene transition (Villagrán 2001; Heusser 1984). Finally, given that the arillus of *Prumnopitys andina* is very palatable, it is possible to hypothesize that following the megafauna extinction the likelihood of seed dispersal may have decreased. However, further research is needed to support this idea.

Major vegetation and sedimentological changes occurred during the transition of layer PB-8 to PB-9 at ca. 12,800 cal yr BP. First, pollen (Figs. 9.3 and 9.5) and seeds concentrations decline reaching minimum values (Lobos 2013; Pino et al. 2019). At the same time, high fire activity is indicated by the peak concentration of micro and macro-charcoal particles in the same layer (Figs. 9.3 and 9.5). Additionally, both layers present important, but short-term lithological changes. Thus, the increase of inorganic mud and gravels at expenses of the sand fraction, and a mild decrease of organic matter support the open tree-less landscape hypothesis (Fig. 9.2). The very low pollen concentration would be related to an increase in the sedimentation rate consequence of the input of colluvial sediments derived from bare soils after fire events. Despite the pollen concentration is very low, there are several big pieces (>30 cm) of burnt woods spread in layer PB-9 supporting the occurrence of intense local-set fires at Pilauco. The presence of pollen and seeds of *Aristotelia chilensis* associated with *Maytenus*, *Gunnera*, Poaceae, and other open landscape indicators may respond to the well-known pioneer behavior of the species after fire disturbances (Fig. 9.5; Muñoz et al. 2014).

Several authors propose that the hypothesis of warming at the end of the last glaciation as a key factor involved in the massive extinction of megafauna recorded around the world at this time (e.g., Prescott et al. 2012 and references therein). However, there is substantial evidence of repeated glacial advances and retreats during the Late Pleistocene (Petit et al. 1999), suggesting that the climatic factor was not the sole cause for the disappearance of the megafauna. For instance, considering there is a strong evidence of early human presence at Pilauco (see Moreno et al. 2019; Navarro-Harris et al. 2019; Chap. 16, this volume) and Monte Verde (~100 km south of Pilauco, Fig. 9.1) (i.e., Dillehay and Pino 1989), is certainly possible that the development and expansion of human groups contemporaneous with changing climate, and anthropogenic-set fires may have triggered the disappearance of megafauna globally (Wroe et al. 2004; Miller et al. 2005; Cosgrove et al. 2010; Villavicencio et al. 2016).

Herbivorous megafauna and/or human activity may have influenced the prevalence of open landscape in the area of Pilauco, whereas forest expansion (*Nothofagus*, Myrtaceae, and Podocarpaceae) was occurring in the rest of Lake District in response to the abrupt warming at the beginning of the last glacial termination at ca. 16,000 cal yr BP (Moreno 1997, 2000; Abarzúa et al. 2004; Pesce and Moreno 2014; Moreno et al. 2015). The gomphotheres inhabited from open grasslands in north-central Chile to open and close forested areas in northwestern Patagonia. Its dietary behavior was generalist, including leaf-browsing feeding habits (González-Guarda et al. 2018). Consequently, the high occurrence of these and other mammals in Pilauco (suggested by bone remains and the dung spore *Sporormiella* sp., Chap. 6, Pino et al. 2013, 2019), may have caused serious disturbances on a local scale, including damages by browsing and soil compaction negatively impacting the forest regeneration. Similar results were evidenced in southern Patagonia when the grasslands disappeared, as a result of climate change and the herbivorous megafauna became extinct. Thus, the *Nothofagus* forests may have had the possibility to expand as observed in other regions (Barnosky et al. 2016).

Johnson (2009) propose three scenarios that result in changes in the vegetation after large megafauna become extinct that also occur in some grade at Pilauco. Increased fire activity and declined coevolved plants are the most evident (could be linked to decline of *Prumnopitys andina, Gomortega keule* and others, to the modern and vulnerable conservation state). The third scenario is the loss of open vegetation and habitat mosaics that is difficult to infer due to the lack of sedimentary record post PB8/PB9 event. From ~12,800 cal yr, BP humans and megafauna evidences disappear at Pilauco (Pino et al. 2019). The erosional unconformity recorded between the layers PB-8/PB-9 (~12,800 cal yr BP) is characterized by the decline of pollen concentration, high concentration of charcoal particles, and high concentrations of Pt, Au, and Pd, and high-temperature impact-related spherules (Pino et al. 2019). Following this major peak, charcoal concentrations decline, but values remain higher than in the underlying unit PB-8. This charcoal peak is the highest in the sequence analyzed and indicates an anomalously large biomass-burning episode beginning near the onset of YD climate change at ~12,800 cal yr BP.

High unstable climate conditions related to the high variability and/or seasonality of precipitations can be related to the Younger Dryas Chronozone (YDC) and Huelmo/Mascardi Cold Reversal, which have been suggested by other paleoclimate records in Patagonia (Moreno et al. 2015; Massafferro et al. 2014). Our results suggest that the unconformity development between the layers PB-8 and PB-9 document the effect of YDC, as is suggested by the low vegetation cover, and strong precipitations that originated the colluvial sediments of the layer PB-9. At the YDC onset, drier conditions are indicated by the disappearance of aquatic ferns (*Isoetes* sp.), following their persistent presence throughout the pre-YDC sequence. These changes were also accompanied by increased biomass burning fostered by the onset of a marked seasonality of rainfall. Furthermore, beginning at the YDC layer the vegetation records show a distinct upward increase (from 5 to 20%) of pioneering/colonizing taxa indicative of more disturbed open habitats. These taxa appear to have colonized the area in response to the abrupt onset of alternating seasonal drying and precipitation (humidity) that also promoted an increase in biomass burning. All these factors indicate that the post-YDC vegetation at Pilauco is closely affiliated with the more northern Valdivian rainforest that favors seasonally drier and warmer conditions. Thus, the vegetation history at Pilauco records a sudden shift at the YDB from cooler, wetter conditions characteristic of the North Patagonian forest to seasonally drier and warmer conditions of the Valdivian rainforest. Coinciding with the termination of the Antarctic Cold Reversal, this shift in vegetation at the YDC continued for at least 100 years suggesting persistent climate change and disturbed landscape (Chap. 15, this volume). This change is associated with the termination of the Antarctic Cold Reversal episode, anti-phased with the shift in the northern hemisphere from the warm Bølling-Allerød period to the colder YD. The Pilauco record shows that this climate shift occurred rapidly within just a few years or less, based on sedimentary deposition rates. It appears that the changes in southern Chilean vegetation (cool/wet north Patagonian forest to warmer/dryer Valdivian and deciduous forests), marking the YDC onset and the termination of the Antarctic Cold Reversal, were caused by the abrupt southward shift of the southern westerly wind belt toward the Southern Ocean and Antarctica (Abarzúa et al. 2004; Moreno et al. 2015; Pino et al. 2019). Multisite comparison based on quantitative climate reconstructions will allow us to develop a regional to the global synthesis of climate changes in order to test this hypothesis and improve data model comparisons.

The interpretation of pollen, seeds, charcoal, and sediments in Pilauco provide insights to refine our understanding of the biological and climatic changes occurred during a period of environmental change, which is not yet well understood in areas of the Southern Hemisphere: the last glacial–interglacial transition. Our investigation provides evidence of a significant decrease in the concentration of arboreal pollen, and the abrupt increase in the concentration of charcoal at around 12,800 cal yr BP. This abrupt environmental change is concomitant with the local extinction of megafauna and the disappearance of the human artifacts, and the onset of high variability and/or seasonality of precipitations during the asteroid-impact documented in Pilauco.

References

Abarzúa AM, Moreno PI (2008) Changing fire regimes in the temperate rainforest region of southern Chile over the last 16,000 yr. Quat Res 69:62–71

Abarzúa AM, Villagrán C, Moreno PI (2004) Deglacial and postglacial climate history in east-central Isla Grande de Chiloé, southern Chile (43°S). Quat Res 62:49–59

Abarzúa AM, Lobos V, Martel-Cea A (2016) ¡Pequeño testigo de grandes cambios!: Polen, semillas y carbón escondidos bajo tierra, In: Pino M (ed) El Sitio Pilauco. Osorno, Patagonia Noroccidental de Chile. Universidad Austral de Chile-TAQUACH, Valdivia, pp 49–54

Abarzúa AM, Pinchicura GA, Jarpa L, Martel-Cea A, Sterken M, Vega R, Pino M (2014) Environmental responses to climatic and cultural changes. In: Dillehay TD (ed) The teleoscopic polity: Andean patriarchy and materiality. Springer International Publishing, Cham, pp 123–136

Amigo J, Rodríguez-Guitián MA, Ramírez C (2011) The lleuque forests of south central Chile: a phytosociological study and syntaxonomical classification within South American temperate forests. Lazaroa 31:85–98

Ariztegui D, Bianchi MM, Masaferro J, Lafargue E, Niessen F (1997) Interhemispheric synchrony of late-glacial climatic instability as recorded in proglacial Lake Mascardi, Argentina. J Quat Sci 12:333–338
Barnosky AD, Lindsey EL, Villavicencio NA, Bostelmann E, Hadly EA, Wanket J, Marshall CR (2016) Variable impact of late-quaternary megafaunal extinction in causing ecological state shifts in North and South America. Proc Natl Acad Sci USA 113:856–861
Blaauw M, Christen JA (2013) Bacon manual-v2.2, p 11. http://chrono.qub.ac.uk/blaauw/manualBacon_2.2.pdf
Blunier T, Brook EJ (2001) Timing of millennial-scale climate change in Antarctica and Greenland during the last glacial period. Science 291:109–112
Blunier T, Schwander J, Stauffer B, Stocker T, Dällenbach A, Indermühle A, Tschumi J, Chappellaz J, Raynaud D, Barnola JM (1997) Timing of the Antarctic cold reversal and the atmospheric CO_2 increase with respect to the Younger Dryas event. Geophys Res Lett 24:2683–2686
Charles CD, Lynch-Stieglitz J, Ninnemann US, Fairbanks RG (1996) Climate connections between the hemisphere revealed by deep sea sediment core/ice core correlations. Earth Planet Sci Lett 142:19–27
Clark PU, Shakun JD, Baker PA, Bartlein PJ, Brewer S, Brook E, Carlson AE, Cheng H, Kaufman DS, Liu Z, Marchitto TM, Mix AC, Morrill C, Otto-Bliesner BL, Pahnke K, Russell JM, Whitlock C, Adkins JF, Blois JL, Clark J, Colman SM, Curry WB, Flower BP, He F, Johnson TC, Lynch-Stieglitz J, Markgraf V, McManus J, Mitrovica JX, Moreno PI, Williams JW (2012) Global climate evolution during the last deglaciation. Proc Natl Acad Sci USA 109:E1134–E1142
Cosgrove R, Field J, Garvey J, Brenner-Coltrain J, Goede A, Charles B, Wroe S, Pike-Tay A, Grün R, Aubert M, Lees W, O'Connell J (2010) Overdone overkill—the archaeological perspective on Tasmanian megafaunal extinctions. J Archaeol Sci 37:2486–2503
Denton GH, Heusser CJ, Lowel TV, Moreno PI, Andersen BG, Heusser LE, Schlühter C, Marchant DR (1999) Interhemispheric linkage of paleoclimate during the last glaciation. Geogr Ann A 81:107–153
Dillehay TD, Pino M (1989) Stratigraphy and chronology. In: Dillehay TD (ed) Monte Verde: a late pleistocene settlement in Chile. Smithsonian Institution Press, Washington, DC, pp 133–145
Faegri K, Iversen J (1989) Textbook of pollen analysis. Wiley, London
González-Guarda E, Petermann-Pichincura A, Tornero C, Domingo L, Agustí J, Pino M, Abarzúa AM, Capriles JM, Villavicencio NA, Labarca R, Tolorza V, Sevilla P, Rivals F (2018) Multiproxy evidence for leaf-browsing and closed habitats in extinct proboscideans (Mammalia, Proboscidea) from Central Chile. Proc Natl Acad Sci USA 115:9258–9263
Grimm EC (1987) CONISS: a Fortran 77 program for stratigraphically constraint cluster analysis by the method of incremental squares. Comp Geosci 13:13–35
Grimm EC (2011) Tilia software 1.7.16. Illinois State museum, research and collections center. Springfield, USA
Hajdas I, Bonani G, Moreno PI, Ariztegui D (2003) Precise radiocarbon dating of late-glacial cooling in mid-latitude south America. Quat Res 59:70–78
Hechenleitner P, Gardner MF, Thomas PI, Echeverría C, Escobar B, Brownless P, Martínez, C (2005) Plantas amenazas del Centro-Sur de Chile: Distribución, Conservación y Propagación. Universidad Austral de Chile-Royal Botanic Garden Edinburgh, Valdivia
Heusser CJ (1966) Late-Pleistocene pollen diagrams from the Province of Llanquihue, southern Chile. Proc Am Philos Soc 110:269–305
Heusser CJ (1971) Pollen and spores of Chile: modern types of the Pteridophyta, Gymnospermae, and Angiospermae. University of Arizona Press, Tucson
Heusser CJ (1983) Quaternary pollen record from Laguna de Tagua Tagua, Chile. Science 219:1429–1432
Heusser CJ (1984) Late-glacial-Holocene climate of the lake district of Chile. Quat Res 22:77–90
Hickey RJ, Macluf C, Taylor WC (2003) A re-evaluation of *Isoetes savatieri* Franchet in Argentina and Chile. Am Fern J 93:126–136
Hoek WZ (2008) The last glacial-interglacial transition. Episodes 31:226–229

Hogg AG, Hua Q, Blackwell PG, Niu M, Buck CE, Guilderson TP, Heaton TJ, Palmer JG, Reimer PJ, Reimer RW, Turney CSM, Zimmerman SRH (2013) SHCal13 Southern Hemisphere Calibration, 0–50,000 Years cal BP. Radiocarbon 55:1889–1903

Jara IA, Moreno PI (2014) Climatic and disturbance influences on the temperate rainforests of northwestern Patagonia (40° S) since ~14,500 cal yr BP. Quat Sci Rev 90:217–228

Johnson CN (2009) Ecological consequences of late quaternary extinctions of megafauna. Proc R Soc B276:2509–2519

Jouzel J, Vaikmae R, Petit JR, Martin M, Duclos Y, Stievenard M, Lorius C, Toots M, Mélières MA, Burckle LH, Barkov NI, Kotlyakov VM (1995) The two-step shape and timing of the last deglaciation in Antarctica. Clim Dynam 11:151–161

Lobos V (2013) Analisis paleovegetacional basado en semillas y frutos fósiles del sitio Paleontológico Pilauco, Pleistoceno Tardío, Osorno, centro-sur de Chile. Undergraduate Dissertation, Universidad Austral de Chile

Macluf CC, Morbelli MA, Giudice GE (2003) Morphology and ultrastructure of megaspores and microspores of *Isoetes savatieri* Franchet (Lycophyta). Rev Palaeobot Palyno 126:197–209

Malhi Y, Doughty CE, Galetti M, Smith FA, Svenning JC, Terborgh JW (2016) Megafauna and ecosystem function from the Pleistocene to the Anthropocene. Proc Natl Acad Sci USA 113:838–846

Massaferro J, Larocque-Tobler I, Brooks SJ, Vandergoes M, Dieffenbacher-Krall A, Moreno P (2014) Quantifying climate change in Huelmo mire (Chile, Northwestern Patagonia) during the last glacial termination using a newly developed chironomid-based temperature model. Palaeogeogr Palaeoclimatol Palaeoecol 399:214–224

Miller GH, Fogel ML, Magee JW, Gagan MK, Clarke SJ, Johnson BJ (2005) Ecosystem collapse in Pleistocene Australia and a human role in megafaunal extinction. Science 309:287–290

Monnin E, Indermühle A, Dällenbach A, Flückiger J, Stauffer B, Stocker TF, Raynaud D, Barnola JM (2001) Atmospheric CO_2 concentrations over the last glacial termination. Science 291:112–114

Montade V, Peyron O, Favier C, Francois JP, Haberle SG (2019) A pollen–climate calibration from western Patagonia for palaeoclimatic reconstructions. J Quat Sci 34:76–86

Moreno PI (1997) Vegetation and climate near Lago Llanquihue in the Chilean Lake District between 20200 and 9500 ^{14}C yr BP. J Quat Sci 12:485–500

Moreno PI (2000) Climate, fire, and vegetation between about 13,000 and 9200 ^{14}C yr B.P. in the Chilean Lake District. Quat Res 54:81–89

Moreno PI, Villa-Martínez R, Cárdenas ML, Sagredo EA (2012) Deglacial changes of the southern margin of the southern westerly winds revealed by terrestrial records from SW Patagonia (52° S). Quat Sci Rev 41:1–21

Moreno PI, Denton GH, Moreno H, Lowell TV, Putnam AE, Kaplan MR (2015) Radiocarbon chronology of the last glacial maximum and its termination in northwestern Patagonia. Quat Sci Rev 122:233–249

Moreno PI, Videla J, Valero-Garcés B, Alloway BV, Heusser LE (2018) A continuous record of vegetation, fire-regime and climatic changes in northwestern Patagonia spanning the last 25,000 years. Quat Sci Rev 198:15–36

Moreno K, Bostelmann E, Macías C, Navarro-Harris X, De Pol-Holz R, Pino M (2019) A late Pleistocene human footprint from the Pilauco archaeological site, northern Patagonia, Chile. PLoS One 14(4):e0213572

Muñoz A, González G, Díaz-Hormazabal I, Riquelme N (2014) *Aristotelia chilensis* (Mol) Stuntz. In: Donoso C (ed) Las espécies arbóreas de los bosques templados de Chile y Argentina. Autoecología, María Cuneo Ediciones, Valdivia, pp 166–180

Navarro-Harris X, Pino M, Guzmán-Marín P, Lira MP, Labarca R, Corgne A (2019). The procurement and use of knappable glassy volcanic raw material from the late Pleistocene Pilauco site, Chilean Northwestern Patagonia. Geoarchaeology 1–21. https://doi.org/10.1002/gea.21736

Pahnke K, Zahn R, Elderfield H, Schulz M (2003) 340,000-year centennial-scale marine record of southern hemisphere climatic oscillation. Science 301:948–952

Peel MC, Finlayson BL, McMahon TA (2007) Updated world map of the Köppen-Geiger climate classification. Hydrol Earth Syst Sci Discuss 4:439–473
Pesce OH, Moreno PI (2014) Vegetation, fire and climate change in central-east Isla Grande de Chiloé (43° S) since the Last Glacial Maximum, northwestern Patagonia. Quat Sci Rev 90:143–157
Petit JR, Jouzel J, Raynaud D, Barkov NI, Barnola JM, Basile I, Bender M, Chappellaz J, Davis M, Delaygue G, Delmotte M, Kotlyakov VM, Legrand M, Lipenkov VY, Lorius C, Pepin L, Ritz C, Saltzman E, Stievenard M (1999) Climate and atmospheric history of the past 420,000 years from the Vostok ice core, Antarctica. Nature 399:429–436
Pino M, Chávez-Hoffmeister M, Navarro-Harris X, Labarca R (2013) The late Pleistocene Pilauco site, Osorno, south-central Chile. Quat Int 299:3–12
Pino M, Martel-Cea A, Astorga G, Abarzúa AM, Cossio N, Navarro X, Lira MP, Labarca R, Lecompte MA, Adedeji V, Moore CR, Bunch TE, Mooney C, Wolbach WS, West A, Kennett JP (2019) Sedimentary record from Patagonia, southern Chile supports cosmic-impact triggering of biomass burning, climate change, and megafaunal extinctions at 12.8 Ka. Sci Rep 9:4413
Prescott GW, Williams DR, Balmford A, Green RE, Manica A (2012) Quantitative global analysis of the role of climate and people in explaining late quaternary megafaunal extinctions. Proc Natl Acad Sci USA 109:4527–4531
Rutter NW, Weaver AJ, Rokosh D, Fanning AF, Wright DG (2000) Data-model comparison of the Younger Dryas event. Can J Earth Sci 37:811–830
San Martin C, Ramírez C, Figueroa H, Ojeda N (1991) Estudio sinecológico del bosque de roble-laurel-lingue del centro-sur de Chile. Bosque (Valdivia) 12:11–27
Stockmarr J (1971) Tablets with spores used in absolute pollen analysis. Pollen Spores 13:615–621
Villagrán C (1985) Análisis palinológico de los cambios vegetacionales durante el Tardiglacial y Post-glacial en Chiloé, Chile. Rev Chil Hist Nat 58:57–69
Villagrán C (2001) Un modelo de la historia de la vegetación de la Cordillera de La Costa de Chile central-sur: la hipótesis glacial de Darwin. Rev Chil Hist Nat 74:793–803
Villavicencio NA, Lindsey EL, Martin FM, Borrero LA, Moreno PI, Marshall CR Barnosky AD (2016) Combination of humans, climate, and vegetation change triggered late quaternary megafauna extinction in the Última Esperanza region, southern Patagonia, Chile. Ecography 39:125–140
Whitlock C, Larsen C (2001) Charcoal as a proxy fire. In: Smol JP, Birks HJB, Last WM (eds) Tracking environmental change using lake sediments: terrestrial, algal, and siliceous indicators. Kluwer Academic Publishers, Dordrecht, The Netherlands, pp 75–97
Wroe S, Field J, Fullagar R, Jermin LS (2004) Megafaunal extinction in the late quaternary and the global overkill hypothesis. Alcheringa 28:291–333

Chapter 10
Diatoms: Microalgae from Pilauco

Leonora Jarpa-Mateluna

Abstract Fossil diatoms from the site of Pilauco were studied to reconstruct the hydrologic changes in the lentic system between ~16,000 and 14,200 cal yr BP. We studied 13 samples from the sedimentary layers PB-7 and PB-8 and counted a minimum of 400 diatom valves per sample. The taxonomic analysis identified 33 taxa at the species level and 11 taxa at the genus level. The results using the CONISS analysis revealed two notoriously different periods in the diatom succession. The PIL-1 zone was dated between 16,000 and 14,800 cal yr BP and marked by the dominant presence of benthic diatom species with an affinity for cold waters such as *Fragilaria construens* and *Fragilaria capucina*. These results indicate a cold environment assigned to the end of the last glaciation and is characterized by shallow waters and reduced humidity. The subsequent period (PIL-2) corresponds to the time frame from 14,800 to 14,200 cal yr BP and is marked by more warmer conditions favouring the growth of macrophytes serving as substrate for epiphytic diatoms such as *Lemnicola hungarica* and *Nitzschia* sp. Furthermore, the presence of *Aulacoseira granulata* has been considered indicative for concurrent repeated flooding of a wetland. In turn, the occurrence of prolonged dry periods longer than a season during the studied time span can be rather discarded. The diatom record at Pilauco emerged as a clear indicator for environmental conditions reflected by relative variations in temperature and humidity. The sediment layers PB-7 and PB-8 display high capacity to preserve diatom frustules and provide valuable evidence for the late Pleistocene palaeo-environmental conditions at the Pilauco site.

Keywords Fragilaria · Lemnicola · Nitzschia · Aulacoseira

L. Jarpa-Mateluna (✉)
Instituto de Ciencias de la Tierra, Universidad Austral de Chile, Valdivia, Chile
e-mail: leojarpa@gmail.com

M. Pino and G. A. Astorga (eds.), *Pilauco: A Late Pleistocene Archaeo-paleontological Site*, The Latin American Studies Book Series,
https://doi.org/10.1007/978-3-030-23918-3_10

10.1 Background

Diatoms are unicellular algae surrounded by a cell wall made of silica. According to their lifestyle, the diatoms can be classified as either planktonic or benthic (Battarbee 2000; Round et al. 1990). The latter is distributed across at least three different habitats and also known as epiphytic, epipelic and episamon depending on the type of substrate they are attached to Battarbee (1986), Round (1964). Their distribution within the water column is apparently not arbitrary but rather depends on the spatio-temporal interactions with abiotic factors (Garzón and Rondón 2011). In small water bodies such as creeks and swamps, the diatom flora is principally composed of benthic diatoms (Round 1964). For environmental reconstructions, it is necessary that the sediment is free of perturbations across temporal horizons (Battarbee 1986, 2000; Schönfelder and Steinberg 2004). This type of studies particularly focuses on deep-water lake deposits although other lentic systems such as palaeo-meanders or flood plains might also condition a continuous sedimentation that favours the preservation of diatom frustules (Schönfelder and Steinberg 2004). Fossil diatoms can also be used for the reconstruction of palaeo-swamp environments (Kiss et al. 2012).

10.2 Paleohydrological Reconstruction of Pilauco

The present study encompasses qualitative and quantitative analyses of fossil diatoms in order to reconstruct environmental changes affecting the lentic system of Pilauco between ~16,000 and 14,200 cal yr BP; Pino et al. (2019). A total of 13 samples of sediment were collected from PB-7 and PB-8 layers (grid 14AC). The sediments were processed according to the cleaning protocol for valves from Battarbee (1986), which implies the oxidation of organic matter with 30% H_2O_2. After oxidation, the coarse fraction was removed by sieving samples down to 500 μm. Density separation with sodium polytungstate (SPT 2.2 g/cm^3) was applied to remove the clay fraction. Permanent triplicate preparations were embedded in Naphrax resin (I.R. = 1.7). A minimum of 400 diatom valves per sample were counted on an optical microscope Carl Zeiss Axion at a (1000x)-magnification.

The taxonomic identification is based on microscope images and morphometric analyses implying length and width measurements as well as by counting the number of areola striae within 10 μm of each photographed species. Some samples were mounted for SEM analysis (Scanning Electron Microscope) to observe characteristic structural details of the frustule to identify particular species. The identification was made according to Krammer and Lange-Bertalot (1986–1991) and Round et al. (1990).

The relative abundance (%) of all taxa was calculated with the software TILIA 1.7.16 (Grimm 2011). A CONISS analysis (Constrained Incremental Sums of Squares) was performed with TILIA on the taxa characterized by >2% of relative abundance in order to detect significant changes in the diatom assemblage (Fig. 10.1).

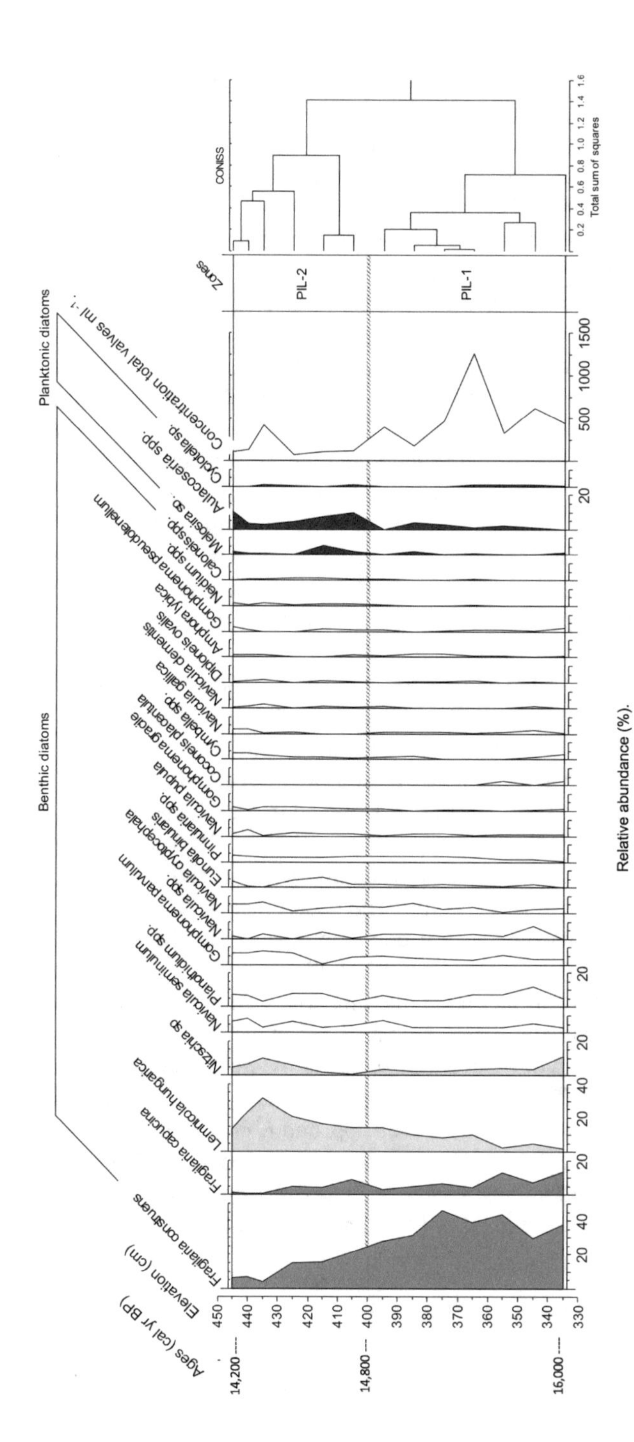

Fig. 10.1 TILIA diagram of vertical (temporal) distribution of diatoms in Pilauco. The CONISS analysis identifies the dominant species in PIL-1 (grey) and PIL-2 (green). The dark blue signature corresponds to planktonic species

The taxonomic analysis identified 44 different taxa; 33 of them were characterized on the species level whereas the remaining 11 on the genus level. The CONISS analysis identified one significant change in the diatom succession exactly in the contact between layers PB-7 and PB-8. The two zones denominated here as PIL-1 and PIL-2 are interpreted to reflect different depositional environments at the site between ~16,000 and 14,200 cal yr BP (Fig. 10.1).

The PIL-1 zone (335–400 cm of local elevation equivalent to the base of PB-7) was developed between ~16,000 and 14,800 cal yr BP. It is dominated by benthic diatom species such as *Fragilaria construens* (Fig. 10.2) and *Fragilaria capucina* with maximum relative abundances of 46% at 375 cm and 14% at 335 cm respectively. Following these maxima, the abundance of both species gradually decreases. Close to the base of the sediment record at approximately 345 cm local elevation, both species show a local decrease of 30% for *F. construens* and 7% for *F. capucina.* At the same level, other species reach their maximum abundances of 7% for *Navicula* spp, 2% for *Eunotia bilunaris*, 5% for *Navicula seminulum* and 12% for *Planothidium* spp. Subsequently, the abundances of *Navicula* spp. and *Eunotia bilunaris* decrease and remain stable throughout the PIL-1 zone. In turn, *Navicula seminulum* and *Planothidium* spp. reach another local maximum close to the limit with the PIL-2 zone of 7% and 6%, respectively. Central and planktonic taxa are scarce throughout the record.

The PIL-2 zone (400–445 cm of local elevation equivalent to base of PB-8 layer) corresponds to the time frame between 14,800 and 14,200 cal yr B.P. The dominant species from the PIL-1 zone decrease significantly reaching minimum relative abundances of 4% for *F. construens* at 435 cm and 1% for *F. capucina* at 440 cm local elevation. The PIL-2 zone is generally dominated by the epiphytic and epipelic taxa *Lemnicola hungarica* (Figs. 10.2 and 10.3) and *Nitzschia* spp. The relative abundances of both taxa gradually increase above 405 cm reaching their maxima at 435 cm local elevation. The occurrence of these taxa is inversely proportional to the abundance of the dominant species in the underlying PIL-1 zone *F. construens* and *F. capucina* (Fig. 10.1). Similar to PIL-1, *Navicula* spp. and *Eunotia bilunaris* exhibit further local maxima at 415 cm with relative abundances of 4% and 6%, respectively. These increases are slight and followed by gradual decreases towards the top of the record except for the third intermittent increase of *Navicula* spp. reaching 3% of relative abundance at 435 cm. At the same elevation, the taxa *Navicula seminulum Planothidium* spp. and *F. contruens* decrease reaching their local minima with 3% and 4% of relative abundance, respectively (Fig. 10.1). At the same time, planktonic taxa such as *Aulacoserira* spp. and *Melosira* sp. are more abundant at this level with 10% (405 cm) and 5% (415 cm), respectively.

10.3 Discussion and Conclusions

The CONISS analysis and the ecological implications of fossil diatoms from the sediment record indicate that between ~16,000 and 14,200 cal yr BP, the Pilauco wetland was characterized by variable environmental conditions reflected by two significantly different periods. The PIL-1 zone covers the time span between ~16,000 and 14,800 cal yr BP and corresponds to the coldest period within the record, which can be inferred from the dominant presence of *F. construens* and *F. capucina*. These benthonic species known from lacustrine and fluvial environments (Alveal et al. 2008; Lotter and Bigler 2000; Polyakova et al. 2006; Urrutia et al. 2007) have been considered indicative for freezing conditions (Lotter and Bigler 2000) and their elevated concentration hints toward rather stable cold climate conditions (Douglas and Smol 1999). The prolonged cold periods likely inhibited the growth of aquatic plants, which is consistent with the low concentration of epiphytic taxa such as *Lemnicola hungarica*. Furthermore, this interpretation is concordant with the documented abundance of *Prumnopitys andina* seeds at Pilauco from the base of PB-8 layer (Lobos 2014). Based on its present spatial distribution, this conifer species has been considered indicative for the cold climate conditions at the end of the last glaciation in south-central Chile (Lobos 2014).

Increased abundances of planktonic diatoms have been generally correlated with high water levels (Urrutia et al. 2010) as well as sufficient turbulence to keep them in suspension (Sterken et al. 2008; Hassan et al. 2013). In contrast, the cold period assigned to the older part of the Pilauco record is characterized by rather

Fig. 10.2 SEM images **a** and **b** of *Lemnicola hungarica* with internal (upper left) and external (upper right) view of the valve with raphe. Scale bars measure 10 μm and 3 μm, respectively. **c** and **d** External views of *Fragilaria construens*. Scale bar measures 2 μm (left) and 1 μm (right)

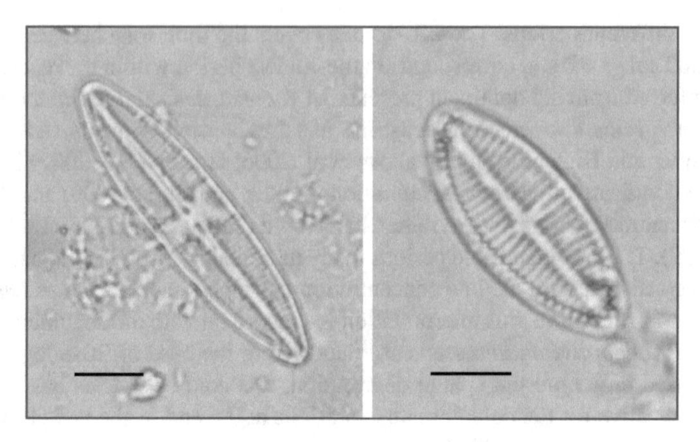

Fig. 10.3 Optical microscope images of *Lemnicola hungarica*. A valve with (left) and without (right) raphe. The black bar is 5 μm long

a scarce occurrence of planktonic diatoms such as *Aulacoseira* spp *Melosira* sp. and *Cyclotella* sp. and was likely less humid. The abundances of *Aulacoseira* sp. and *Melosira* sp. increase towards the subsequent more recent period between 14,800 and 14,200 cal yr BP when the wetland was possibly affected by frequent and probably more permanent flooding. Kiss et al. (2012) suggest a positive correlation between the size of the water current and the number of planktonic diatom species. The Pilauco record indicates only slight increases of the abundances of these species, which indicates a rather stable depth of the water body.

The lower abundances of *F. construens* y *F. capucina* near the base of PB-8 reflect a gradual increase in temperatures favouring the growth of aquatic plants along the shores of the system that would offer the potential substrate for epiphytic diatoms. This suggestion is supported by the increased abundance of *Lemnicola hungarica* and *Nitzschia* sp. The former is a specialized epiphytic species associated with the Lemnaceae family and living is attached to the roots of this macrophyte (Buczkó 2007). However, seeds and pollen from the Lemnaceae family have not been registered among the aquatic plant record from the Pilauco sediments (Lobos 2014; Abarzúa and Gajardo 2008). These results suggest that *Lemnicola hungarica* might have either been associated with other plant species or characterized by a different lifestyle. Notably, other species from the same genus can live attached to sand grains leaves branches in shallow lakes small creeks or swamps (García and Souza 2006).

In turn, the ecological conditions required by species from the genus *Nitzschia* have been described as corresponding to eutrophic habitats and turbid waters (Round et al. 1990). An increased turbidity of the water in the studied fossil system might be explained by the presence of macrophytes and their decomposition products (Salomoni et al. 2006) as well as the apparently high content of fine material indicated by the sedimentological analysis (Chap. 3, this volume).

The continuous presence of water-dependant benthic species throughout the diatom record indicates the lack of dry periods within the wetland. The *Navicula* spp. and *Pinnularia* spp. species might exhibit epiphytic or subaerial living habitat and resist extremely dry conditions (Round 1964; Round et al. 1990; Stoermer and Smol 1999). Nonetheless, their mere presence is not indicative enough for drought because their abundance is never exclusively high throughout the record. Considering the results from the present study, the presence of dung beetles (Tello 2012) that are dependent on dry periods can then rather be assigned to seasonal variabilities.

Nitzschia spp. and *Eunotia binularis* species show a constant presence throughout the studied record. These taxa prefer eutrophic environments (Sterken et al. 2008; Round et al. 1990; Salomoni et al. 2006) which implies high-nutrient content within the system during the entire period. This conclusion is also supported by the elevated content of reduced organic matter in the sediments (Chap. 3, this volume). Based on the results from the diatom record, we infer two periods in the evolution of the wetland at Pilauco marked by contrasting climatic conditions. Accordingly, the older period was characterized by cold climate shallow water and reduced humidity and corresponds to the time frame between 16,000 and 14,800 cal yr BP. In turn, the subsequent period bracketed between 14,800 and 14,200 cal yr B.P. was apparently warmer and more humid which favoured the growth of macrophytes that served as living substrate for epiphytic diatoms such as *Lemnicola hungarica* and *Nitzschia* sp. Furthermore, there is no evidence for the occurrence of prolonged dry periods within the studied time frame.

References

Abarzúa A, Gajardo A, (2008) ¿Y qué nos cuentan los polen? Reconstruyendo la historia climática y vegetacional del sitio Pilauco. In: Pino M (ed) Pilauco: Un sitio complejo del Pleistoceno tardío. Osorno Norpatagonia chilena. Universidad Austral de Chile. Imprenta América, Chile. pp 49–53

Alveal IE, Cruces FJ, Araneda AE, Grosjeans M, Urrutia RE (2008) Estructura comunitaria de diatomeas presentes en los sedimentos superficiales de ocho lagos andinos de Chile central. Rev Chil Hist Nat 81:83–94

Battarbee RW (1986) Diatom analysis. In: Berglund BE (ed) Handbook of holocene palaeoecology and palaeohydrology. Wiley, pp 527–570

Battarbee RW (2000) Palaeolimnological approaches to climate change with special regard to the biological record. Quat Sci Rev 19:107–124

Buczkó K (2007) The occurrence of the epiphytic diatom Lemnicola hungarica on different European Lemnaceae species. Fottea Olomouc 7(1):77–84

Douglas MS, Smol JS (1999) Freshwater diatoms as indicators of environmental change in the high arctic. In: Stoermer EF, Smol JP (eds) The diatoms: application for the environmental and earth sciences. Cambridge University Press, Cambridge, pp 227–244

García M, Souza VF (2006) Lemnocola hungarica (Grunow) Round & Basson from southern Brazil: Ultra-structure plastid morphology and ecology. Diatom Res 21(2):465–471

Garzón EP, Rondón JD (2011) Diversidad y distribución de diatomeas en un arroyo de montaña de los Andes Colombianos. Limnology 33(1):177–191

Hassan GS, Francesco CG, Dieguez S (2013) The significance of modern diatoms as paleoenvironmental indicators along an altitudinal gradient in the Andean piedmont of central Argentina. Palaeogeogr Palaeoclimatol Palaeoecol 369:49–360

Kiss K, Klee R, Ector L, Ács É (2012) Centric diatoms of large rivers and tributaries in Hungary: morphology and biogeographic distribution. Acta Botanica Croat 71(2):311–363

Krammer K, Lange-Bertalot H (1986–1991) Bacillariophyceae 4 Vols. in: Süsswasserflora von Mitteleuropa. Stuttgart/Jena: Gustav Fischer Verlag: Volume 2/1 (1986): Naviculaceae. Volume 2/2 (1988): Bacillariaceae Epithemiaceae Surirellaceae. Volume 2/3 (1991): Centrales Fragilariaceae Eunotiaceae. Volume 2/4 (1991): Achnanthaceae kritische Ergänzungen zu Navicula (Lineolatae) und Gomphonema

Lobos V (2014) Análisis paleovegetacional basados en semillas y frutos fósiles del Sitio Pelontológico Pilauco Pleistoceno Tardío Osorno centro- sur de Chile. Undergraduate Dissertation, Universidad Austral de Chile

Lotter AF, Bigler C (2000) Do diatoms in the swiss alps reflect the length of ice-cover? Aquat Sci 62:125–141

Pino M, Martel-Cea A, Astorga G, Abarzúa AM, Cossio N, Navarro X, Labarca R, Lecompte MA, Adedeji V, Moore CR, Bunch TE, Mooney C, Wolbach WS, West A, Kennett JP (2019) Sedimentary record from Patagonia southern Chile supports cosmic-impact triggering of biomass burning climate change and megafaunal extinctions at 12.8 ka. Sci Rep 29:1–29

Polyakova YI, Klyuvtkina TS, Novichkova EA, Bauch HA, Kassens H (2006) High-resolution reconstruction of Lena River discharge during the late holocene inferred from microalgae assemblages. Polarforschung 75(2–3):83–90

Round FE (1964) The ecology of benthic algae. In: Jackson DP (ed) Algae and man. Plenum Press, New York, pp 138–184

Round FE, Crawford RM, Mann DG (1990) The diatoms biology & morphology of the genera. Cambridge University Press

Salomoni SE, Rocha O, Callegaro VL, Lobo EA (2006) Epilithic diatoms as indicators of water quality in the Gravataí river Rio Grande do Sul Brazil. Hydrobiologia 559:233–246

Schönfelder I, Steinberg CW (2004) How did the nutrient concentrations change in northeastern German lowland rivers during the last four millennia?—a paleolimnological study of floodplain sediments. Studia Quaternaria 21:129–138

Sterken M, Verleyen E, Sabbe K, Terryn G, Charlet F, Bertrand S, Boes X, Fagel N, de Batist M, Vyverman W (2008) Late Quaternary climatic changes in southern Chile as recorded in a diatom sequence of Lago Puyehue. J Paleolimnol 39:219–235

Stoermer EF, Smol JP (eds) (1999) The diatoms: applications for the environmental and earth sciences. Cambridge University Press, Cambridge

Tello F (2012) Coleópteros fósiles y reconstrucción paleoambiental de la capa PB-7 del Sitio Paleontológico Pilauco Osorno Norpatagonia de Chile. Undergraduate Dissertation, Universidad Austral de Chile

Urrutia U, Araneda A, Cruces F, Torres L, Chirinos L, Treutlerc H, Fagel N, Bertrand S, Alvial I, Barra R, Chapron E (2007) Changes in diatom pollen and chironomid assemblages in response to a recent volcanic event in Lake Galletue (Chilean Andes). Limnologica 37:49–62

Urrutia R, Araneda A, Torres L, Cruces F, Vivero C, Torrejón F, Barra R, Fagel N, Scharf B (2010) Late Holocene environmental changes inferred from diatom chironomid and pollen assemblages in an Andean lake in Central Chile Lake Laja (36S). Hydrobiologia 648:207–225

Chapter 11
Phytolith Analysis from Coprolites of Pilauco

Valentina Álvarez-Barra

Abstract The analysis of microfossils such as phytoliths may provide strong evidence regarding the characteristics of the dominant vegetation in ancient landscapes. The phytolites recovered from the coprolites found at the Pilauco site were analyzed to identify what types of plants were consumed by the fauna that inhabited the site. Phytolith remains evidenced mainly the presence of C3 grasses from the Poaceae family, which is in accordance with previous pollen analysis at Pilauco. Two out of the four coprolite samples analyzed likely correspond to *Equus* (andium). This was concluded based on the similar phytolith morphology associated with C3 grasses that match the eating habit of the genus in the southern hemisphere as described in the literature, and the presence of parasites eggs specific to equines. The remaining samples showed that the coprolite producer had an herbivorous diet with parasites indicating likely a ruminant animal.

Keywords C3 · Ruminant · Equus · Pilauco

11.1 Background

The beginning of phytolith studies dates back to the mid-nineteenth century when German biologists analyzed samples collected by Darwin in his expedition to Cabo Verde during the journey of the Beagle in 1846 (Power-Jones 1992).

Phytoliths are generated as a result of a silica mineralization process. The plant incorporates silica into its tissues during the process of absorption of water from the soil. The silica is absorbed in its soluble state, monosilicic acid (H_4SiO_4), and is transported through the xylem, a water-transporting tissue present in vascular plants, along with other elements found in the water (Bowdery 1989). The incorporation

V. Álvarez-Barra (✉)
Department of Palynology and Climate Dynamics. Albrecht-von-Haller-Institute for Plant Sciences, Georg-August-Universität Göttingen, Untere Karspüle 2, 37073 Göttingen, Lower Saxony, Germany
e-mail: valentina.alvarez@biologie.uni-goettingen.de

M. Pino and G. A. Astorga (eds.), *Pilauco: A Late Pleistocene Archaeo-paleontological Site*, The Latin American Studies Book Series,
https://doi.org/10.1007/978-3-030-23918-3_11

of silica is made via the aerial organs of the plant by the process of transpiration, in those organs where phytoliths are found in greatest proportion (Piperno 2006). Once the silica arrives at the aerial parts of the plant and saturates it with the solution that contains it, it is deposited as silica dioxide ($SiO_2 + 4–9\% H_2O$) (Epstein 1994).

The SiO_2 is deposited in plants in the cellular lumen, outer surface of the cells, and intercellular spaces. Several species of pteridophytes, gymnosperms, and angiosperms produce phytoliths, but not all plants produce them, and their quantities are variable within each taxon (Pearsall 1989). In various taxa and specific tissues of aerial organs, there have been identified zones where the deposition of solid silica occurs (Blackman and Perry 1968; Bozarth 1987; Dorweiler and Doebley 1997; Iriarte 2003), such as for instance, plant epidermis, covering the seeds and fruits of numerous trees and grasses (Ernst et al. 1995; Kealhofer and Piperno 1998; Runge and Runge 1997); subepidermal tissues of orchids and palm leaves (Tomlinson 1961); epidermis of bracts and seeds of grasses (Mulholland 1993); and bulliform cells, related to the regulation of water loss during transpiration (Gajardo-Pichincura 2011).

The function of silica in plants has been thoroughly studied for specific groups. In general, it is thought that the presence of silica provides structural support to plants; mitigates toxic effects of aluminum and other heavy minerals present in the soil (Carnelli et al. 2002) and provides protection from herbivores and pathogens (McNaughton and Tarrants 1983; McNaughton et al. 1985; Marshner 1995).

The analysis of phytolites generally includes three approximations, (1) Taxonomic, (2) Typologic, and (3) Taxonomic–typologic. The first and partly the third emphasizes the morphology in relation to the structure of the plant. This approach is often used when working with modern plant material or articulated plant material present in archeological sites. Thus, the base for analysis of phytoliths is to understand how the different forms of the phytoliths recovered from the ancient material are related with the anatomy of the plant in some taxonomic level (Carter and Lian 2000). Because the phytoliths present a variety of forms, depending on where they are produced, it is possible to identify the type of plant structure consumed or used in different contexts at the same site. Thus, it is important to know the morphology, which is determined by the anatomy of the parental plant. This makes essential a sound taxonomic and anatomic knowledge of different plant groups. This is mainly because, the resulting morphology of the phytolith depends on the site of deposition within the plant, having been molded by the space that contains it (Fig. 11.1). The most common deposition sites within plants are: (a) intracellular, corresponding to the cellular lumen where the phytolith acquires the exact shape of the cell, presenting therefore a diagnostic form; (b) extracellular, where the silica is deposited on the cell membrane acquiring its external form; and (c) intercellular, where the silica refills spaces between the cells (Piperno 2006).

Because it is possible to identify differences in the internal anatomy of plant structures for each taxon (e.g., Kranz-type anatomy for C4 plants). This indicates evidence of morphotypes of different phytoliths, which makes it possible to identify families, genera and in some cases, species (Table 11.1; Fig. 11.2). Both the production and deposition of silica in different plant taxa is under genetic and metabolic

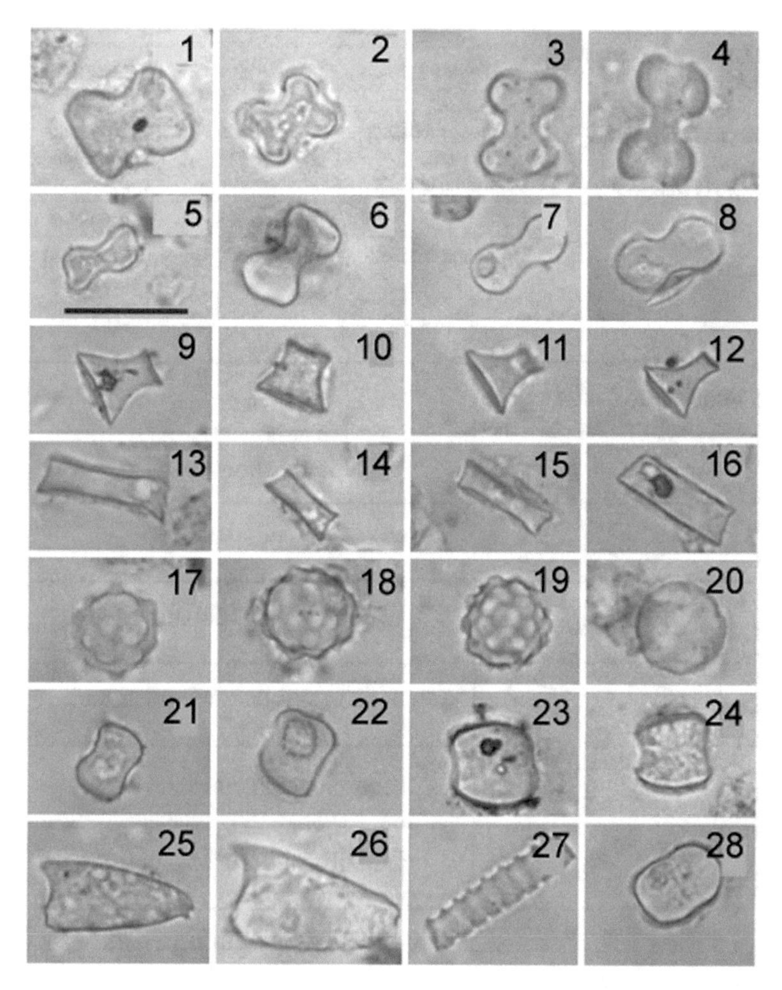

Fig. 11.1 Pilauco phytoliths and their morphologies (size < 40 μm) 1–8: bilobate; 9–16: rondel; 17–19: globular granulate; 20: globular psilate; 21–24: saddle; 25–26: acicular; 27: elongate echinate; 28: oblong phytolith. Adapted from Erra (2010)

control (Commoner and Zucker 1953; Blackman 1969; Dorweiler and Doebley 1997; Piperno et al. 2002).

However, it should be noted that not all vascular plants produce phytoliths, and moreover, not all contain diagnostic morphologies allowing their identification. Many different phytolith morphologies can be found in one taxon; and the same shape can be present among many other taxonomic groups (Rovner 1971). According to Piperno (2006), there are five patterns of phytoliths production of taxonomic significance in plants, which are subdivided into pteridophytes, basal angiosperms, monocotyledons, and dicotyledons. In Chilean species native, phytolith production

Table 11.1 Association of the different phytolith morphologies with a particular taxonomic level and their paleoecological significance (Twiss 1992)

Morphotype	Subfamily	Paleoecological indicator
Rondel, trapezoids, conical and pyramidal shapes	Pooideae C3	Temperate environments
Saddle	Chloridoideae C4	Arid environments
Cross and bilobed	Panicoidea C4	Humid and temperate environments

Schematic drawings	Name
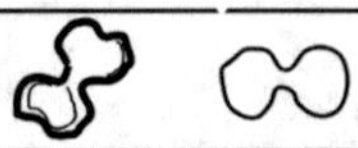	Bilobate short cell
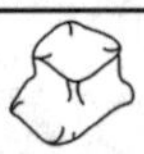	Trapeziform short cell
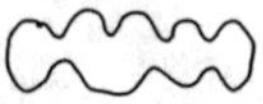	Cylindrical polilobate
	Trapeziform polylobate and sinuate
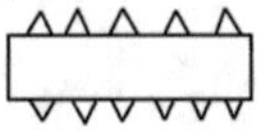	Elongate echinate long cell
	Cuneiform bulliform cell
	Parallepipedal bulliform cell
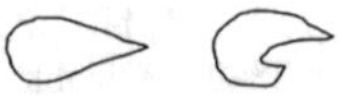	Acicular and unciform hair cell
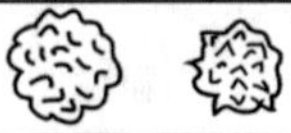	Globular granulate and echinate
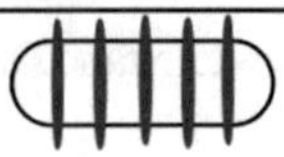	Cylindric sulcate tracheid

Fig. 11.2 Some examples of classification according to the "International Code for Phytolith Nomenclature 1.0" (Madella et al. 2005)

patterns are led by families such as Equisetaceae, Cyperaceae, and Cucurbitaceae, among others (Gajardo-Pichincura 2011).

The study of phytoliths extracted from coprolites is a relatively novel approach and most of the researches have focused on the study of prehistoric human diet (Horrocks et al. 2002; Horrocks and Irwin 2003). It has also provided a record of the possible causes of extinction of some vertebrates (Presad et al. 2005; Piperno and Sues 2005). In this case, the collection of phytoliths extracted from coprolites from the paleontological site Pilauco was analyzed in order to infer the diet of the biological coprolite producers, as well as provide information of the main characteristics of the vegetation in which these animals lived during the late Pleistocene.

11.2 Materials and Methods

Fresh subsamples between two and five grams were taken from four coprolites (MHMOP/PI/585; MHMOP/PI/586; MHMOP/PI/587; and MHMOP/PI/588). The four coprolite samples used were recovered from the Pilauco layer PB-7. The sample MHMOP/PI/587 was dated at the University of Arizona AMS laboratory (sample AA-81812, GEOUACH-57) providing a radiocarbon age of 11,004 ± 186 A.P. (see Chap. 3, this volume).

The recovery of phytoliths from coprolites requires the elimination of all organic material, so several chemical attacks were carried out, where at each stage, the different portions that comprise them are eliminated. With the application of KOH, the humic acids contained in the material are eliminated. In the second place, 10% of HCI 10% applied to remove carbonates. Finally, H_2O_2 is used to remove carbon organic matter. Sodium polytungstate solution (SPT solution) was used to separate by flotation the phytoliths of other particles, whose density is higher than the density of the solution. The assembly of the phytoliths is carried out on slides and sealed with Entellan©. The phytoliths are observed, classified, and counted using an optic microscope. For the identification of the phytoliths' morphologies recovered from the study material, the nomenclature code proposed by Madella et al. (2005) was used. Classification of phytolith morphotypes recovered from the material is carried out in accordance with the classification proposed by Twiss 1992 (Table 11.1; Fig. 11.3).

11.3 Phytoliths Recorded in the Coprolites

The most frequent morphology in sample MHMOP/PI/585 corresponds to cuneiform bulliform cells with rondel shapes predominating in the short cells. In sample MHMOP/PI/566, cuneiform bulliform shapes are very abundant, and long cells predominate over short cells. Samples MHMOP/PI/587 and MHMOP/PI/585 present a high frequency of bulliform cells, in both parallelepiped and cuneiform shapes. The

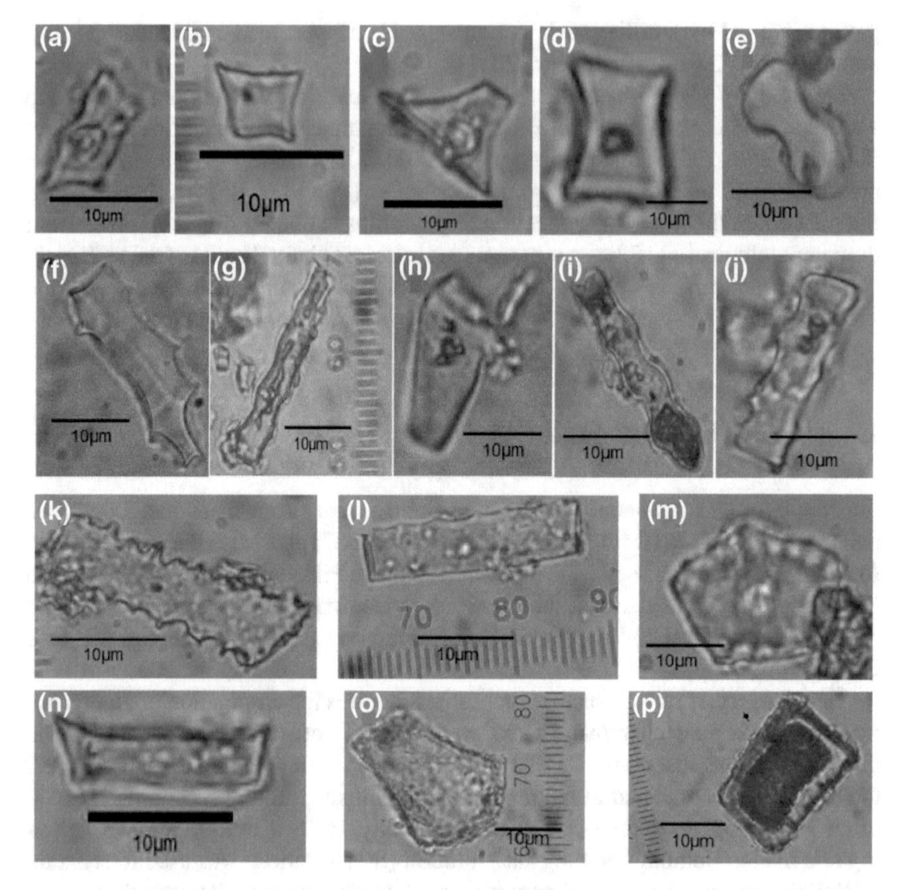

Fig. 11.3 Some examples of the morphotypes observed in coprolite samples from the Pilauco site, Chile: **a–b** trapezoid; **c–d** rondel; **e** bilobated; **f** polyhedral; **g** elongate, sinuous; **h** unknown short type; **i–j** elongate; **k–l** elongate, spiny; **m** Cyperaceae phytolith; **n** trapezoid; **o–p** bulliform shapes

rondel form stands out between the short cells and the three long cell forms are at a similar frequency. Finally, bulliform shapes dominate sample MHMOP/PI/588.

Phytoliths were grouped according to characteristic subfamily forms: C3 Pooidea, C4 Chloridoidea, and Panicoidea as indicated in Tables 11.2 and 11.3. Regarding short cells, the most abundant shapes within the samples were rondel, bilobed, and psilate trapezoid. It was possible to determine that for samples 585, 586 y 588, the percentage of C3 short cells are the highest. C4 forms predominate in sample 587 (Fig. 11.4).

Table 11.2 Classification of phytoliths according to C3 and C4 morphologies and their respective frequency in each coprolite sample

Form	585	586	857	587	C3	C4
Bilobed	18	8	4	4	–	X
Cross	5	0	1	0	–	X
Saddle	2	0	0	6	–	X
Rondel	59	30	2	51	X	–
Polylobed trapezoid	12	3	2	6	X	–
Psilated trapezoid	14	23	0	3	X	–
Sinuated trapezoid	8	18	0	0	X	–

Table 11.3 Percentage of short cells compared to total, and the percentage of C3 and C4 cells for each coprolite sample

Sample	585	586	587	588
Total phytoliths	550	407	223	696
Total short cells	118	82	9	70
% short cells	22	20	4	11
Total C3 cells	93	74	4	60
Total C4 cells	25	8	5	10
% C3 cells	79	90	44	86
% C4 cells	21	10	56	14

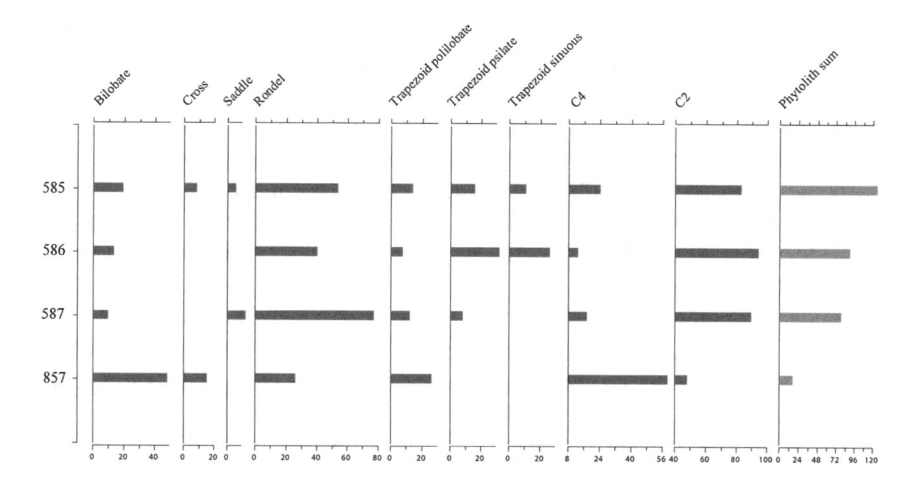

Fig. 11.4 Phytolith types in each coprolite analyzed (percentage), and total of counted phytolith. Green and red bars indicate C4 and C3 shapes, respectively

11.4 Discussion and Conclusions

According to the described morphotypes, it is possible to identify and relate two phytoliths to plant subfamilies (Figs. 11.2, 11.3, and 11.4; Tables 11.2 and 11.3). The "rondel" morphotypes correspond to short cells formed in the epidermal tissue of grasses of the subfamily Pooidea (C3) (Mulholland 1989; Bremond et al. 2008). This morphology is abundant in samples 585 and 586. Additionally, bilobed and cross phytoliths types are characteristic of C4 grasses (subfamily Panicoideae). Elongated forms correspond to phytoliths formed in the epidermis of stems and leaves of Poaceae. Finally, phytoliths that represent dicotyledons plants are those of polyhedral form, although this morphology was not represented in the analyzed coprolite samples.

The results observed in Fig. 11.3 indicate a high percentage of C3 grasses in samples 586, 588 and 585 (Table 11.3). Like the pollen analysis, these results show that Poaceae were part of the vegetation surrounding Pilauco, although important differences exist between both reconstructions. Pooideas (C3 grasses) reflect temperate environments; while the Panicoideas and Chloridoideas provide evidence of humid temperate and arid climatic conditions respectively (Table 11.1; Kellogg 2001; Piperno 2006). The frequencies of phytolith forms associated with type C3 plants (Fig. 11.4) dominate, which concurs with the distribution of C3 plants in the Southern Hemisphere during the Late Pleistocene, when this type of plant predominated at higher latitudes. Pilauco is located at approximately 40° S, within the range of C3 abundance (Prado et al. 2011). However, coprolite 587 presents 56% of C4 phytolith grasses with only nine short cells present in the sample. The presence of type C4 phytoliths is discordant with what is indicated in the literature, where they are associated with low latitudes in the southern hemisphere during the Late Pleistocene. Also, in middle latitudes, there is a transition of C3 and C4 grasses (McFadden et al. 1999). Regarding the total number of phytoliths corresponding to Pooideas, Chloridoideas, and Panicoideas, Twiss (1992) indicates that the high rates of Pooideas (C3) imply temperate environments. This might indicate the transition toward the Holocene, which implied ecological adaptations and strategies developed by forest and meadows communities in response to the changes in temperature and precipitation (Abarzúa et al. 2004; Solari 2007). Nevertheless, natural and anthropogenic disturbances processes might have contributed to these change as well (see Chap. 9, this volume)

Pollen studies carried out in the stratigraphic column 14AD (PB-7 layer) describe a landscape composed by forests and meadows with a predominance of Poaceas, Asteraceas (see Chap. 9, this volume). The phytoliths recovered from coprolites indicate abundance of type C3 Poaceaae, according to the results of the pollen analysis. However, Asteraceas were not recorded within phytoliths, which could be related to the palatability of the group (e.g., Katz et al. 2014). The small quantity of forms corresponding to dicotyledon shrubs or trees can indicate that the animal that produced the coprolite did not browse this group of plants. Regarding the biological producer of the coprolites, the percentage of C3 phytoliths for samples 585 and 586 is very high (Fig. 11.4), which is concordant with diet studies of the Pleistocene horse in the Southern Hemisphere (McFadden et al. 1999).

The percentages of forms associated with C3 and C4 grasses are very similar in sample 587, where the percentage of C4 grasses is more than 50%. Although the number of short cell phytoliths is very small (9), a possible explanation for this is that the coprolite-producing animal consumed C4 grasses developed as a result of the change in climate produced by the transition from Pleistocene to Holocene. As the difference in percentages of C3 and C4 grasses in this sample is very small, it could be inferred that the animal was an opportunistic feeder. *Equus (andium)* is described as such in the literature, and in many cases, its diet was mixed in terms of the type of plant consumed (Asevedo et al. 2011; Sánchez et al. 2003).

The analyzed coprolite samples (MHMOP/PI/585 and MHMOP/PI/856), could correspond to *Equus* (andium), mainly based on the morphology type associated with C3 grasses. That type of plants matches the feeding habits described in the literature for this genus in the Southern Hemisphere and is supported by the presence of parasite eggs specific to equids. The other samples show that the producer had an herbivorous diet, and the parasites found indicate that it may correspond to a ruminant animal. However, dicotyledon phytoliths are very scarce in the four samples analyzed, so there is not enough evidence to infer whether some of the samples were produced by a gomphothere.

References

Abarzúa AM, Villagrán C, Moreno PI (2004) Deglacial and postglacial climate history in east-central Isla Grande de Chiloé, Southern Chile (43ªS). Quat Res (Orlando) 62:49–59

Asevedo L, Winck GR, Mothé D, Avilla LS (2011) Ancient diet of the Pleistocene gomphothere Notiomastodon platensis (Mammalia, Proboscidea, Gomphotheriidae) from lowland mid-latitudes of South America: stereomicrowear and tooth calculus analyses combined. Quat Int 30:1–11

Blackman E (1969) Observations on the development of the silica cells of the leaf sheath of wheat (Triticum aestivum). Can J Bot 47:827–838

Blackman E, Perry DW (1968) Opaline silica deposition in rye (Cecale cereale L). Ann Bot (Lond) 32:199–206

Bowdery D (1989) Phytoliths analysis: introduction and applications. In: Beck W, Clarke A, Head L(eds) Plants in Australian Archaeology, St. Lucia: archaeology and material culture studies in anthropology. Tempus 1, pp 161–196

Bozarth S (1987) Diagnostic opal phytoliths from rinds of selected Curcubita species. Am Antiq 52:607–615

Bremond L, Alexandre A, Matthew JW, Hély C, Williamson D, Schâfer P, Majule A, Guiot J (2008) Phytolith indices as proxies of grass subfamilies on East African tropical mountains. Glob Planet Change 61:209–224

Carnelli A, Madella M, Theurillat JP, Ammann B (2002) Aluminium in the Opal silica reticule of phytoliths: a new tool in palaeoecological studies. Am J Bot 89(2):346–351

Carter JA, Lian BO (2000) Palaeoenvironmental reconstruction from the last interglacial using phytolith analysis, Southeastern North Island, New Zealand. J Quat Sci. 15:733–743

Commoner B, Zucker ML (1953) Cellular differentiation: an experimental approach. In: Loomes WE (ed) Growth and differentiation in plants. Ames, Iowa, pp 339–392

Dorweiler JE, Doebley J (1997) Developmental analysis of teosinte glume architecture 1: a major locus in the evolution of maize (Poaceae). Am J Bot 84:1313–1322

Epstein E (1994) The anomaly of silicon in plant biology. Proc Natl Acad Sci USA 91:11–17

Ernst WOH, Vis RD, Piccoli F (1995) Silicon in developing nuts of the sedge Schoenus nigricans. J Plant Physiol 146:481–488

Erra G (2010) Asignación sistemática y paleocomunidades inferidas a partir del estudio fitolitico de sedimentos cuaternarios de Entre Ríos-Argentina, Boletin de la Sociedad Argentina de Botanica. 45:309–319

Gajardo-Pichincura A (2011) Posibilidades del análisis de fitolitos en pequeños humedales del centro-sur Chile: El caso del humedal El Valle, Araucanía-Chile Dissertation, Universitat Rovira i Virgili

Horrocks M, Irwin GJ (2003) Pollen, phytoliths and diatoms in prehistoric coprolites from Kohila, Bay of Plenty, New Zeland. J Archaeol Sci 30:13–20

Horrocks M, Jones MD, Beever RE, Sutton DG (2002) Analysis of plants microfossils in prehistoric coprolites from Harataonga Bay. Great Barrier Island, New Zeland. J R Soc N Z 32:617–628

Iriarte J (2003) Emergent cultural complexity in the wetlands of uruguay during the middle holocene. Dissertation, University of Kentucky

Katz O, Lev-Yadun S, Bar P (2014) Do phytoliths play an antiherbivory role in Southwest Asian Asteraceae species and to what extent? Flora 209(7):349–358

Kealhofer L, Piperno DR (1998) Opal phytoliths in Southeast Asian flora. Smithsonian Contributions to Botany, No 88. Smithsonian Institution Press, Washington, DC

Kellogg EA (2001) Evolutionary history of the grasses. J Plant Physiol 125:1198–1205

Madella M, Alexandre A, Ball T (2005) International code for phytolith nomenclature 1.0. Ann Bot (Lond) 96:253–260

Marshner H (1995) Mineral nutrition of higher plants. Academic Press, London

McFadden JB, Cerling TE, Harris J, Prado J (1999) Ancient latitudinal gradients of C3/C4 grasses interpreted from stable isotopes of new world Pleistocene horse (Equus) teeth. Glob Ecol Biogeogr 8:137–149

McNaughton SJ, Tarrants JL (1983) Grass leaf silicification: natural selection for an inducible defense against herbivores. Proc Natl Acad Sci USA 80(3):790–791

McNaughton SJ, Tarrants JL, Mc Naughton MM, Davis RH (1985) Silica as defense against herbivory and a growth promoter in Africans grasses. Ecology 66(2):528–535

Mulholland SC (1989) Phytoliths shape frequencies in North Dakota grasses: a comparison to general patterns. J Archaeol Sci 16:489–511

Mulholland SC (1993) A test of phytoliths analysis at Big Hidatasa, North Dakota. In: Pearsall M, Piperno DR (eds) Current research in phytoliths analysis: applications in archaeology and paleoecology, MASCA research papers in science and archaeology, vol 10. MASCA, The University Museum of Archaeology and Anthropology, University of Pennsylvania, Philadelphia, pp 131–145

Pearsall D (1989) Paleoethnobotany: a handbook of procedures. Academic Press, San Diego

Piperno DR (2006) Phytoliths: a comprehensive guide for archaeologist and paleoecologist. AltaMira Press, UK

Piperno DR, Sues HD (2005) Dinosaurs dined on grass science 310:1126–1128

Piperno DR, Holst L, Wessel-Beaver L, Andres TC (2002) Evidence for the control of phytoliths formation in Curcubita fruits by the hard rind (Hr) genetic locus: archaeological and ecological implications. Proc Natl Acad Sci USA 99:10923–10928

Power-Jones A (1992) Great expectations: a short historical review of European phytoliths systematics. In: Rapp G, Mulholland S (eds) Phytoliths systematics: emerging issues. Advances in archaeological and museum science 1. Plenum Press, New York, pp 15–35

Prado JL, Sánchez B, Alberdi MT (2011) Ancient feeding ecology inferred from stable isotopic evidence from fossil horses in South America over the past 3 Ma. BMC Ecol 11(1):15

Presad V, Stromberg CAE, Alimohammadian H, Sahni A (2005) Dinosaurs coprolites and the early evolution of grasses and grazers. Science 310:1177–1179

Rovner I (1971) Potential of opal phytoliths for use in paleoecological reconstruction. Quat Res (Orlando) 1(3):343–359

Runge F, Runge J (1997) Opal phytoliths in East African plants and soils. In: Pinilla A, Juan-Tresserras J, Machado MJ (eds) The state-of-the-art of phytoliths in soils and plants. Monografías del Centro de Ciencias Medioambientales, Consejo Superior de Investigaciones Científicas, Madrid, pp 71–81

Sánchez B, Prado JL, Alberdi MT (2003) Paleodiet, ecology, and extinction of Pleistocene gomphotheres (Proboscidea) from Pampean Region (Argentina). Coloq Paleontol 1:617–625

Solari ME (2007) Historia Ambiental Holocénica de la Región sur-austral de Chile (X–XII Región). Rev Austral Cienc Soc 13:79–92

Tomlinson PB (1961) Anatomy of monocotiyledons II: palmae. Oxford University Press, London

Twiss PC (1992) Predicted world distribution of C3 and C4 grass phytoliths. In: Rapp G, Mulholland S (eds) Phytolith systematics. Plenum Press, New York, pp 113–128

Chapter 12
Fossil Coleoptera from the Pilauco Site: An Approach to Late Pleistocene Microenvironments

Francisco Tello and Fernanda Torres

Abstract Quaternary fossil beetles have been widely used as environmental indicators and reconstructions of past climate. In Chile, records of fossil Coleoptera range from ~28,000 to 5200 cal yr BP with the higher concentration of fossil sites in Patagonia. This chapter presents an overview of the aforementioned studies, with an emphasis on the taxonomic beetle composition and some environmental inferences of the Pilauco site. In Pilauco, 13 families and 21 species of fossil beetles have been described with age constraints ranging between 15,750 and 14,700 cal yr BP. *Enochrus vicinus* and *Germainiellus dentipennis* are the most abundant species at Pilauco. Based on the record of Coleoptera, was possible infer that the surrounding vegetation around Pilauco was dominated by forest of *Nothofagus* sp., grasslands and wetlands. Beetle assemblages also provide information regarding the relationship between megafauna and coleopterans. This is mainly because; large herbivorous mammals may have provided substrates for the subsistence of dung beetles and scavengers, acting at the same time as modifiers of the site conditions and promoting greater environmental heterogeneity. These results highlight the need to incorporate modern information of intraspecific species dynamics and disturbances to studies of environmental and climatic paleo reconstructions based on fossil beetles.

Keywords Fossil beetles · Quaternary entomology · Paleoenvironmental reconstructions · Pilauco

---

F. Tello (✉)
Facultad de Ciencias Forestales y Recursos Naturales, Programa de Doctorado en Ecosistemas Forestales y Recursos Naturales, Universidad Austral de Chile, Valdivia, Chile
e-mail: ftelloa@yahoo.cl

Center for Climate and Resilience Research "CR2", FONDAP, CONICYT, Santiago, Chile

F. Tello · F. Torres
Transdisciplinary Center for Quaternary Research "TAQUACH", Valdivia, Chile

M. Pino and G. A. Astorga (eds.), *Pilauco: A Late Pleistocene Archaeo-paleontological Site*, The Latin American Studies Book Series,
https://doi.org/10.1007/978-3-030-23918-3_12

12.1 Background

Fossil beetles are widely used for the reconstruction of Pleistocene and Holocene environments (e.g., Coope 1970; Whitehouse 2007; Ashworth 2007; Massaferro et al. 2008; Elias 2010; Tello et al. 2017). The Order Coleoptera contains about 350,000–400,000 described species and is a group with the greater richness and abundance of species on Earth (Stork et al. 2015). The diversity of these insects is probably related to the ability of adults to inhabit the terrestrial or aerial space.

These qualities are promoted by the following morphological characteristics (Fig. 12.1): (a) The head is the structure containing an important number of sensory organs, such as the antennae and eyes. In addition, the chewing-mandibular apparatus in some cases form defenses and courtship structures; (b) the thorax includes the pronotum, which articulates the rest of the thorax. Generally, it has a shield shape that provides protection to the Coleoptera. The six legs are assembled in the ventral thoracic section, with each pair of legs being articulated in a specific thoracic section (prothorax, mesothorax, and metathorax). These articulated segments are highly sclerotized but allow for mobility of the legs. The elytra emerge in the dorsal thoracic section and are also highly sclerotized, forming a shell that protects the abdomen and wings. Although there are beetles that actually use elytra to fly, these structures are mainly used as a protection mechanism, usually covering the entire abdomen or an important part of it. The wings are membranous structures, located below the elytra emerging from the second thoracic section that are frequently used for flight; and (c) the interior of the abdomen contains the viscera and reproductive organs. It is highly sclerotized to protect these organs while remaining flexible for the utilization of them (i.e., storing a large amount of eggs).

In Chile, approximately 4000 species have been described (Elgueta 2000), which are widely distributed in continental and insular territories, excluding Antarctica. The beetle fauna of Chile is characterized by a high degree of endemism. This is mainly a result of the biological isolation imposed by geographic barriers, such as extreme aridity of the Atacama Desert in northern Chile, the high peaks of the Andean Range, as well as being a product of the Australasian connection 14 Ma years ago. In taxonomic terms, most well-represented families are Staphylinidae, Tenebrionidae, Curculionidae, and Carabidae. Other important groups include Scarabaeidae, Elateridae, Cerambycidae, and Chrysomelidae, although they have a distinctly smaller number of species (Elgueta 2000, 2008).

In direct relation to the taxonomic diversity of the order, coleopterans are associated with a wide variety of habitats, inhabiting terrestrial, continental aquatic, and coastal environments. However, the highest beetle diversity is found in forests, living on tree trunks, branches, roots, foliage, flowers, and fruits. On the ground, they are found among leaf litter and under dead or decaying logs, where they forage or feed on organic matter. Other more specialized species feed on feces, dead animals, and other food remains or inhabit bird nests where they parasitize. Additionally, there are other that inhabit continental aquatic environments, such as ponds, rivers, streams, and all types of wetlands.

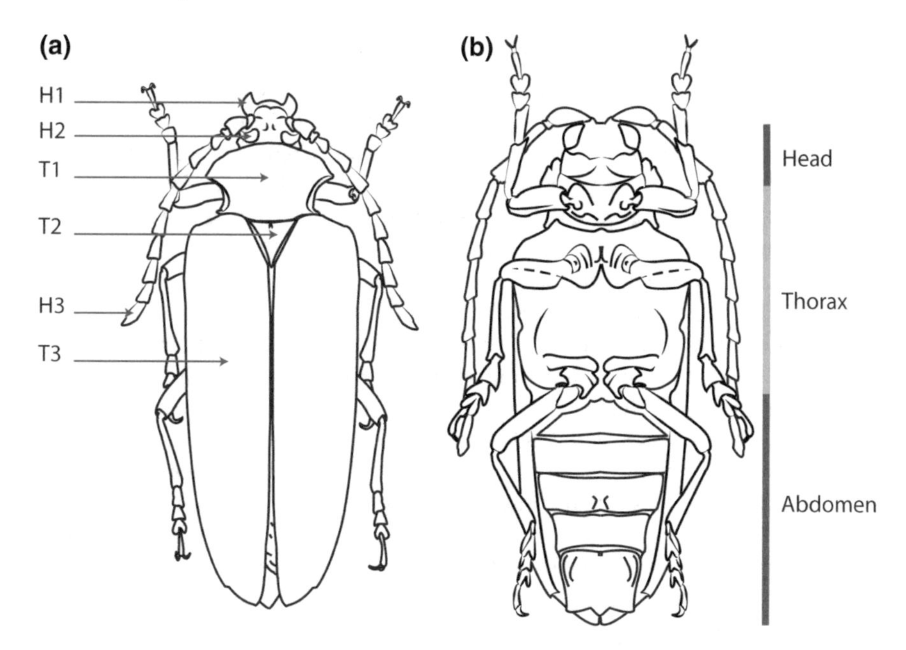

Fig. 12.1 General morphology of Coleoptera. **a** In dorsal view, H1 = mandible, H2 = eye, H3 = antennae, T1 = pronotum, T2 = scutellum, T3 = elytra. **b** Ventral view

Coleoptera order has been of particular importance for a wide spectrum of scientific disciplines, both within natural and applied sciences. For example, this group is frequently studied in agriculture, biomedicine, forensics, ecology, genetics, and even innovation and technology sciences, among others. In particular, several investigations have used fossil beetles as modern biological indicators, as well as tools for the interpretation of Quaternary environmental/climate conditions, early humans dwelling habits and preexisting megafauna (Ashworth 2001, 2007; Ashworth and Hoganson 1987, 1993; Elias 2010; Hoganson et al. 1989; Hoganson and Ashworth 1992).

According to Elias (2006), beetles share several characteristics that make them particularly useful in past climate and environmental reconstructions; (a) Coleoptera have an enormous diversity of species that are distributed in specific niches; (b) this group of insects are sufficiently sensitive to changes in their environment; (c) they have a good capacity for dispersal and colonization by rapidly migrating to newly available niches; (d) their fossil record is abundant in a variety of depositional settings; (e) beetles have usually a hard exoskeleton and important body parts are preserved in sediments, which is useful for the subsequent identification of species; (f) most of the species found in Pleistocene records have modern analogues; (g) most Coleoptera have niche specificity and therefore, limited to a specific environment or climate; and (h) Coleoptera are the best studied group within entomology making possible to obtain detailed information of species recorded in Quaternary sediments.

Since beetles are dependent on temperature, climate variations constitute a force of change at the mesoscale (i.e., hundreds to thousands of years) in the distribution of population of Coleoptera (Ashworth 2001). On the other hand, geological process (e.g., volcanic eruptions, glacier activity, mountain range uplift) promote changes on a macro-evolutionary scale (thousands of years or more). Both forcing factors have generated substantial changes on beetle distribution contributing to the evolutionary process they have sustained up to the present. Therefore, in order to use Coleoptera as biological proxies in climatic and environmental interpretations, it is necessary to understand the natural history of their modern analogous.

In South America, the use of fossil beetles has mainly been focused on understanding the climatic fluctuations during the late glacial and Holocene (the last ~19,000 years ago), both in Chile and Argentina. These investigations have focused on indicating how climatic variations, glacial activity, and the development of human settlements have influenced the distribution of beetle communities. Specifically, south of the 39° S, fossil entomofaunas ranging from ~24,000 to 4500 cal yr BP have been analyzed by Hoganson et al. (1989), Hoganson and Ashworth (1992), Ashworth and Hoganson (1993), Ashworth (2007), Tello et al. (2017) providing valuable information regarding past environmental conditions. Many of the aforementioned studies correspond to peat deposits that have facilitated the preservation of fossil material. In these sites, several species of fossil Coleoptera have been recorded associated with Valdivian temperate rain forests, including, for instance, the species *Germainiellus dentipennis* (Coleoptera: Curculionidae) from 11,000 to 12,000 cal yr BP. These records indicate cold temperate conditions (temperatures 4–6 °C lower than present) and environments dominated by tree species of the genus *Nothofagus*, about 13,000–12,500 cal yr BP. The coldest indicators correspond to Coleoptera associated with North Patagonian forests and Magellanic moorlad, for example, carabids associated with open environments and cold climates during the past 18,000 to 14,000 cal yr BP (Ashworth and Markgraf 1989; Hoganson and Ashworth 1992; Ashworth 2007). Another record at Puerto Edén (49° 08′ S; Ashworth and Hoganson 1993) is represented by fossil beetles associated with young forests growing in the vicinity of glaciers, which have been interpreted as a glacial refuge (Fig. 12.2).

Fossil beetles recorded at Pilauco spanning 15,750 and 14,700 cal yr BP. They were deposited under transitional climatic and environmental conditions, probably affected by natural disturbance events ranging from intense volcanism to wildfires, as suggested by Jara and Moreno (2014) for northwestern Patagonia (40° S). The vegetation in Pilauco was dominated by a mixed variety of taxa corresponding to the Chilean North Patagonian forest and Valdivian temperate rain forest (Abarzúa and Moreno 2008) under humid and cold climate conditions (Villagrán 2001). These conditions allowed the development of the genus *Nothofagus* trees to which multiple species of Coleoptera are presently associated. Additionally, in Pilauco, flooding episodes may have generated the availability of environments such as banks, swamps, and meadows inhabited by Coleoptera (Pino et al. 2013).

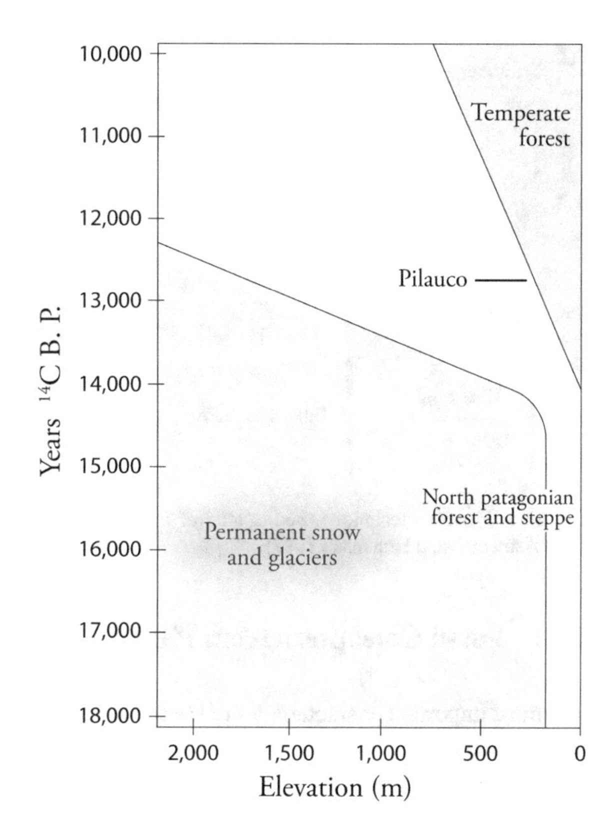

Fig. 12.2 Summary of major geological and biological changes occurring between 39 and 44° S during the last 18,000 years BP Modified from Ashworth and Hoganson (1993)

12.2 Methods

The fossil remains were recovered using the flotation technique, especially adapted to the conditions of the Pilauco site. This technique consists of using a vertical water flow to separate the organic matter from the sediment (step 1, Fig. 12.3). The organic remains (fossil seeds, wood fragments, insects, mites, etc.) were then washed and the fossil remains of beetles are recovered with kerosene or manually under a stereomicroscope (steps 2 and 3, Fig. 12.3). Once beetle remains were separated, ultrasound was used to remove the remaining sediments. The remains were mounted on entomological slides and stored for their later study (step 4, Fig. 12.3). In Quaternary fossil deposits, as is the case of Pilauco, is common to find disarticulated parts of fossil beetles entombed in sediments. For practical purposes, the studies often concentrated on the head, elytra, pronotum, and to a lesser extent on the legs, because such characters provide a most useful taxonomic diagnostic for the identification of species. In some cases, chitin has remained stable and has preserved the original color of the insect, facilitating the determination.

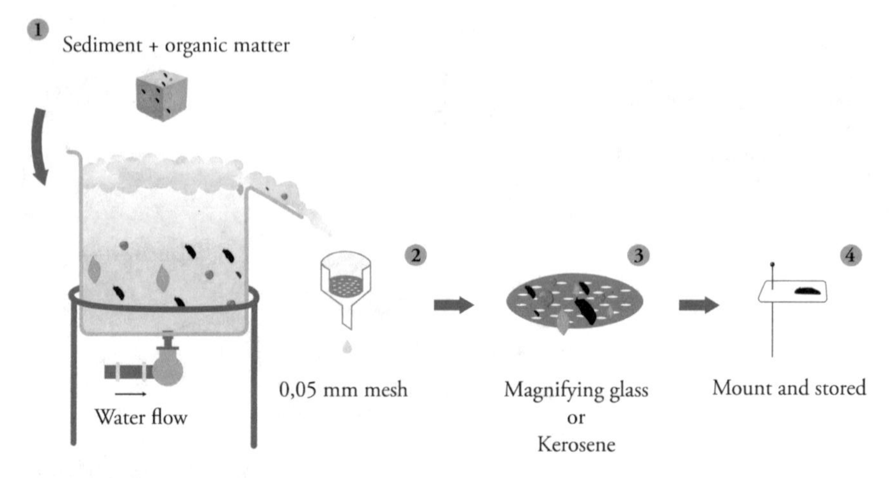

Fig. 12.3 Flotation technique used to recover fossil beetle remains from the sediments (adapted from Ashworth and Hoganson 1987)

12.3 Fossil Coleoptera from Pilauco

The most important characteristics of the fossil beetles recorded at Pilauco have been described by Tello et al. (2017) and are presented below. The diagnoses presented here represent key elements that allowed an interpretation of the fossil remains, so the most frequent structures in the fossil record are detailed.

Suborder ADEPHAGA
Family CARABIDAE Latreille (1802)

Diagnosis. Small to large body size, ranging from 1 to 85 mm. Body of variable shape, sometimes oval, oblong or elongated, flat or convex. This group has a characteristic metallic body coloration that is usually black and rarely extremely colorful. The body is often glabrous, with usually large, non-faceted, and circular eyes and prominent jaws. The pronotum is narrower than the elytra having a well-marked lateral edge. On the edges, there are a variable number of bristles, one of which is markedly visible in the basal area. The elytra present a variety of striae, frequently between 8 and 9 longitudinal grooves (Solervicens 2014).

In terms of taxonomic diversity, the Carabidae is the fourth most important family in Chile. This group includes 95 genera and 365 species of which 18% are endemic to Chile (Leal 2001). They are characterized by being outstanding predators of larvae, snails, and worms, among others. Their legs move rapidly and are adapted to hunt. In Chile, they have a wide distribution and are frequently found in prairie environments, forests, and riversides.

The fossil record of Pilauco indicates the presence of several species of the Bembidiini tribe, which are known to be associated with water bodies and temperate climates. Most species live on riverbanks or in stagnant waters, where they feed on

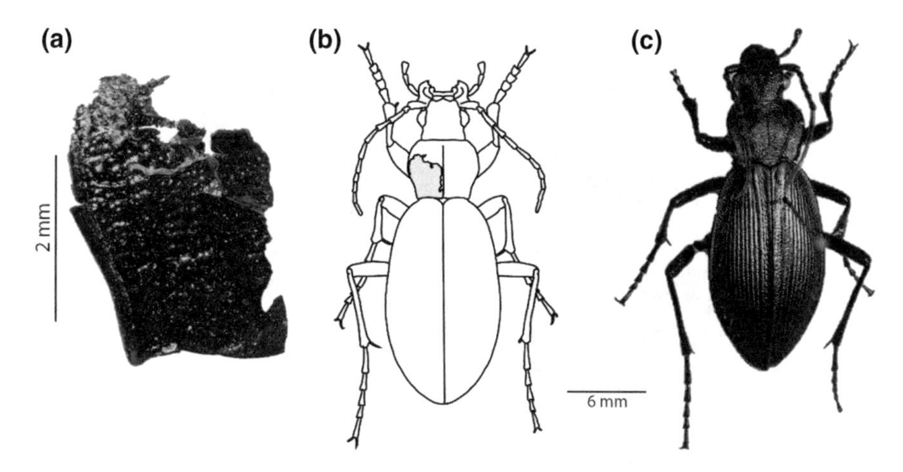

Fig. 12.4 **a** Fossil pronotum fragment of *Ceroglossus* sp. **b** The area painted in gray indicates the segment of the body to which the fossil corresponds. **c** Modern specimen of *Ceroglossus* sp.

other arthropods (Maddison 2012). Additionally, present also in the fossil record of Pilauco is *Ceroglossus* cf. *chilensis* (Fig. 12.4), which inhabit forest formations, usually as a part of *Nothofagus* understory (Briones et al. 2012). They are active predators of other invertebrates and can be associated with decaying organic matter (Briones and Jerez 2007). This genus is highly striking due to its metallic coloration.

Family DYTISCIDAE Leach (1815)

Diagnosis. Beetles of medium size ranging from 2.5 to 27 mm. Elongated and oval body, dorsally and ventrally convex, smooth, and glabrous. Legs are elongated and modified in the form of paddles for swimming. In the posterior section of the legs, the tarsi are densely populated with bristles.

Members of the Dytiscidae family are present in continental lentic water environments, such as permanent and temporary lakes, ponds, and wetlands. There are also species adapted to extreme conditions and potamon areas (Illies and Botosaneanu 1963). In lotic environments, they may be present in shallow overflowing areas. Larvae and adults are often aquatic and carnivorous; adults have a high capacity for flight (Jerez and Moroni 2006). In Chile, 33 species have been reported (Elgueta 2008), and members of the family have been recorded at Pilauco; however, the fossil remains did not allow for higher taxonomic classification.

Family TRACHYPACHIDAE Thomson (1857)

Diagnosis. Small-sized coleoptera from 3.8 to 7 mm long. The body is generally dark and relatively glossy or with the metallic surface. They are similar to the Carabidae family. Well-developed eyes. Pronotum and elytra have a faint, irregular dotted surface (Beutel and Arndt 2005). The Trachypachidae family is a small group of adephagous Coleoptera with six modern described species, only two of them occurring in Chile. These small beetles were much more diverse during the first half of the Mesozoic Era (Ponomarenko 1977), followed by a decline of the group due to the

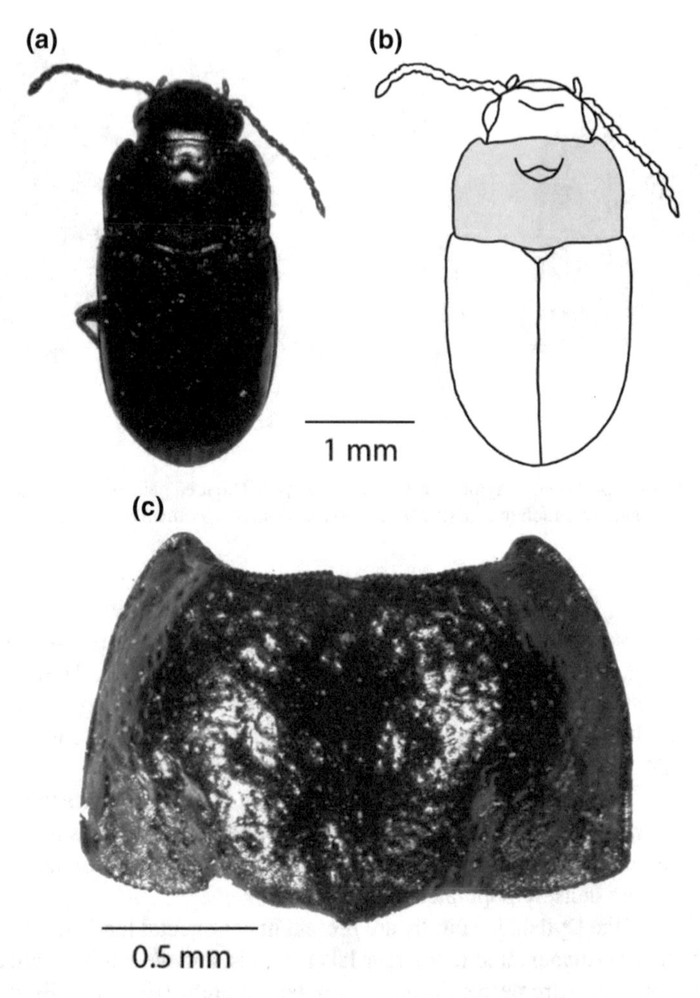

Fig. 12.5 **a** Modern specimen of *Systolosoma lateritium*. **b** The area painted in gray indicates the segment of the body to which the fossil corresponds. **c** Pronotum fossil of *S. lateritium*

expansion of the Carabidae family (Beutel 1998). The fossil remains of the species *Systolosoma lateritium* have been recorded at Pilauco (Fig. 12.5), and modern individuals of the species have been collected in forests dominated by the evergreen species *Nothofagus dombeyi* and *Nothofagus pumilio*. It is believed that these beetles are predators, however, there are no major precedents in this regard (Beutel and Arndt 2005).

Suborder POLYPHAGA
Family BRENTIDAE Billberg (1820)

Diagnosis. Small-sized Coleoptera, from 1.5 to 3 mm that present pear-shaped body and convex oval elytra. Hairs or scales can sometimes be observed. Members of

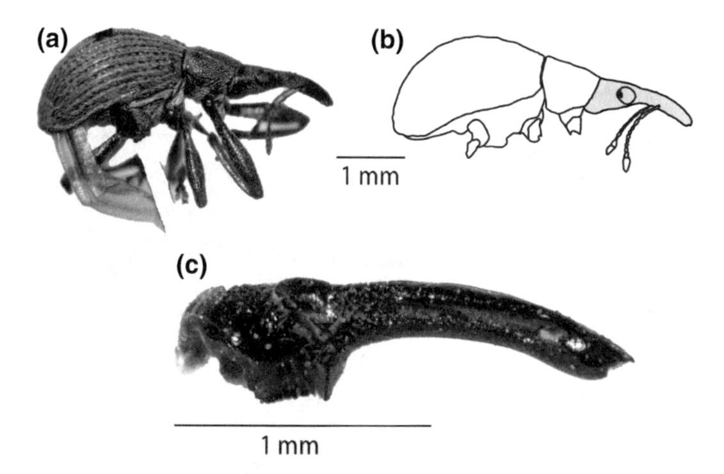

Fig. 12.6 **a** Modern specimen of the genus *Mythapion*. **b** Gray-colored area indicates the segment of the body to which the fossil corresponds. **c** Fossil head remains of *Mythapion trifolianum*

this group often present a long and narrow head forming a "trunk". The antennae are inserted proximally to the base of the face (Solervicens 2014). The family is typically associated with plants of the Fabaceae and Asteraceae families (Elgueta and Marvaldi 2006).

In Chile, 13 species members of the Brentidae family have been recorded within 8 genera (Elgueta 2008). In Pilauco, fossil remains corresponding to individuals of the species *Mythapion trifolianum* have been recorded (Fig. 12.6). Modern individuals of this species are typically associated with herbaceous plant species such as *Senecio smithii* (Asteraceae) and *Trifolium* sp. (Fabaceae) (Elgueta and Marvaldi 2006).

Family CURCULIONIDAE Latreille (1802)

Diagnosis. Medium sized coleopterans with body ranging from 1 to 40 mm in length. Their body shape is often oval, slightly elongated, and strongly convex dorsally. They may have bristles or scales attached to the body and/or obvious hairs.

The head is characterized by having an elongated face; sometimes short, forming a "horn", eyes of variable size, usually large, sometimes diminished or missing. The antennae, in some cases, are inserted distally on the face or proximally/medially in relation to the eyes with variations within subfamilies (Solervicens 2014). The pronotum is generally cylindrical and narrower than the elytra. Elytra often present variable characteristics; they are strongly convex, with tubercles and shallow or deep groove. These beetles have highly variable habits and are typically associated with plant species (Elgueta and Marvaldi 2006). In general, the larvae are found underground or in leaf litter, where they feed on roots, stems, or organic matter. Similarly, it is possible to find larvae that feed on the soft parts, buds, and shoots of trees and shrubs, or on other hard parts of trees, such as bark or xylem. Adults are usually phytophagous. In Chile, 525 species have been described within 171 genera (Elgueta and Marvaldi 2006; Elgueta 2008).

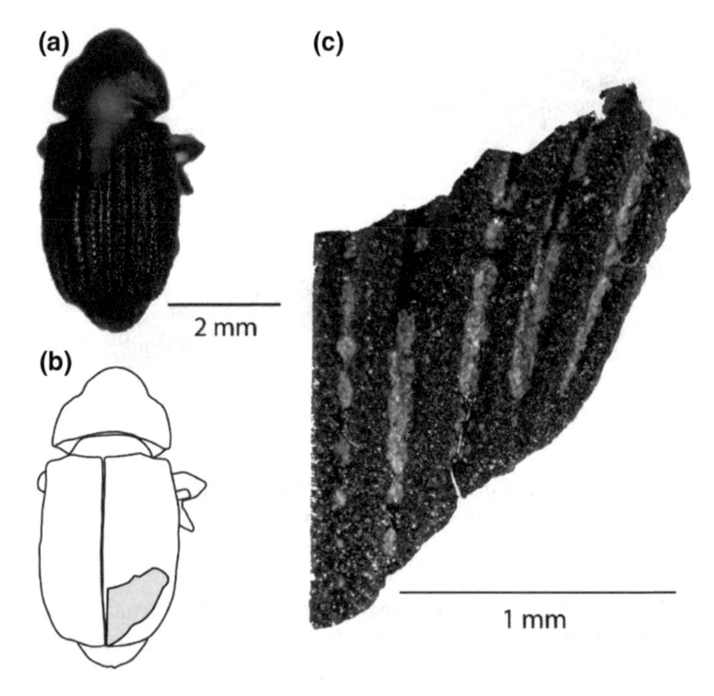

Fig. 12.7 **a** Modern specimen of *Psepholax dentipes*. **b** The area painted in gray indicates the segment of the body to which the fossil corresponds. **c** Fossil elytra fragment of *P. dentipes*

Fossils of Curculionidae are the most abundant and diverse at Pilauco. Of these, *Germainiellus dentipennis* associated with Magallanic forest and Valdivian temperate forests (Morrone 1993), is the most abundant species at Pilauco. Furthermore, the species *Aegorhinus schoenherri, A. bulbifer*, and *Psepholax dentipes* (Fig. 12.7) associated with *Nothofagus* forests (Elgueta et al. 2008) have also been recorded.

Rhyephenes maillei (Fig. 12.8) is a generalist species, of which two fossilized elytra have been found at Pilauco. In the larval stage this species feeds on *Maytenus boaria*, *Nothofagus antarctica*, *N. dombeyi*, *N. nervosa*, *Lithraea caustica*, and *Laureliopsis philippiana*, among others (Elgueta et al. 2008; Lanfranco et al. 2001; Lanfranco and Ruiz 2010; Rojas and Gallardo 2004). *Puranius australis*, *Puranius fasciculiger*, and *Dasydema* sp. are three of the species recorded whose modern analogous are associated with soil in a larval stage, either in grasslands or *Nothofagus* forests (Elgueta and Marvaldi 2006, MNHN record).[1]

Family HYDROPHILIDAE Latreille (1802)

Diagnosis. Body of variable size, ranging from 0.7 to 50 mm. Body shape usually oval, acutely dorsally convex and ventrally flat, smooth, and usually glabrous. The eyes are clearly visible dorsally, while the head has a marked V-shaped fronto-parietal suture (Klimaszewski and Watt 1997; Solervicens 2014).

[1]Information obtained from the National Museum of Natural History.

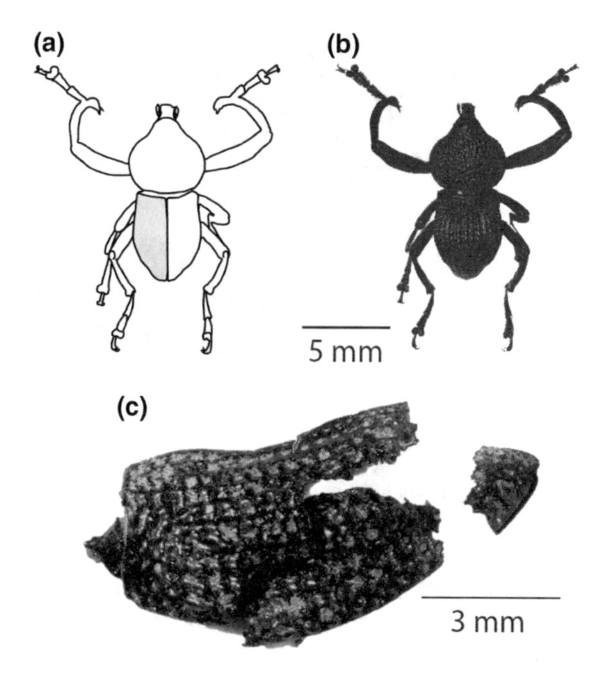

Fig. 12.8 **a** Gray-colored area indicates the segment of the body to which the fossil corresponds. **b** Modern specimen of *Rhyephenes maillei*. **c** Elytra fossil from *R. maillei*

Most species of the group are characterized by being aquatic. Others inhabit the fresh dung of mammals or decomposing substrates. In Chile, around 30 species have been described (Elgueta 2000; Solervicens 2014).

In Pilauco, two genera and two species from the subfamily Hydrophilinae have been recorded: *Enochrus vicinus* and *Tropisternus setiger*, characterized as such because they are aquatic organisms. Larvae are predators while adult individuals are saprophagous (Jerez and Moroni 2006).

Family LEIODIDAE Fleming (1821)

Diagnosis. Small body ranging from 1 to 8 mm in length usually elongated, convex and flattened. Generally dark to brownish-gray colored, slightly setose to glabrous. The elytra are convex, longer than wider, with distinct epipleura. The transverse striation is clearly visible, with separated striations parallel to the suture. Species from the Leiodidae family can be found in a wide diversity of habitats and present different life forms. The most common within the group correspond to saprophagous species (Newton and Thayer 2005). In Chile, 58 species have been reported (Elgueta 2000), which are distributed mainly in southern Chile (Seago and Newton 2009).

The fossil record of Pilauco includes specimens of *Chiliopelates nigrus* (Fig. 12.9), the genera *cf. Nemadiopsis*, and *Orchesia*. Both modern genera *Orchesia* and *Nemadiopsis* have a wide variety of habitat preferences and are associated with mosses, fungi, or simply feeding on organic waste in the soil (Lawrence 1991).

Family PTINIDAE Latreille (1802)

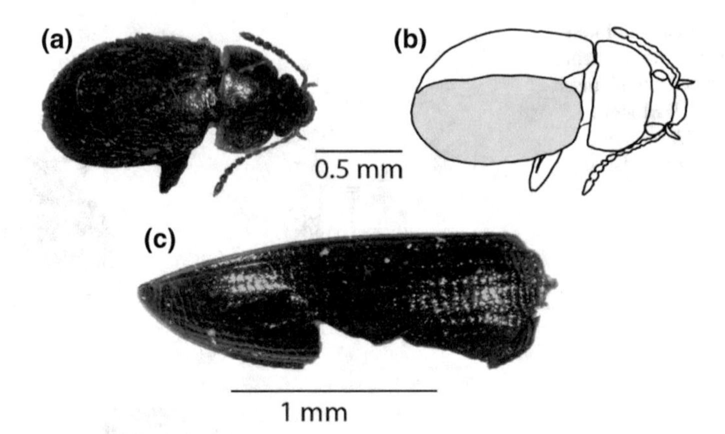

Fig. 12.9 **a** Modern specimen of *Chiliopelates nigrus*. **b** Gray-colored area indicates the segment of the body to which the fossil corresponds. **c** Fossil elytron of *Chiliopelates nigrus*

Diagnosis. Coleoptera of small size, usually ranging from 0.9 to 10 mm. Dark, brown or metallic colored body, strongly to moderately convex. The head is sometimes hidden in dorsal view. Most of these Coleoptera feed on dead wood, whereas others inhabit or feed on fungi (Solervicens 2014). In Chile, 95 species have been reported for the Ptinidae family.

At Pilauco, a fossil remains of the species *Stichtoptychus tenuivittatus* has been recorded, which corresponds to an endemic genus of the Dorcatominae subfamily associated with dead wood and fungi (White 1980).

Family SCARABAEIDAE Latreille (1802)

Diagnosis. Group presenting highly variable body size, commonly between 3 and 30 mm, but sometimes reaching 160 mm in length. The body shape is frequently oblong to oval and convex. Short face, sometimes with tubers or horns around the clipeous, eyes partially divided by the face. The group presents convex elytra with or without longitudinal striae.

This family is extremely diverse and abundant in tropical climates and they are associated with a wide variety of habitats, being frequently phytophagous. They are also associated with dung, carrion, humus, fungi, dead wood, and pollen (Solervicens 2014). In Chile, 180 species have been reported (Elgueta 2008).

In Pilauco, three subfamilies of the Scarabaeidae family have been recorded: Aphodiinae, represented by the species *Aphodius (Calamosternus) granarius*; Scarabaeinae with *Homocopris torulosus* (Fig. 12.10), and the subfamily Melolonthinae represented by specimens of the genus *Sericoides* (Fig. 12.11). The first two species of this family correspond to organisms associated with dung. *A. granarius* is an endocoprid species widely distributed in Chile, with a preference for temperate zones. It is mainly associated with herbivorous animal farms, sometimes being found in decaying wood and also associated with carrion (Ashworth and Hoganson 1987; Smith and Skelley 2007). In contrast, *H. torulosus* is a paracoprid dung beetle

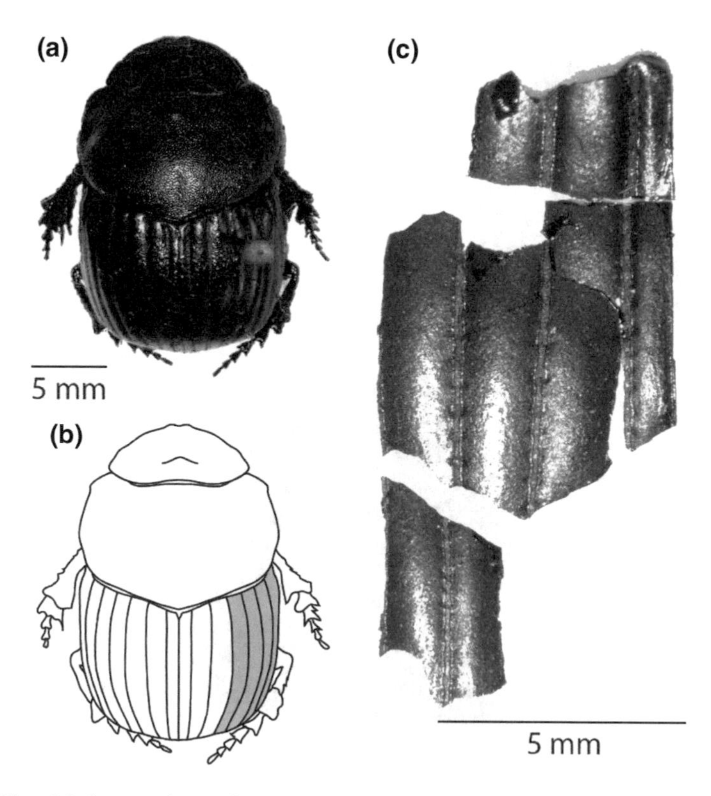

Fig. 12.10 **a** Modern specimen of *Homocopris torulosus*. **b** Gray-colored area indicates the segment of the body to which the fossil corresponds. **c** Elytra fossil remains of *Homocopris torulosus*

associated with herbivorous mammal dung. Multiple species of the genus *Sericoides* are described as defoliator insects in *Nothofagus* species, while feeding on roots during the larval stage (Lanfranco and Ruiz 2010).

Family STAPHYLINIDAE Latreille (1802)

Diagnosis. Group of coleoptera of small size, between 0.6 and 25 mm in length. The body is usually black or brown, often bright or iridescent. Has an elongated appearance, glabrous or fuzzy.

It is characterized by having exposed a variable number of abdominal segments, meaning that elytra are truncated (Newton and Thayer 2005). This family is one of the most abundant and diverse of the Order Coleoptera. In Chile, around 1000 species widely distributed in different ecoregions have been recorded (Elgueta 2000).

In Pilauco, both the subfamilies Aleocharinae and Oxytelinae are represented. The latter includes the species *Anotylus sulcicollis*. Modern individuals of *A. sulcicollis* are considered saprophagous, associated with dead wood and dung (Ashworth and Hoganson 1987; Newton and Thayer 2005). Thus, it is possible to consider this species as a generalist in terms of habitat, as is usually found associated to different substrates.

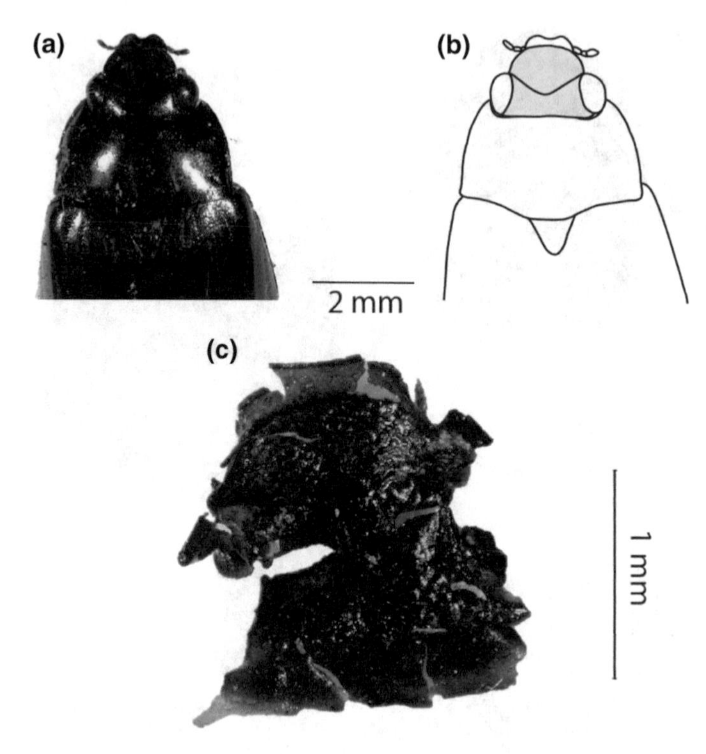

Fig. 12.11 **a** Modern specimen of *Sericoides* sp. **b** Gray-colored indicates the segment of the body to which the fossil corresponds. **c** Fossilized rostrum remains of *Sericoides* sp.

12.4 Discussion and Conclusions

Fossil Coleoptera provide an opportunity to interpret climatic and environmental characteristics of the past. Thus, in order to generate these inferences in sites with the presence of humans and megafauna, it is necessary to know the intrinsic characteristics of modern analogous insects, and the interactions that arise from natural and/or anthropogenic disturbances.

The presence of aquatic beetles at Pilauco indicates the prolonged occurrence of waterlogging conditions in some areas, with shallow and stagnant water (Tello et al. 2017). Furthermore, the presence of Carabidae fossils is commonly associated with water bodies (e.g., Ashworth and Hoganson 1993; Kuzmina et al. 2008). Perennial or ephemeral streams descending from the slope of the hill, to the north of the site could explain the co-occurrence of these environments.

In contrast, the curculionid record of Pilauco is related to *Nothofagus* forests, and in particular to the species *N. antarctica*, *N. dombeyi* and *N. obliqua* (Morrone 1993; Lanfranco et al. 2001; Rojas and Gallardo 2004; Elgueta and Marvaldi 2006; Elgueta et al. 2008; Lanfranco and Ruiz 2010; Zavala et al. 2011; Werenkraut and Ruggiero 2012). Additionally, based on the curculionid species recorded at Pilauco, it is also possible to infer the presence of other shrub and tree species, such as *Aristotelia chilensis*, *Discaria serratifolia*, *Maytenus boaria*, and *Austrocedrus chilensis*.

Other beetle families recorded at Pilauco, such as Melandryidae and Ptinidae are mainly associated with fungi and ferns (Lawrence 1991), which are an important component in North Patagonian forests (Ramírez and Dillehay 1989). These include *Lophosoria quadripinnata, Blechnum* sp., and *Isoetes* sp., which have been identified in the fossil pollen record of Pilauco (see Chap. 9, this volume).

Overall, the fossil beetle reconstruction is concordant with the analyses of pollen and fossil seeds at Pilauco. The vegetation reconstruction indicates the abundance of non-arboreal species including different types of ferns, *Gunnera magellanica* and plant taxa from the families Poaceae, Asteraceae, Ranunculaceae, and Cyperaceae.

Additionally, the presence of dung fossil beetles at Pilauco can be associated with herbivorous megafauna that inhabited this area (e.g., Ashworth and Nelson 2014). These beetles have a strong association with large mammals, since a large number of them rely on their excrement to feed and nest (Halffter and Favila 1993). At Pilauco, 11 coprolite samples have been identified, including parasites that facilitate a direct link with horses (see Chap. 13, this volume). The assemblage of manure-dependent beetles requires fecal material exposed to the atmosphere and not submerged under water. It can therefore be interpreted that places endured at the site, at least seasonally, which were sufficiently free of water. For example, some diatoms such as *Navicula dicephala*, *N. lengana*, and *Pinnularia divergens*, together with the presence of aquatic and marsh plant species, indicate the presence of wetlands and/or possible lagoons (see Chap. 10, this volume).

Based on abovementioned fossil beetle evidences, the environment at Pilauco (between 15,750 and 14,700 cal yr BP) can be confidently interpreted as a mosaic of *Nothofagus* forests interspersed with wetlands, and/or, lacustrine environments alongside with non-waterlogged grassland and shrub areas. This landscape differs from the general environment in which the site is situated. The analyses of pollen and seed have described Pilauco as mainly dominated by North Patagonian Forest, with elements, such as *N. dombeyi*-type, *Podocarpus andina*, *Podocarpus nubigena*, *Saxegothaea conspicua*, *Fitzroya/Pilgerodendron*-type, Myrtaceae, *Weinmannia trichosperma*, *Hydrangea serratifolia*, *Maytenus* sp., *A. chilensis*, and *Aextoxicon punctatum*. These results coincide with those recorded in other contemporary sites in the area such as Huelmo (Moreno and León 2003) and Lago Condorito, where the cold–humid conditions related to the forests of Northern Patagonia are those at the end of the last glaciation (see Chap. 9, this volume).

Monte Verde II was the first archeological site in Chile where fossil beetles have been studied. With the exception of Trachypachidae, all other beetle families are common at both sites (Hoganson et al. 1989; Tello et al. 2017). Approximately, 70% of the Pilauco fossil beetle taxa coincide with the assemblages of Monte Verde II. The latter is represented by beetles that were associated with trees of the genus *Nothofagus* (forests and understory), grassland areas, lacustrine environments, high and low energy streams, ferns, fungi and wood (Hoganson et al. 1989). According to this author, these conditions and species are strongly related to Valdivian temperate rain forest and species that live under similar warm-temperate modern climatic conditions at Monte Verde. These environmental conditions may have favored the early establishment of human settlements. However, the environmental conditions at Monte Verde II may have been modified by human activities at the time, mainly because the site was used as campsite. In contrast to Monte Verde II, Pilauco was a place that humans seem to have used sporadically for scavenging and the fabrication and/or retouching of tools, and not as a permanent campsite. These activities may have been the result of the availability of large mammals in the area (Recabarren et al. 2014). The local fossil evidence of gomphotheres and horses at Pilauco reinforces this hypothesis (see Chaps. 4 and 5, this volume).

Gomphotheres, on the other hand, share similarities with modern elephants in terms of size and possibly similar behavior, so that they have been called ecosystem engineers of the past (Barnosky et al. 2016), because they likely have the capacity to modify their environment. This could mean that the fossil beetle environmental reconstruction at Pilauco may have a strong disturbance component, caused by the presence of humans in combination with the impact of large mammals (Tello et al. 2017). However, more research is necessary to gain a better understanding of the effects of climate, megafauna and natural/anthropic disturbances on past ecosystems (Tello et al. 2017).

References

Abarzúa A, Moreno P (2008) Changing fire regimes in the temperate rainforest region of Southern Chile over the last 16,000 yr. Quat Res (Orlando) 69(1):62–71

Ashworth A (2001) Perspectives on quaternary beetles and climate change. In: Gerhard L, Harrisony WE, Hanson BM (eds) Geological perspectives of global climate change. American association of petroleum geologists studies in geology #47, Tulsa, Oklahom, pp 153–168

Ashworth A (2007) Late pleistocene of South America. In: Elias S (ed) Encyclopedia of quaternary science. Elsevier, Amsterdam, pp 212–221

Ashworth A, Hoganson J (1987) Coleoptera bioassociations along an elevational gradient in the Lake Region of Southern Chile, and comments on the postglacial development of the fauna. Ann Entomol Soc Am 80(6):865–895

Ashworth A, Hoganson J (1993) Magnitude and rapidity of the climatic change marking the end of the pleistocene in the mid-latitudes of South America. Palaeogeogr Palaeoclimatol Palaeoecol 101:263–270

Ashworth A, Markgraf V (1989) Climate of the chilean channels between 11,000 and 10,000 yr B.P. based on fossil beetle and pollen analyses. Rev Chil Hist Nat 62:61–74

Ashworth A, Nelson A (2014) The paleoenvironment of the olympia beds based on beetle fossils from Discovery Park, Seattle, Washington, USA. Quat Int 341:243–254

Barnosky A, Lindsey E, Villavicencio N, Bostelmann E, Hadly E, Wanket J, Marshall C (2016) Variable impact of late-quaternary megafaunal extinction in causing ecological state shifts in North and South America. Proc Natl Acad Sci USA 113(4):856–861

Beutel R (1998). Trachypachidae and the phylogeny of Adephaga (Coleoptera). In: Ball G, Casale A, Vigna Taglianti A (eds). Phylogeny and classification of Caraboidea, XX I.C.E. Firenze. Museo Regionali di Science Naturali, Torino, pp 81–106

Beutel R, Arndt E (2005). Trachypachidae C.G. Thomson, 1857. In: Beutel R, Leschen R (eds). Morphology and systematics (Archostemata, Adephaga, Myxophaga, Polyphaga partim), vol 1, pp 116–118

Briones R, Jerez V (2007) Efecto de la edad de la plantación de Pinus radiata en la abundancia de Ceroglossus chilensis (Coleóptera: Carabidae) en la Región del Biobío Chile. Bosque 28(3):207–214

Briones R, Gárate F, Jerez V (2012) Insectos de Chile nativos, introducidos y con problemas de conservación, Guía de, Campo edn. Corporación Chilena de la Madera, Concepción, Chile

Coope GR (1970) Interpretations of quaternary insect fossils. Ann Rev Entomol 15(1):97–121

Elgueta M (2000). Estado actual del conocimiento de los coleópteros de Chile (Insecta: Coleoptera). PrIBES-2000: Proyecto para Iberoamérica de Entomología Sistemática 17:145–154

Elgueta M (2008) Holometábolos. Orden Coleóptera. In: CONAMA Biodiversiad de Chile, Patrimonio y Desafíos. Ocho Libros Editores, Santiago, Chile, pp 144–150

Elgueta M, Marvaldi A (2006) Lista Sistemática de las especies de Curculionoidea (Insecta: Coleoptera) presentes en Chile, con su sinonimia. Bol Mus Nac Hist Nat, Chile 55:113–153

Elgueta M, Arias E, Will K (2008) Curculionoidea (Coleoptera) en follaje de árboles del centro-sur de Chile. Contribuciones Taxonómicas en órdenes de Insectos Hiperdiversos, pp 177–200

Elias S (2006) Quaternary beetle research: the state of art. Quat Sci Rev 25:1731–1737

Elias S (2010) Advances in quaternary entomology, developments in quaternary science. Elsevier Science

Halffter G, Favila M (1993) The Scarabaeinae (Insecta: Coleoptera) an animal group for analysing, inventorying and monitoring biodiversity in tropical rainforest and modified landscapes. Biol Intern 27:15–21

Hoganson J, Ashworth A (1992) Fossil beetle evidence for climatic change 18,000–10,000 years B.P. in South-Central Chile. Quat Res (Orlando) 37:101–116

Hoganson J, Gunderson M, Ashworth A (1989) Fossil-beetle analysis. In: Dillehay TD (ed) Monte Verde—a late pleistocene settlement in Chile. Smithsonian Institution Press, Washington

Illies J, Botosaneanu L (1963) Problemes et métodes de la classfication et de la zonation de eaux courantes considerées surtout du point de vue faunistique. Mitt Int Verein Theor Angew Limnol 12:1–57

Jara I, Moreno P (2014) Climatic and disturbance influences on the temperate rainforests of Northwestern Patagonia (40 °S) since ~14,500 cal yr BP. Quat Sci Rev 90:217–228

Jerez V, Moroni J (2006) Diversidad de coleopteros acuaticos en Chile. Gayana 70(1):72–81

Klimaszewski J, Watt J (1997) Coleoptera: family-group review and keys to identification. Fauna N Z 37:199 pp

Kuzmina S, Elias S, Matheus P, Store J, Sher J (2008) Paleoenvironmental reconstruction of the last glacial maximum, inferred from insect fossils from a tephra buried soil at Tempest Lake, Seward Peninsula, Alaska. Palaeogeogr Palaeoclimatol Palaeoecol 267:245–255

Lanfranco D, Ruiz C (2010) Entomología Forestal en Chile. Ediciones Universidad Austral de Chile

Lanfranco D, Ide S, Ruiz C, Peredo H, Vives I (2001) Razón sexual de Hylurgus ligniperda (F.), Hylastes ater (Paykull) y Gnathotrupes spp. (Coleoptera: Scolytidae). Bosque 22(2):85–88

Lawrence A (1991) Melandryidae (Tenebrionoidea) (=Serropalpidae). In: Stehr F (ed) Immature Insects: (2). Kendall/Hunt Publisher Company, Saint Louis USA, pp 505–508

Leal R (2001) Diversidad de la familia Carabidae (Coleoptera) en Chile. Niemelä, pp 549–571

Maddison D (2012) Phylogeny of Bembidion and related ground beetles (Coleoptera: Carabidae: Trechinae: Bembidiini: Bembidiina). Mol Phylogenet Evol 63(3):53–76

Massaferro J, Ashworth A, Brooks A (2008) Quaternary fossil insects from Patagonia. In: Rabassa J (ed) Late Cenozoic of Patagonia and Tierra del Fuego. Developments in quaternary science. Elsevier, Frankfurt, Germany, pp 393–410

Moreno PI, León AL (2003) Abrupt vegetation changes during the last glacial to Holocene transition in mid-latitude South America. J Quat Sci 18(8):787–800

Morrone J (1993) Revisión sistemática de un nuevo género de Rhytirrhinini (Coleoptera: Curculionidae), con un análisis biogeográfico del domino Subantártico. Bol Soc Biol Concepc, Chile 64:21–145

Newton A, Thayer M (2005) Catalog of higher taxa of Staphyliniformia and genera and subgenera of Staphylinoidea. Chicago, Field Museum of Natural History. https://www.fieldmuseum.org/peet-program. Accessed November 2016

Pino M, Chávez-Hoffmeister M, Navarro-Harris X, Labarca R (2013) The late Pleistocene Pilauco site, Osorno, South-Central Chile. Quat Int 299:3–12

Ponomarenko A (1977) Suborder Adephaga. In: Arnoldy V, Jeriki V, Nikritin L, Ponomarenko A (eds) Mesozoic Coleoptera, Trudy Paleontol Inst Akad Nauk SSSR, vol 161, pp 1–183

Ramírez C, Dillehay T (1989) Research area-background information. In: Dillehay T (ed) Monte Verde, vol I. Palaeoenvironment and site context. Smithsonian Press, Washington, DC, USA, pp 53–87

Recabarren O, Pino M, Alberdi MT (2014) La Familia Gomphotheriidae en América del Sur: evidencia de molares al norte de la Patagonia chilena. Estud Geol (Madr) 70(1):1–12

Rojas E, Gallardo R (2004) Manual de insectos asociados a maderas de la zona sur de Chile. Servicio Agrícola y Ganadero. División Protección Agrícola. Proyecto vigilancia y control de plagas forestales. 64p

Seago A, Newton A (2009) A New genus of Leiodid beetle from Chile with generic key and species checklist of described Neopelatopini (Coleoptera: Leiodidae: Camiarinae). Ann Zool (Wars) 59(3):297–304

Smith A, Skelley P (2007) A review of the Aphodiinae (Coleoptera: Scarabaeidae) of Southern South America. Zootaxa 1458:1–80

Solervicens J (2014) Coleópteros de la Reserva Nacional Río Clarillo, en Chile central: taxonomía, biología y biogeografía. CONAF, Chile

Stork NE, McBroom J, Gely C, Hamilton AJ (2015) New approaches narrow global species estimates for beetles, insects, and terrestrial arthropods. Proc Natl Acad Sci USA 112:7519–7523

Tello F, Elgueta M, Abarzúa A, Torres F, Pino M (2017) Fossils beetles from Pilauco, South-Central Chile: an upper pleistocene paleoenvironmental reconstruction. Quat Int 449:58–66

Villagrán C (2001) Un modelo de la historia de la vegetación de la Cordillera de La Costa de Chile central-sur: la hipótesis glacial de Darwin. Rev Chil Hist Nat 74(4):793–803

Werenkraut V, Ruggiero A (2012) Altitudinal variation in the taxonomic composition of ground-dwelling beetle assemblages in NW Patagonia, Argentina: environmental correlates at regional and local scales. Insect Conserv Divers 6:82–92

White R (1980) A taxonomic study of the new world genus Stichtoptychus Fall (Coleoptera Anobiidae). Technology Bull 1602, Department of Agric, USA, pp 1–35

Whitehouse N (2007) The study of fossil insect remains in environmental and archaeological investigations. In: Murphy E, Whitehouse N (eds) Environmental Archaeology in Ireland. Oxbow Books, pp 136–163

Zavala A, Elgueta M, Abarzúa B, Aguilera A, Quiroz A, Rebolledo R (2011) Diversity and distribution of the Aegorhinus genus in the La Araucanía Region of Chile, with special reference to A. superciliosus and A. nodipennis. Cien Inv Agr 38:367–377

Chapter 13
Coprolites as Proxies for Paleoparasitology at Pilauco

Leonora Salvadores-Cerda and Felipe Ramírez-Mercado

Abstract Coprolites are trace fossils. One of the ways to determine the presence of coprolites is to determine the presence of parasites in them. The four coprolites analysed here were collected in situ from the fossil-bearing layer PB-8. The paleoparasitological analyses were performed in two steps, the first is to confirm the classification of the sampled material as coprolites, and in the second place, to develop new analyses to add new parasite taxa. We used the diagnostic parasitological sedimentation–flotation technique to carry out the microscopic analyses. The same protocol was used in some sediment samples tacked in adjacent levels to the coprolites. Only the studied fossils and not the sediments contain parasitic eggs, which indicate faecal or intestinal origin and can be therefore considered as coprolites. The occurrence of eggs from *Anoplocephala* sp. indicate that the host was a horse, whereas *Moniezia* sp. can be correlated with deer and camelids.

Keywords *Anoplocephala* · *Moniezia* · Pilauco

13.1 Background

Coprolites represent fossilized dung from the digestive system of marine and terrestrial animals (Häntzschel et al. 1968). The term derives from the Greek words '*kopros*' meaning 'excrements' and '*lithos*', meaning 'stone' or 'rock' (Amstutz 1958). Coprolites are classified as trace fossils, which correspond to sedimentary structures that record or evidence the existence of an organism, such as footmarks, eggs, perforations, nests, etc. The main challenge of studying coprolites is to define their origin, as most of the material in question could be easily confused with organic

L. Salvadores-Cerda (✉)
Instituto de Ciencias de la Tierra, Palaeontology Master Program, Universidad Austral de Chile, Valdivia, Chile
e-mail: leonorasalvadores@gmail.com

F. Ramírez-Mercado
LabNat Pilauco, Valdivia, Chile

M. Pino and G. A. Astorga (eds.), *Pilauco: A Late Pleistocene Archaeo-paleontological Site*, The Latin American Studies Book Series,
https://doi.org/10.1007/978-3-030-23918-3_13

concretions or with organic sediments. Starting with the pioneer work of Amstutz (1958), the literature provides numerous references to coprolites from all continents (exception Antarctica), which are associated with invertebrates, fish and reptiles in paleontological contexts since the Paleozoic, as well as with mammals and humans in Cenozoic paleontological and/or archaeological sites in Europe, Asia, Africa and America (Hunt et al. 1994).

The coprolites often represent the first-hand evidence of the past presence of animals and humans in archaeological and/or paleontological sites providing the basis for their further study and characterization (Jouy-Avantin et al. 2003). They contain a great variety of information about the existence and conduct of their producers (Fugassa 2007), allowing for a better understanding and more robust interpretation of the relation between organisms and their substrate, as well as their past life conditions. Most of the information stored in fossilized faeces originates from the preservation of microfossils and macroscopic plant remnants. Quaternary coprolites are particularly suited for detailed analysis of pollen particles and parasitic remnants due to their excellent state of preservation (Araújo et al. 2003). Nevertheless, these fossils are scarcely found in Quaternary deposits of South America, as bearing sediments are commonly unconsolidated, which makes the recognition and recovery of coprolites particularly challenging. Generally, coprolites can be obtained from sedimentary layers or directly from the mummified or conserved bodies of humans and animals. Even if the source organisms cannot be identified, the content of coprolites can be indicative for their diet and ecology (Hunt et al. 1994).

The most common mineral elements in ruminant faeces include spherulites, phytoliths and detrital dust, as well as diatoms, ostracods and small gastropods, all of which offer valuable information about characteristics of the pasture areas (Badal and Atienza 2005). These elements recovered from coprolites allow for a variety of inferences about the diet of corresponding animals, their prevailing paleo-environment and the associated invertebrate fauna, which also serves as a paleo-ecological bioindicator. The pollen record recovered from coprolites serves mainly as a general proxy for the regional characteristics of vegetation, whereas the macroscopic remnants (mostly seeds and phytoliths) provide insights into dietary habits (Carrión 2002). The macroscopic analysis of coprolites of desiccated or mineralized faeces allows for the identification of developmental parasitic stages, remnants from acarians and insects, plant epidermis, phytoliths, pollen particles, spores and diatoms (Chimento and Rey 2008; van Geel et al. 2008).

The discipline of paleoparasitology developed as an independent branch of parasitology with the first discovery of parasitic eggs in archaeological material (Araújo and Ferreira 2000). The parasitic relationship of organisms depends on their behaviour and living environment. Paleoparasitological studies can therefore be evaluated from the medical perspective, but also in order to understand past events in human evolution (Bouchet et al. 2003a).

Parasitic behaviour implies a system of ecological relationships between parasite, host, and their environment, and provides further insights into sanitary, cultural and ecological aspects. Parasites are considered as biological indicators for zoonosis, environmental impacts, phylogenetic relationships between host and parasitic species, human migrations, occupation of new ecological niches, amongst others (Fugassa et al. 2010).

There are two scientific concepts in paleoparasitological studies adopted in the Americas. North American archaeological studies typically involve biologists, who are trained in archaeological techniques and are able to interpret their findings in terms of dietary habits, ecology and cultural complexity. As a consequence, Reinhard (1990) introduced the term of archaeo-parasitology referring to the study of parasites in archaeological contexts, a concept, which years later was applied in areas in Peru and Chile (Bouchet et al. 2003a, b). In turn, paleoparasitological studies from Brazil followed a different path. Towards the end of the 70s, Ferreira and his collaborators were the first to denominate the study of parasites recovered from archaeological and paleontological material as paleoparasitology (Ferreira et al. 1989). These parasitologists were mainly interested in parasitic diseases from the past, the time duration of the relationship between parasites and their hosts, and the prehistoric distribution of parasites (Araújo and Ferreira 2000). As a consequence, they began to participate in archaeological studies in order to investigate coprolites, mummies, and other sources for parasitological information. Consequently, the Brazilian parasitologists focused their interest mainly on diseases from the Old World and the Pre-Colombian New World and on prehistoric migrations based on paleoparasitological data (Araújo et al. 1981).

The first macro and micropaleontological studies in Pilauco date from 2008 and include the paleoparasitological analysis of presumed coprolites (Salvadores-Cerda and Abarzúa 2008), where the parasite eggs corresponding to nematodes and cestodes were described (Salvadores-Cerda et al. 2008). On the other hand, the analyzed bone material evidences the existence of a wide variety of extinct and existing mammals (Pino et al. 2013 and their references) with which relationships can be inferred.

13.2 Materials and Methods

The analysed material from Pilauco was collected in situ from the fossil-bearing layer PB-8 (see Chap 3, this volume). The three samples (MHMOP/PI/585, MHMOP/PI/586 and MHMOP/PI/587; 551, 256 y 620 gr, respectively) were excavated from grid 7AC (see Chap. 1, this volume). A fourth sample (MHMOP/PI/588, 2458 gr) was extracted from the exploration trench No. 5, which was associated with the first geological evaluation of Los Notros site (see Chap. 14, this volume). Samples MHMOP/PI/585 and MHMOP/PI/587 were subdivided into 10 subsamples, MHMOP/PI/588 into 20 subsamples and MHMOP/PI/586 in 5 subsamples. Four pseudo replicas were obtained from each of these subsamples giving a total of 180 pseudo-replicas.

Sample MHMOP/PI/587 yielded a radiocarbon age of 11,004 $\pm$ 186 BP, which corresponds to the youngest mammal fossil date obtained from the PB-8 layer in Pilauco (see Chap. 2, this volume).

The paleoparasitological analyses were performed in two stages: (1) to confirm the definition of the sampled material as coprolites and (2) to perform new analyses confirming earlier findings and adding new data evidencing a possible new nematode in the record (Ramírez 2012). During the first stage of the study, we followed a protocol including the observation, weighting and macroscopic description of the material documenting the colour, remnants from plants or insect chitin and seeds (Badal and Atienza, 2005). According to the protocol of the Parasitological Veterinary Laboratory at Universidad Austral de Chile, we performed the diagnostic parasitological sedimentation–flotation technique, microscopic analyses, and obtained photographs of the samples. The same protocol was used in some sediment samples, which were obtained in the same levels where the coprolites were found. The paleoparasitological analysis encompassed a veterinary technique for parasitological diagnosis of coprolite samples to obtain a set of images from each parasitic form recovered from the pseudo replicas of the original coprolite specimens. The results were first compared to samples of actual parasites from the Parasitological Veterinary Laboratory at Universidad Austral de Chile. The taxonomic comparative study was completed based on data from literature and the record of actual and fossil parasites and applying a descriptive qualitative approach based on morphological and biometric characteristics to assign parasitic genera.

Samples have been stored in the Paleontological Laboratory of the Instituto de Ciencias de las Tierra at Universidad Austral de Chile.

13.3 The Coprolites from Pilauco

No parasites were recorded in the sediments analysed at the Pilauco site. The paleontological sampled material is characterized by a wide gamma of dark brownish variegated colours, but lighter and with less olive colouring that the host sediment:

MHMOP/PI/585: brownish-dark chocolate colour, with peaty texture; high content of plant remnants, such as branch fragments, grasses, seeds and elytra from insects.

MHMOP/PI/586: brownish-black colour grading into brownish-orange towards the centre; brownish-green in aerial contact; high content of plant remnants.

MHMOP/PI/587: brownish-black colour, with peaty-muddy texture; medium content of plant remnants of different sizes and colours.

MHMOP/PI/588: brownish-black grading into brownish-dark chocolate colour towards the centre; low content of plant remnants, such as branch fragments, grasses, seeds and elytra from insects. The record and measurements were performed on eggs from nematodes as follows:

- *Trichuris* sp., 8 eggs with lengths of 40–50 μm, 2 eggs with lengths of 60 μm; symmetric and lemon-shaped, brownish-yellow colouring, no internal segmentation, thick shell consisting of two flat layers and outstanding polar plugs (Fig. 13.1). The measurements and description correspond to those associated with *Trichuris* spp. reported in the literature (Fig. 13.2).

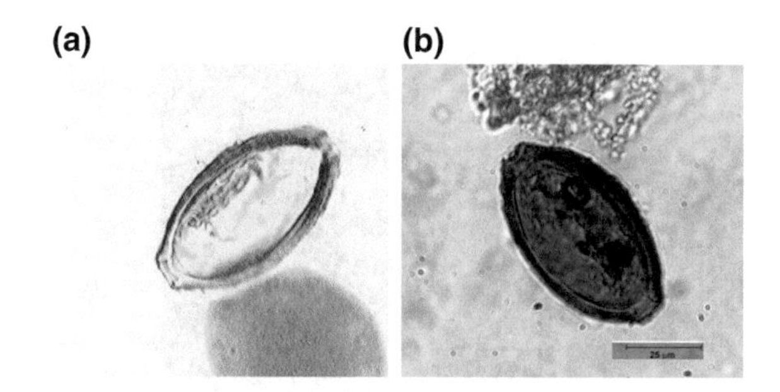

Fig. 13.1 **a** An egg from *Trichuris* sp. recovered from coprolite MHMOP/PI/587, 40 × 50 μm of length. Scale of observation corresponds to 400x; **b** An egg from *Trichuris* sp. Recovered from coprolite MHMOP/PI/585, 60 μm of length. Scale bar measures 25 μm

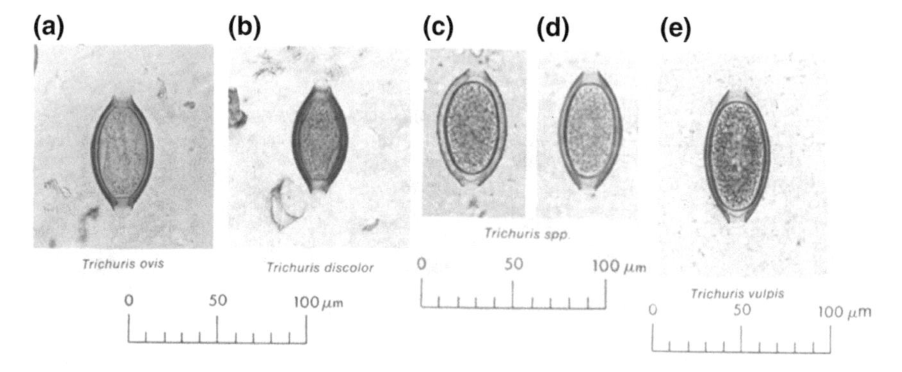

Fig. 13.2 **a** and **b** Eggs from *Trichuris* sp. recognized as common parasites in ruminants with variable size around 60 μm; **c** and **d** Eggs from *Trichuris* sp. found in cats; **e** Eggs from *Trichuris vulpis*, lemon-shaped, with polar plugs and average size around 75 μm (Bowman 2004)

- *Capillaria* sp., 2 eggs with lengths of 50 μm and width of 30 μm, symmetric, sub-parallel lateral margins, shell with an internal hyaline layer and external layer with irregular surface margins and slightly outstanding mucous polar plugs (Fig. 13.3). Figure 13.4 shows another example of *Capillaria* sp. in an archaeological context.

We also obtained eggs from cestodes assigned to the Anoplocephalidae family, and probably corresponding to *Anoplocephala* sp. The record encompasses eight asymmetric eggs with lengths between 50 and 80 μm, triangular shape or one semi-circular margin and thick refracting outer shell (Fig. 13.5). One egg (length/width 54/48 μm) displays darker interior impeding the identification of internal structures. Considering their shape and dimensions, the eggs from Pilauco assigned to the genera *Anoplocephala* correspond to eggs from the same living specie (Fig. 13.6). They are characterized by a thick and irregular external shell providing them a semicircular form with one flattened extremity.

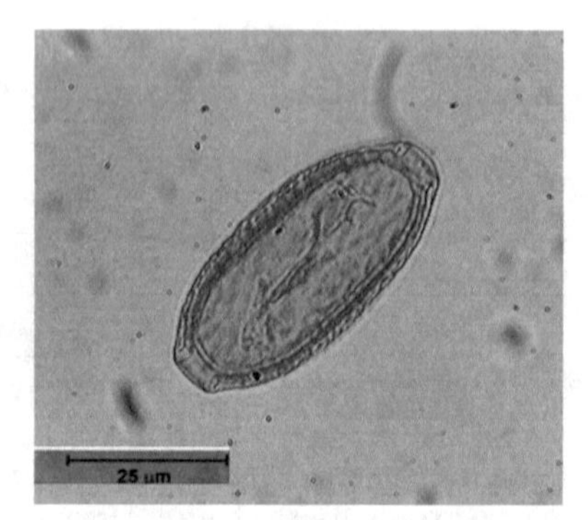

Fig. 13.3 An egg from *Capillaria sp.* recovered from coprolite sample MHMOP/PI/587 from the Pilauco site. Length 50 μm, width 30 μm. Scale of observation corresponds to 250x. Scale bar measures 25 μm

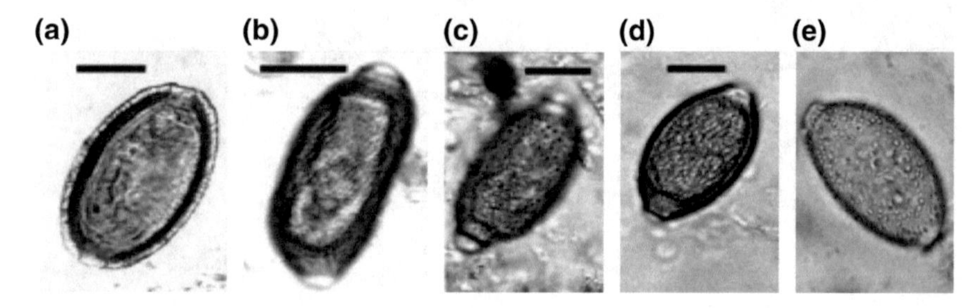

Fig. 13.4 Different morphotypes among eggs from *Capillaria* sp. (**a**–**d**) (Fugassa et al. 2008). Scale bar measures 20 μm; **e** Length/width 57/34 μm (Harter and Bouchet 2002)

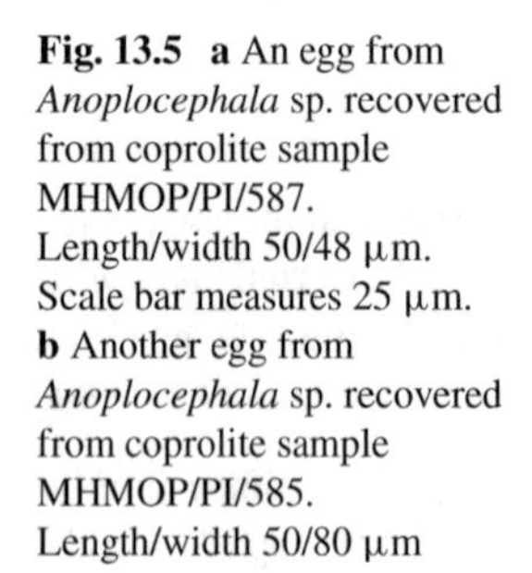

Fig. 13.5 **a** An egg from *Anoplocephala* sp. recovered from coprolite sample MHMOP/PI/587. Length/width 50/48 μm. Scale bar measures 25 μm. **b** Another egg from *Anoplocephala* sp. recovered from coprolite sample MHMOP/PI/585. Length/width 50/80 μm

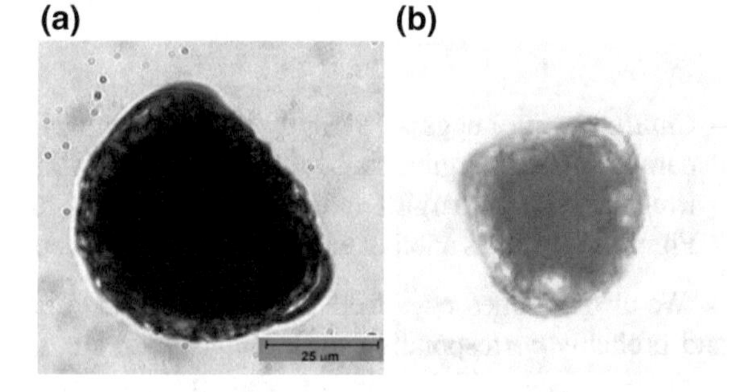

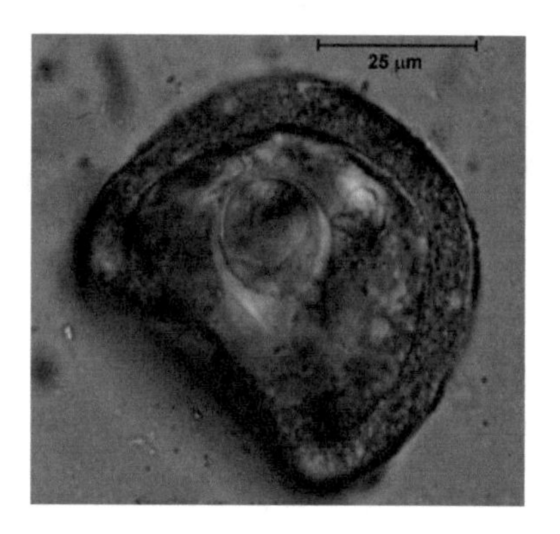

Fig. 13.6 An egg from *Anoplocephala perfoliata*. The piriform apparatus can be observed in the interior (Merijo 2001). Scale bar measures 25 µm

Additional findings include a cubic element with dimensions of approximately 50–60 µm exhibiting similarity with eggs from the cestode *Moniezia benedeni*, which is currently found in beef (Fig. 13.9), as well as other two elements with pyramidal and semi-cubic shape, respectively. As for eggs assigned to *Anoplocepha*, it is impossible to recognize particular structures, such as the oncosphere or the hexacanth embryo (with three pairs of hooks) and the embryophore corresponding to the piriform apparatus. The outer shell is thick and prominent at the edges (Figs. 13.7 and 13.8)

The last egg specimen exhibits at least two peripheral shells and can be therefore likely assigned to parasites from the Ascariddiae family (Fig. 13.9). The first shell is clearly delimited, and exhibits displays irregularities. The underlying shell is thicker and lacks a clear limit towards the internal granular texture of the egg. The egg exhibits a brownish colouring and a diameter of ~100 µm. These characteristics allow the classification of the egg specimen on the family taxonomic rank. The assignment of generic and specific classification is inhibited due to the degree of disintegration of the specimen.

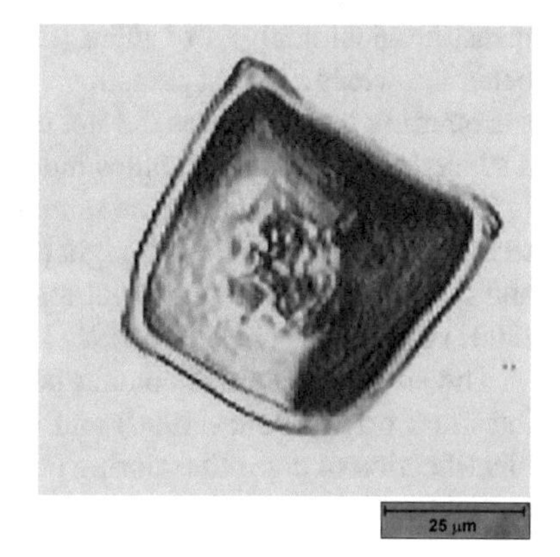

Fig. 13.7 A cubit element similar to an egg from *Moniezia benedeni* recovered from coprolite sample MHMOP/PI/587. Length/width 50 × 60 µm. Scale bar measures 25 µm

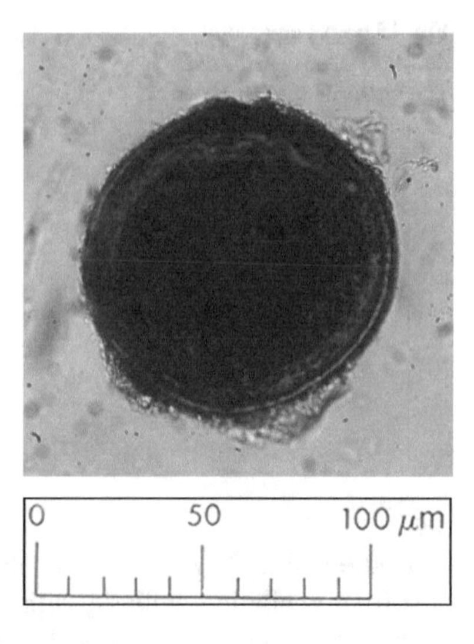

Fig. 13.8 **a** An egg from *Moniezia benedeni*, collection of the Parasitological Veterinary Laboratory, Universidad Austral de Chile). Scale bar measures 25 μm. Scale of observation corresponds to 400x. **b** An egg from *Moniezia expansa* (Varcárcel 2010)

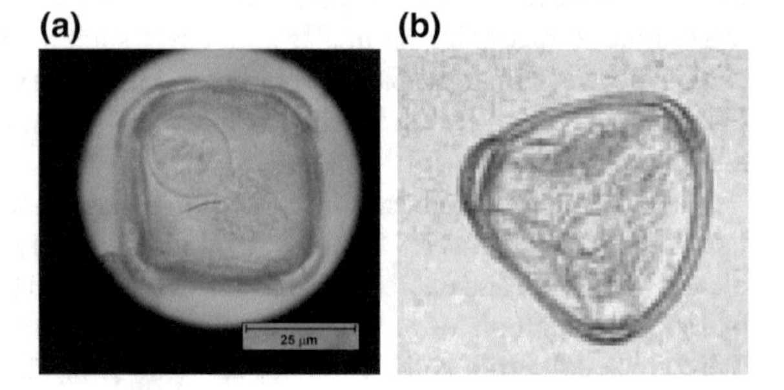

Fig. 13.9 An egg assigned to the Ascarididae family. Recovered from coprolite sample MHMOP/PI/585. It presents a sub-spherical form and a thick shell

13.4 Discussion and Conclusions

The analysis of sediments contiguous to coprolites is of great help to support the parasitological analysis of them (Bouchet 1995). Sometimes sediments provide a better knowledge of past parasitosis that contaminate soils. In Pilauco, the sediments surrounding the coprolites did not contain parasites. That situation confirm that the analysed samples are coprolites indeed.

Generally, the shapes of the sampled faeces are unknown and they are not characterized by certain forms that might allow for the definition of morphometric patterns and the assignation to terrestrial mammals, as it was done in the studies of Chame (2003) and Häntzschel et al. (1968).

The studied materials contain parasitic eggs, which strongly suggests faecal or intestinal origin (Jones 1990) and can be therefore considered as coprolites. The identification of coprolites during paleontological excavations is generally challenging and usually occurs when typical characteristics of faecal origin are conserved

(Fugassa and Guichón 2005). Unfortunately, the preservation of such features is not very common, because of the readily occurring disintegration of coprolites by decomposition of organic matter, especially due to bacterial activity (Häntzschel et al. 1968). Coprolites are frequently deformed by footsteps or found as smaller fragments, which are difficult to distinguish (Fugassa and Guichón 2005). Additionally, increased humidity at the site of deposition hinders their compaction. In the case of the Pilauco site, there are several outcrops of phreatic water within the fossil-bearing layer, which particularly impedes the in situ recognition of coprolites. Successful paleoparasitological analyses are therefore challenging and provide a useful tool to confirm the fossil faecal character of the sampled material and possibly identify the producers.

Considering the different parasitic genera registered at Pilauco, it can be argued that the cestodes from the Anoplocephalidae family are typical parasites for a wide variety of vertebrates including ruminants, horses, primates, elephants (McAloon 2004), and domestic donkeys (Getachew et al. 2010). Fugassa (2006) describes eggs from Anoplocephalidae from the Late Pleistocene and associates them with rodent faeces, although admitting that this is a rather novel parasite–host relationship that has not been reported before. Anoplocephalidae parasites require both a secondary and a definitive host to complete their biological life cycle (Denegri et al. 1998). The biological life cycle of the genus *Anoplocephala* starts when the cestode releases the proglottids in the interior of the host, which then disintegrate and liberate the parasitic eggs in the faeces of the definitive host (Reinemeyer and Nielsen 2009). In the open environment, the eggs are being consumed by a variety of oribatid mites, such as those present in Pilauco (Tello 2013, Tello et al. 2017), which represent the intermediate host for the parasitic cisticercoid life stage (McAloon 2004). Horses are usually infected due to the occasional pasture and ingestion of infectious acarians (Reinemeyer and Nielsen 2009).

The eggs from this genus are irregular and display a basic pattern, typical for all cestodes, that allows their straightforward identification. The oncosphere or hexacanth embryo is located in the centre and surrounded by the embryophore. For the genus *Anolocephala*, the embryophore is represented by a distinctive pear-shaped two-tailed formation, referred to as 'pyriform apparatus' (Bowman 2004). The anoplosephalid eggs recovered from Pilauco display the typical size and shape characteristics but lack the pyriform apparatus. This might be due to the long period of time elapsed since deposition of the eggs and the increased humidity at the site.

Currently, the equids maintain their species-typical association with the parasitic species *Anoplocephala perfiolata, Anoplocephala magna* and *Paranoplocephala mamillana.* The occurrence of eggs from *Anoplocephala* sp. (Fig. 13.6) and their association with fossil equids are represented by the subspecies *Equus (Amerhippus)* typical for this (Recabarren et al. 2011), let us infer that equids from the end of the Llanquihue glaciation in Northern Patagonia hosted the same cestodes as in the present. The coprolites, or some of them, were then likely produced by *Equus (Amerhippus)* sp., which agrees with the study of phytoliths in coprolite samples (see Chap. 11, this volume).

The analysed material from samples MHMOP/PI/585, MHMOP/PI/586 and MHMOP/PI/588 is characterized by increased content of C3-phytoliths, which is concordant with alimentary studies of Pleistocene horses from the Southern Hemisphere. The scarce occurrence of dicotyledons indicates that the consumers were nibbling on these types of plants, but those did not form a fundamental part of their diet (Álvarez 2011).

The parasite eggs assigned to the genus *Anoplocephala* are characterized by differing metrics: 48/54 μm (length/width) from sample MHMOP/PI/587 and 45 μm (diameter) from sample MHMOP/PI/585, whereas most of the remaining eggs are between 50 and 80 μm. The number of species assigned to this genus and the biometric measurements on their eggs are rather limited and basically reduced to two species—*Anoplocephala perfoliata* with 65/80 μm and *Anoplocephala magna* with 50/60 μm length and width, respectively. Furthermore, the species *Anoplocephala manubriata* hosted by *Elephas maximus* and *Loxodonta Africana* and *Anoplocephala gigantean* hosted by the African white rhinoceros exhibit length/width dimensions of 70/80 μm and 77/95 μm, respectively. However, these metrics fall within the range of eggs from *Anoplocephala perfoliata* and *Anoplocephala magna* and can be therefore related to *Equus (Amerhippus)* sp. as well.

On the other hand, the genus *Moniezia* is represented by two species (*Moniezia benedeni* and *Moniezia expansa*) from the Anoplocephalidae family. These are characterized by the same biological life cycle as *Anoplocephala* sp. with oribatid mites as intermediate hosts and belong to the group of ruminant-specific parasites. *Moniezia benedeni* (80/90 μm) is specialized on cattle and has irregular, triangular or cubic eggs with pyriform apparatus. *Moniezia expansa* (50–65/60–75 μm) parasitize ruminants (preferably sheep) and camelids and has irregular, triangular or pyramidal eggs with pyriform apparatus.

The morphological characteristics and biometrics of fossil eggs from *Moniezia* sp. correspond to those found in current ruminants, suggesting that these Late Pleistocene parasites might also have been specialized on this particular host. In Late Pleistocene South America, the ruminants are represented by Lamini (Artiodactyla: Camelidae) (genus *Hemiauchenia* Gervais and Ameghino; Labarca et al. 2013) and the small deer pudú (*cf. Pudu* Molina) (González et al. 2014).

The species from the genus *Trichuris* have a direct life cycle and reach their mature stage within a single host, which becomes infected by the ingestion of embryonated eggs in the open environment that eclose in the small intestine. The adult nematodes can be found in the blind and adjacent part of the large intestine and liberate their eggs through the faeces (Spickler 2005).

The brownish-yellowish eggs are elliptic and exhibit a thick two-layered shell, discontinuous at the distal extremities (Atías 1999; Table 13.1), which bear the typical polar mucous plugs providing the lemon-shaped egg form. The eggs remained intact during the processes of desiccation and re-hydration allowing for the identification of species based on biometric data (Ferreira et al. 1991). Importantly, the metrics of parasite eggs are only occasionally useful, as for most of the species, these vary between 50 and 60 μm (length) and 25 and 35 μm (width) (Chandler 1930). Ferreira et al. (1991) identify some species of *Trichuris* in South American mammals, such as *Trichuris felis* Diesing, 1851 in *Felis tigrina* (tiger cat), *Trichuris reesali* Wolfgang, 1951 in *Didelphis marsupialis* (common opossum), *Trichuris urichi* Wolfgang, 1951 in *Didelphis* sp. (opossums) and *Trichuris miocastoris* Enigk, 1933 in *Myocastor coypus* (coipo).

Some of the parasite eggs recovered from coprolite samples MHMOP/PI/585, MHMOP/PI/587 and MHMOP/PI/588 exhibit lengths of 60 μm including the polar plugs. All species mentioned above correspond to these metrics except for *Trichuris suis* that stands out in the biometric record with lengths between 72 and 90 μm. Notably, some of the species are represented by both normal-sized and oversized eggs, at which the latter are generally better preserved in the coprolite samples from Pilauco. None of the species from this genus affects equids. We can therefore discard *Equus (Amerhippus)* sp. as a possible host, but still consider the camelids, deer and rodents.

In fact, micromammals form a significant component in the predator's diet and play an important role within the ecosystem by dispersing seeds and spores, hosting a variety of parasite species, and providing reservoir for numerous zoonoses (Sardella and Fugassa 2010).

After Fugassa et al. (2010), the presence of *Capillaria* nematode eggs from archaeological sites could be related to high-latitude locations, the hunter-gatherer life style strategies or with taphonomic factors.

Fugassa and his collaborators (e.g. Fugassa et al. 2010) document parasite eggs from *Trichuris* sp., *Capillaria* sp., anoplocephalids and other cestodes and nematodes from archaeological contexts in Argentinean Patagonia dated between 6500 and 8000 cal yr BP. These studies broadened our understanding of the temporal distribution of species and genera in Patagonia (Fugassa 2006), which is now extended even more by the findings from Pilauco dated at 12,870 cal yr BP.

The parasite eggs from the nematode genus *Capillaria* vary between species in terms of their size, colour, shape and plug type. These parasites affect a wide variety of vertebrates including fish, birds (Figueiroa et al. 2002), rodents (Robles et al. 2012), carnivores (Fugassa et al. 2008), camelids (Rodríguez et al. 2012), ruminants (Moreno and Gómez 1991) and humans (Fugassa et al. 2008). The study of Rodríguez et al. (2012) realized on *Lama glama* documents parasite eggs from *Capillaria* sp. characterized by 65.3/35.2 μm (length/width) metrics, lemon-shaped form, two slightly outstanding opercula, transparent appearance and a thick shell with a rough surface. As mentioned above, two of the parasite eggs recovered at Pilauco exhibit size metrics of 50/30 μm of length/width. Nevertheless, these can not be unambiguously assigned to the same species, because of the differing ornamentation of their outer shell, which is very prominent on one egg and nearly absent on the other. Therefore, it is not possible to associate the *Capillaria* spp. parasites with a particular host, but *Equus* spp. can at least be discarded.

The ascarids are nematode parasites and their eggs exhibit a distinctive thin, homogeneous and transparent chitin shell. The outer surface is covered with a protein layer lined with lipids forming a narrow space filled with liquids, which separates the capsule from the embryonic content (Bowman 2004). The eggs associated with the genus *Toxocara* contain a single cell and can be identified by their shape, size, and by the thickness of the corrugated shell (Bowman 2004). Their life cycle begins with the deposition of the faeces and after an incubation period in the open environment,

Table 13.1 Morpho-biometric data for the genera *Anoplocephala*, *Trichuris,* and for ascarids
Source Atías (1999), Borgsteede et al. (2012), Chandler (1930), Denegri et al. (1998), Hendrix (1999), Knight (1971), McAloon (2004), Quiroz-Romero (1990), Reinemeyer and Nielsen (2009), Soulsby (1987)

Family	Genus	Species	Host	Eggs	
				Morphology	Size (μm)
Anoplocephalidae	*Anoplocephala*	*manubriata*	Asian elephant		
Anoplocephalidae	*Anoplocephala*	*magna*	Equid	Similar to *A. perfoliata;* pyriform apparatus similar to a heel	50–60
Anoplocephalidae	*Anoplocephala*	*perfoliata*	Equid (greater global prevalence)	Thick shell, D-shaped or semi-triangular	65–80
Anoplocephalidae	*Anoplocephala*	*gigantea*	African rhinoceros	–	–
Ascarididae	*Toxocara*	*canis*	Canine	Sub-globular, thick finely decorated shell	85–90 × 75
Ascarididae	*Toxocara*	*cati*	Feline	Similar to *T. canis*	75 × 70
Ascarididae	*Toxocara*	*mystax*	Lion, leopard, lynx	–	65–75
Ascarididae	*Toxocara*	*vitulorum*	Calfe, buffalo, bison	Sub-spheric, albuminoid, finely decorated external shell	75–98 × 60–75
Trichuridae	*Trichuris*	*discolor*	Cattle, mountain sheep, bighorn sheep	–	55–67 × 26–34
Trichuridae	*Trichuris*	*skrjabini*	Camel, sheep, goat, elk, Persian gazelle, deer and axis deer	–	58–79 × 29–34
Trichuridae	*Trichuris*	*ovis*	Cattle, lamb, goat, and other rumiants	Brown, barrel-shaped, with polar plugs on each pole	70–80 × 30–35
Trichuridae	*Trichuris*	*tenuis*	Camel	–	54–60

(continued)

Table 13.1 (continued)

Family	Genus	Species	Host	Eggs	
				Morphology	Size (μm)
Trichuridae	*Trichuris*	*vulpis*	Dogs	Prominent, but small polar plugs	70–80 × 30–42
Trichuridae	*Trichuris*	*muris*	Rats	–	–
Trichuridae	*Trichuris*	*trichiura*	Humans	Elliptic, brownish, thick, two-layered shell, mucous polar plugs on each side' barrel- or lemon-shaped	50–54 × 22–23

during which the temperature, humidity and the oxygen supply need to be optimal in order to reach the second larval stage within the egg (Quiroz-Romero 1990). The infestation occurs orally, the eggs subsequently eclose in the small intestine and the larvae migrate to various organs without continuing their development. The host needs to be a pregnant female exemplar for the larvae to migrate finally reaching the liver and the lungs, where they remain until their birth (Quiroz-Romero 1990). The egg specimen possibly assigned to the ascarids has a diameter of 126 μm and there are only two species that resemble its morphological characteristics and size. The first one corresponds to *Toxocara vitulorum* in cattle, buffalo (75–93 μm and 60–75 μm for length and width, respectively). Different species of the *Toxocara* genus can infest canines, felines and humans (Fisher 2003). The second parasite *Parascaris equorum.* This is morphologically less similar to the recovered Pilauco egg and reaches a larger diameter (90–100μ). *P. equorum* and is clearly restricted to equids. Therefore, unlike *T. vitulorum, P. equorum* could be associated with the vertebrate record of the site.

Another genus from the Ascarididae family with relevance for the present study is *Parascaris,* as these parasites also infest living host species correlated with those from the vertebrate record at Pilauco and exhibit morphometric characteristics similar to the possible ascarid egg described above. One representative from this genus is *P. equorum*, whose life cycle begins with the ingestion of eggs in the third larval stage by horses when feeding, pasturing or drinking water. Subsequently, the larvae eclose and penetrate the wall of the small intestine and migrate towards the lungs. After migration to the respiratory system, the larvae are repeatedly swallowed and return to the small intestine, reaching their mature stage in the proximal duodenum and jejunum. By completing their migration, the parasites increase in size and the first eggs of *P. equorumm* in the host faeces between 72 and 110 days after infestation (Koudela and Bodecek 2006). In terms of morphology, the *P. equorumm* eggs are less similar to the parasite egg in question from Pilauco, but they can reach diameters up to 90–11 μm and are strictly limited to equid hosts. Hence, in contrast to *Toxocara vitulorum*, the *Parascaris equorum* species could be associated with the record of vertebrates from Pilauco and assigned a possible host.

The particular taphonomic processes affecting coprolites at Pilauco have not been described yet, but we could infer that they were similar to the impacts suffered by other fossil remnants at the site (Chap. 8, this volume).

The preservation of coprolites endured high humidity and anoxic conditions following the deposition of the overlying sedimentary layers. The anoxic conditions allowed for the preservation of microorganisms, but the increased humidity inhibited their consolidation. In turn, Bouchet et al. (2003b) suggest that the humidity stored in deeper layers might have favoured the preservation of organic material, including parasite eggs, but altered the faecal mineralization. Generally, the presence of muddy and sediments as those from Pilauco, can potentially challenge the identification of the eggs as expulsed with the coprolite versus a posterior contamination with current parasites. Nonetheless, this possibility has been discarded, as all parasitic analyses in the immediate surroundings and excavation sites proofed negative.

Finally, there are also several diagenetic factors that might influence the occurrence and degree of preservation of parasitic traces (Bouchet et al. 2003a), which makes the evaluation of the original parasitic load ever more difficult (this load corresponds to the moment of death of the host exemplar or of the deposition of the faeces). The data derived from this archaeological context can only be interpreted in terms of prevalence of certain parasitic forms, emphasizing that a paleo-epidemiological study *sensu lato* is difficult to achieve.

The occurrence of isolated egg specimens additionally hinders their identification, because statistical methods cannot be applied and the dimensions of one single egg specimen often fall within the range of dimensions of several parasitic species that are taxonomically related (Confalonieri et al. 1988). Nonetheless, when the specimens are well preserved, the diagnostic features of some parasitic forms allow for a straightforward identification on the genus level.

The results from the parasitological analyses are the most robust evidence that the collected aggregations of organic material correspond to faeces. Fortunately, the Pleistocene vertebrates from Pilauco are related with current fauna taxa. This allows for the establishment of a relationship between the parasites *Anoplocephala* sp., *Moniezia* sp. and the acarids oribatids with particular vertebrate species that might have participated in the parasite life cycle as definite hosts. With *Trichuris* sp. and *Capillaria* sp., host parasite relationships can also be established. It can be deduced that both the extinct and extant fauna might have been infested by the same parasites. Such correspondence between past and present hosts exists, for example, for *Notiomastodon cf. N. platensis* and elephants, *Equus (Amerhipus)* and modern horses, *Hemiauchenia* sp. and camelids, *Pudu puda* and deer and ruminants, *Conepatus* sp. and skunks and *Myocastor coypus* and *Loxodontomys micropus* and current rodents. The morphological and biometric data on parasite eggs available in the literature allow only for their assignation on the genus level.

The best-conserved parasite eggs from coprolite samples from Pilauco are those assigned to *Trichuris* spp. and *Capillaria* spp. None of the species belonging to these genera infests equids, therefore we can discard *Equus (Amerhippus) andium* Hoffstetter and *N. platensis* as possible hosts, but still consider camelids, deer and rodents. Unfortunately, the volume and morphology of the corresponding coprolites are not suitable as a proxy for potential *Trichuris* sp. hosts, because their shape is not characteristic enough to be recognized during fieldwork.

Notably, the four analysed samples exhibit variable content of plant remnants and sample MHMOP/PI/587 contains parasite eggs from three different genera. For these reasons, we cannot discard a possible contamination or superposition of faeces from different species, particularly considering that some animals (e.g. camelids) make use of designated defecation sites. This might explain the occurrence of cestodes (typical for equids) and nematodes (not found in equids) in the same sample. The coprolite samples from Pilauco cannot be all associated with producers from a single taxon. Nonetheless, based on the actual species-specific parasitological characteristics, the presence of eggs from *Anoplocephala* sp. coincides with evidence for *Equus* sp., whereas *Moniezia* sp. can be correlated with deer and camelids.

The genus *Moniezia* is currently specialized on ruminants, but this condition might have been given in the Late Pleistocene as well. During this time epoch, the only representatives from this group correspond to Lamini (Artiodactyla: Camelidae) represented by the genus *Hemiauchenia* Gervais and Ameghino and the small deer Pudú (*cf. Pudu* Molina).

It can be therefore concluded that most of the coprolites from Pilauco originate from equids and camelids registered on the site, which is partially consistent with the phytolithic analysis of Álvarez (2011) and identifies *Equus (Amerhippus)* sp. as their most probable producer. The parasites from the genera *Trichuris, Capillaria*, the Ascarididae family are characterized by direct life cycles, which implies that the coprolites bearing their eggs originate from definitive hosts.

References

Álvarez V (2011) Análisis de fitolitos recuperados desde coprolitos del sitio Pilauco Osorno-Chile: Aproximaciones sobre paleoecología y paleodieta. Undergraduate Dissertation, Universidad Austral de Chile

Amstutz GC (1958) Coprolites: a review of the literature and a study of specimens from Southern Washington. J Sediment Petrol 28(4):498–508

Araújo A, Ferreira LF (2000) Paleoparasitology and the antiquity of human host-parasite relationship. Mem Inst Oswaldo Cruz 95(1):89–93

Araújo A, Ferreira LF, Confalonieri U (1981) A contribution to the study of helminth findings in archaeological material in Brazil. Rev Bras Biol 41:873–881

Araújo A, Jansen AM, Bouchet F, Reinhard K, Ferreira LF (2003) Parasitism, the diversity of life, and paleoparasitology. Mem Inst Oswaldo Cruz 98(1):5–11

Atías A (1999) Tricocefalosis. In: Atías A (ed) Parasitología Médica, pp 172–177

Badal E, Atienza V (2005) Análisis microscópico de coprolitos de herbívoros hallados en contextos arqueológicos. In: Abstracts of the VI Congreso Ibérico de Arqueometría, Valencia. 16 al 19 de November 2005

Borgsteede FHM, Holzhauer M, Herder FL, Veldhuis-Wolterbeek EG, Hegeman C (2012) *Toxocara vitulorum* in suckling calves in The Netherlands. Res Vet Sci 92:254–256

Bouchet F (1995) Recovery of helminth eggs from archeological excavations of the Grand Louvre (Paris, France). J Parasitol 81(5):785–787

Bouchet F, Guidon N, Dittmar K, Harter S, Ferreira LF, Miranda-Chaves S, Reinhard K, Araújo A (2003a) Parasite remains in archaeological sites. Mem Inst Oswaldo Cruz 98(1):47–52

Bouchet F, Araújo A, Harter S, Miranda-Chaves S, Nascimento A, Laurent J, Ferreira LF (2003b) *Toxocara canis* (Werner, 1782) eggs in the pleistocene site of Menez-Dregan, France (300,000–500,000 years before present). Mem Inst Oswaldo Cruz 98(1):137–139

Bowman DD (2004) Parasitología para veterinarios. Octava edición, Elsevier, España

Carrión JS (2002) A taphonomic study of modern pollen assemblages from dung and surface sediments in arid environments of Spain. Rev Palaeobot Palynol 120:217–232

Chame M (2003) Terrestrial mammal feces: a morphometric summary and description. Mem Inst Oswaldo Cruz 98(1):71–94

Chandler AC (1930) Specific characters in the Genus *Trichuris*, with a description of a new species *Trichuris tenuis* from a camel. J Parasitol 16(4):198–206

Chimento NR, Rey L (2008) Hallazgo de una feca fósil en el Pleistoceno Superior Holoceno Inferior del Partido de General Guido, Provincia de Buenos Aires, Argentina. Rev Mus Argentino Cienc Nat 10(2):239–254

Confalonieri UEC, Ferreira LF, Araujo AJ, Ribeiro-Filho BM (1988) El Uso de un test estadístico para la identificación de huevos de helmintos en coprolitos. PPANewsletter 62:7–8

Denegri GM, Bernardina W, Pérez-Serrano J, Rodríguez-Caabeiro F (1998) Anoplocephalid cestodes of veterinary and medical significance: a review. Folia Parasitol 45:1–8

Ferreira LF, Araújo A, Confalonieri U, Chame M (1989) Acanthocephalan eggs in animal coprolites from archeological sites from Brazil. Mem Inst Oswaldo Cruz, Rios de Janeiro. 84(2):201–203

Ferreira LF, Araujo A, Confalonieri U, Chame M, Gomes DC (1991) *Trichuris* eggs in animal coprolites dated from 30,000 years ago. J Parasitol 77(3):491–493

Figueiroa M, Bianque A, Dowell M, Soares A, Santiago V, Alves R, Evencio A (2002) Parásitos gastrointestinales de aves silvestres en cautiverio en el Estado de Pernambuco. Brasil Parasitol Latinoam 57(1–2):50–54

Fisher M (2003) *Toxocara cati*: an underestimated zoonotic agent. Trends Parasitol 19(4):167–170

Fugassa MH (2006) Examen paleoparasitológico de sedimentos de un sitio arqueológico, Río Mayo, Chubut, Argentina. Parasitol Latinoam 61:172–175

Fugassa MH (2007) Camélidos, parásitos y ocupaciones humanas: registros paleoparasitológicos en Cerro Casa de Piedra 7 (Parque Nacional Perito Moreno, Santa Cruz, Argentina). Intersecciones Antropol 8:265–269

Fugassa MH, Guichón R (2005) Análisis paleoparasitológico de coprolitos hallados en sitios arqueológicos de Patagonia Austral: definiciones y perspectivas. Magallania, Chile, vol 33, no 2, pp 13–19

Fugassa MH, Taglioretti V, Gonçalves MLC, Araújo A, Sardella NH, Denegri GM (2008) *Capillaria* spp. eggs in patagonian archaeological sites: statistical analysis of morphometric data. Mem Inst Oswaldo Cruz, Rio de Janeiro, vol 103, no 1, pp 104–105

Fugassa MH, Sardella NH, Beltrame MO, Cumino A (2010) Parasitología. Abstracts of Jornadas de la Asociación Argentina de Parasitología Veterinaria

Getachew M, Trawford A, Feseha G, Reid SWJ (2010) Gastrointestinal parasites of working donkeys of Ethiopia. Trop Anim Health Prod 42:27–33

González E, Labarca R, Chávez M, Pino M (2014) First fossil record of the smallest deer *cf. Pudu* Molina, 1792 (Artiodactyla, Cervidae) in the Late Pleistocene of South America. J Vert Paleontol 34(2):483–488

Häntzschel W, El-Baz F, Amstutz GC (1968) Coprolites: an annotated bibliography. GSA, 132p

Harter S, Bouchet F (2002) Paléoparasitologie: apports des méthodes de la parasitologie médicale à l'étude des populations anciennes. Bull Mem Soc Anthropol Paris 3–4:363–370

Hendrix CM (1999) Diagnóstico parasitológico veterinario. Segunda edición, Mosby-Doyma, Madrid

Hunt AP, Chin K, Lockley MG (1994) The palaeobiology of vertebrates coprolites. In: Donovan SK (ed) The palaeobiology of trace fossils. Wiley, pp 256–280

Jones A (1990) Coprolites and fecal concretions. Brean down excavations 1983–1987, vol 15, pp 242–245. English Heritage Archaeological Report, London

Jouy-Avantin F, Debenath A, Moigne AM, Moné H (2003) A standardized method for the description and the study of coprolites. J Archaeol Sci 30:367–372

Knight RA (1971) Redescriptions of *Trichuris discolor* (Von Linstow, 1906) y *T. skrjabini* (Baksakov, 1924) from domestic ruminants in the United States and comparisons with *T. ovis* (Abilgaard, 1795). J Parasitol 57(2):302–310

Koudela B, Bodecek S (2006) Effects of low and high temperatures on viability of parascaris equorum eggs suspended in water. Vet Parasitol 142:123–128

Labarca R, Pino M, Recabarren O (2013) Los Lamini (Cetartiodactyla: Camelidae) extintos del yacimiento de Pilauco (Norpatagonia chilena): aspectos taxonómicos y tafonómicos preliminares. Estud Geol (Madr) 69(2):255–269

McAloon FM (2004) Oribatid mites as intermediate hosts of *Anoplocephala manubriata*, cestode of the asian elephant in India. Exp Appl Acarol 32:181–185

Merijo E (2001) Population dynamics of oribatid mited (Acari: Oribatida) on horse pastures of North Central Florida. Dissertation, Universidad of Florida

Pino M, Chávez-Hoffmeister M, Navarro-Harris X, Labarca R (2013) The Late Pleistocene Pilauco site, Osorno, South-Central Chile. Quat Int 299:3–12

Quiroz-Romero H (1990) Parasitología. Editorial Limusa S.A. de C.V. Balderas 95, México

Ramírez FE (2012) Análisis parasitológico de coprolitos del sitio paleontológico Pilauco, Pleistoceno Superior. Dissertation, Universidad Austral de Chile

Recabarren O, Pino M, Cid I (2011) New record of *Equus* (Mammalia: Equidae) from the late pleistocene of Central-South Chile. Rev Chil Hist Nat 84:535–542

Reinemeyer CR, Nielsen MK (2009) Parasitism and colic. Vet Clin North Am Equine Pract 25:233–245

Reinhard KJ (1990) Archaeoparasitology in North America. Am J Phys Anthropol 82(2):145–163

Robles MR, Bain O, Navone GT (2012) Description of a new Capillariinae (Nematoda: Trichuridae) from *Scapteromys aquaticus* (Cricetidae: Sigmodontinae) from Buenos Aires, Argentina. J Parasitol 98(3):627–639

Rodríguez M, Martínez FA, García E, Glesmann V, Portillo C, Oviedo A (2012) *Capillaria* sp. en *Lama glama* (ruminantia, camelidae). Vet Arg. vol XXIX, no 290

Salvadores-Cerda L, Abarzúa AM (2008) Los Coprolitos del Sitio Pilauco. In: Pino M (ed) Pilauco: un sitio complejo del Pleistoceno Tardío. Universidad Austral de Chile. Valdivia, Chile., pp 85–89

Salvadores-Cerda L, Moreno K, Montero I, Recabarren O, Sievers G, Valenzuela G, Torres P, Abarzúa AM, Navarro RX, Pino M (2008) Análisis de coprolitos de mamíferos del Pleistoceno Superior, Sitio Pilauco Bajo, Centro Sur de Chile. In: Abstract of III Congreso Latinoamericano de Paleontología de Vertebrados

Sardella NH, Fugassa MH (2010) Importancia de los parásitos hallados en las heces de roedores en la interpretación paleoparasitológica y zooarqueológica en Patagonia. In: Abstract of Jornadas de la Asociación Argentina de Parasitología Veterinaria, XlX Encuentro Rioplatense de Veterinarios Endoparasitólogos

Soulsby EJL (1987) Parasitología y enfermedades parasitarias en los animales domésticos. Séptima edición, Nueva Editorial Interamericana, México

Spickler A (2005) Trichuriasis: trichocephaliasis, trichocephalosis, whipworm infestation. Center for Food Security and Public Health College of Veterinary Medicine Iowa State University Ames y Institute for International Cooperation in Animal Biologics an OIE Collaborating Center Iowa State University College of Veterinary Medicine

Tello F (2013) Coleópteros fósiles y reconstrucción paleoambiental de la capa PB-7 del sitio Pilauco, Osorno, Norpatagonia de Chile. Undergraduate Dissertation, Universidad Austral de Chile

Tello F, Elgueta M, Abarzúa AM, Torres F, Pino M (2017) Fossil beetles from Pilauco, south-central Chile: an Upper Pleistocene paleoenvironmental reconstruction. Quatern Int 449:58–66

van Geel B, Aptroot A, Baittinger C, Birks HH, Bull ID, Cross HB, Evershed R.P., Gravendeel B., Kompanje EJO, Kuperus PDM, Nierop KGJ, Pals JP, Tikhonov AN, van Reenen G, van Tienderen PH (2008) The ecological implications of a Yakutian mammoth's last meal. Quat Res (Orlando) 69:361–376

Chapter 14
The Site Los Notros: Geology and First Taxonomic Descriptions

María Paz Lira, Rafael Labarca, Daniel Fritte, Hugo Oyarzo and Mario Pino

Abstract The site Los Notros, located just 60 m west of the site Pilauco in Osorno (Chile), was discovered in 2008, with excavations beginning in April 2016. The geology of Los Notros is similar to that of the Pilauco site where strata LN-1 and LN-2 are equivalent to layers PB-7 and PB-8 from Pilauco. However, Los Notros includes an additional stratum (LN-3) of prominent black color, finer texture, and similar age as the discordance between PB-8/PB-9, whereas layer LN-4 is equivalent to PB-9. Additionally, a chronostratigraphic relationship between both sites is suggested based on concordant radiocarbon ages. Excavations at the Los Notros have provided 15 fossil specimens from layers LN-1 and LN-2, 11 of which were taxonomically determined at least up to order level. A rich dental record represents the families Equidae, Gomphotheriidae, and Cervidae, whereas a nearly complete gomphothere tusk lacking torsion and enamel strips allowed its assignment to *Notiomastodon platensis*. Thus, the presence of *N. platensis* in northern Patagonia—the southern limit of the species range—is confirmed together with the identification of *Equus (A.)* sp., and the extension of the distributional range for the extinct genus *Antifer* from Cervidae.

Keywords Los Notros · Notiomastodon · Antifer

M. P. Lira (✉) · D. Fritte · H. Oyarzo · M. Pino
Pilauco LabNat, Facultad de Ciencias, Universidad Austral de Chile, Osorno, Chile
e-mail: mpazlira@gmail.com

R. Labarca · M. Pino
Escuela de Arqueología and Transdisciplinary Center for Quaternary Research in the South of Chile (TAQUACH), Universidad Austral de Chile, Valdivia, Chile

M. Pino
Instituto de Ciencias de la Tierra and Transdisciplinary Center for Quaternary Research in the South of Chile (TAQUACH), Universidad Austral de Chile, Valdivia, Chile

M. Pino and G. A. Astorga (eds.), *Pilauco: A Late Pleistocene Archaeo-paleontological Site*, The Latin American Studies Book Series,
https://doi.org/10.1007/978-3-030-23918-3_14

Fig. 14.1 Location of Los Notros (red) and Pilauco (green) sites, Osorno, Chile

14.1 Background

The discovery of the site Los Notros dates back to the summer of 2008, during the first research campaign at the Pilauco (2007–2008). Different exploratory pits were dug around the Pilauco site in order to gain a broader understanding of the neighborhood geology and, specifically, to investigate why there were so many terraces at Los Notros neighborhood that did not correspond to the general geomorphological model—one having been formed in each period of glaciation. Flat surfaces were explored using a backhoe and their stratigraphic study concluded that almost all of the flat surfaces were not natural terraces but embankments made by the house building Fourcade Company. During one of several random surveys performed at Los Notros neighborhood square—distant only 60 m northwest of the Pilauco site—the exterior of a right gomphothere tusk was observed. This finding promoted the properly excavation of the site in April 2016, after obtaining the permit from the Council of National Monuments.

The site Los Notros site is located in the neighborhood squared at the intersection of the streets Río Aconcagua and Mayor Arturo Bertín (Fig. 14.1), coordinates 40°34′10″S and 73°06′15″W (Datum WGS 84).

14.2 Methods

Wet sieving was used to quantify the percentages of gravel (>2 mm), sand (2–0.063 mm), and mud (<0.063 mm). Hydrogen peroxide was used in the sand and mud fractions in order to burn off organic matter; the mass percentage of organic matter was calculated by differential weight after this procedure.

The dates in Los Notros were determined at the NSF Arizona Accelerator Mass Spectrometry Facility laboratory in Tuscon, AZ. All ages were calibrated with the CALIB 7.1 program, using the SHcal13 curve (Stuiver and Reimer 1993; Hogg et al. 2003). The calibrated ages are presented with 2σ.

Fourteen fossil specimens were analyzed, obtained between 2016 and 2017 from layers LN-1 and LN-2, BP (Table 14.1). The anatomical and taxonomical identification were based mainly on a morphological base considering diagnostic anatomical traits and by using direct comparison between fossil specimens and modern reference materials (e.g., *Equus caballus* Linnaeus, *Lama guanicoe* Müller). We used also fossil materials from the paleontological collection of Pilauco and the paleontology section of the Museum of Natural History in Santiago. Specifically, for Equidae fossil materials, measurements of particular anatomical units were obtained in order to allow for a complementary metric approximation in accordance with Eisenmann et al. (1988). Also, these were compared with literature (e.g., Alberdi et al. 2003; Frassinetti and Alberdi 2000; McFadden and Azzaroli 1987; Prado and Alberdi 2008; Recabarren et al. 2011; Rincón et al. 2006). The terminology used is based on Eisenmann et al. (1988), Alcaraz (2010), Smutz and Bezuidenhout (1987) and Alberdi et al. (2002).

Table 14.1 Summary of fossil remains from layers LN-1 and LN-2 site Los Notros, Chile

Code	Taxon	Specimen	Grid	Layer	Local elevation (cm)
MHMOP/LN/3	*Equus (A.)* sp.	I1–2	BA1	LN-1	538
MHMOP/LN/4	Mammalia indet	Bone fragment	BA7	LN-1	564–553
MHMOP/LN/8	*Notiomastodon platensis*	M2 (left)	BH6	LN-1	520–509
MHMOP/LN/9	*Notiomastodon platensis*	M1 (left)	BI6	LN-1	524–514.5
MHMOP/LN/10	Mammalia indet	Bone fragment	BI5	LN-1	517
MHMOP/LN/11	Mammalia indet	Bone fragment	BI5	LN-1	511
MHMOP/LN/12	Mammalia indet	Bone fragment	BI5	LN-1	505–510
MHMOP/LN/13	Artiodactyla indet	Distal Scapula	BI5	LN-1	515
MHMOP/LN/15	*Notiomastodon platensis*	Tusk	BI5–BI6	LN-1	517–504
MHMOP/LN/1	*Equus (A.)* sp.	m1–2 (left)	BH7	LN-2	584
MHMOP/LN/2	*Antifer ultra*	Antler	BH7–BI7	LN-2	604–596
MHMOP/LN/5	*Equus (A.)* sp.	M1–2 (left)	BH5	LN-2	595
MHMOP/LN/7	Equidae indet	Undetermined molar	BJ6	LN-2	589–587
MHMOP/LN/6	Artiodactyla indet	Undetermined molar	BA7	LN-2/LN-1	574–564

14.3 Stratigraphy and Sedimentology

The site Los Notros is located on a plain at the foothill formed by the San Pablo Sequence (see Chap. 3, this volume). This geomorphological context is very similar to that of Pilauco, but unlike the latter, Los Notros is located around 15 m further from the hill that is the source of the material composing most of the strata. Given the proximity of both sites, it was expected that the stratigraphy would be the same. However, several differences were observed as the excavation progressed.

Four strata (Fig. 14.2) are recorded in Los Notros area, discordantly deposited over a lapilli tuff stratum concordant with the basal layer PB-1 at Pilauco. Unlike the Pilauco site—where the tuff has a slightly bluish to green-gray color—the layer at Los Notros exhibits an intense turquoise color. Furthermore, in the northeastern part of the site, the strata are buttressed by fluvial gravel attributed to a paleo beach formed by the Damas River, concordant with layer PB-6 at Pilauco (Pino and Miralles 2008; Pino et al. 2013, 2016; Pino et al. 2019).

LN-1: Basal peat of 60 cm average thickness that lies sporadically over the PB-1 tuff. In the northeastern area of the site overlies a paleo beach created by the Damas River (PB-6). The fabric of the layer corresponds to a muddy-sandy supported matrix. Wet sieving of sediments from three locations within the site shows average values of sand and mud of 28% and 39%, respectively. The remaining 33% corresponds to gravel clasts ranging from 2 to 5 mm in diameter, with subrounded shapes and low sphericity. The composition of the clasts is mainly basalt (45%); blue-gray tuff (30%), dacite (15%), and the remaining 10% corresponded to granite, andesite, and volcanic scoria. Organic matter is present in both the sand and the mud.

Fig. 14.2 Stratigraphic profile of site Los Notros, western zone. Layer LN-3 display different texture, being much finer than the others younger and older layers

In the western zone, the stratum has grayish yellow color (5/3 25 YR, Munsell Scale), likely as the result of dissolution and subsequent precipitation of iron oxides deposited by fluctuations of the phreatic level. These oxidized aggregates provide consolidation and hardness to this layer, in addition to a slight orange tone. In contrast, the layer of the eastern sector near the northern hill, presents a brownish with yellow hue (5/3 7.5 YR). It oxidizes quickly when in contact with the air, darkening its color. Layer LN-1 present high plasticity, retaining the shape of the clasts in the matrix after separation. A one-meter-wide water pipe made of concrete and iron; created to collect excess and rainwater runs along the boundary between the western and eastern sectors. This acts as an impermeable barrier that restricts the circulation of groundwater, favoring the precipitation of oxides in the western zone. This stratum bears the fossil remains of one gomphothere individual.

LN-2: Peat of 10 to 20 cm thickness, arranged concordantly and in relative flat contact to layer LN-1. The matrix is very similar to that of layer LN-1, although with differences in color, while the clasts show the same composition. The color in the western sector is grayish brown (5/1 5 YR) and, as in layer LN-1, iron oxides are observed, though in smaller quantities. Alternatively, in the eastern sector, the sediment is reddish brown in color (4/3 5 YR). In this zone, determining the contact between LN-1 and LN-2 is complex, since the two layers are similar in color. Wet sieving of two sediment samples obtained at different sample sites yielded average proportions of 28, 25, and 47% of gravel, sand, and mud, respectively. These values are similar to those of the underlying layer, corroborating the similarity observed in the field.

Toward the top of this layer, near the contact with LN-3, the number of clasts increases abruptly by. ~70%. The size of the clasts is ~5 mm; they are spherical and well rounded, formed mainly by white tuff. Until now, dental remains representing the Equidae family and a complete antler of Cervidae, conferred to extinct genus *Antifer* sp., have been found in this layer.

LN-3: A contrastingly darker and thinner layer, when compared to other layers. Its average thickness is between 8 and 12 cm, reaching 20 cm in some areas. The fabric is a supported matrix. Wet textural analyses from three sample sites indicate average values of 6, 24 and 70% for gravel, sand, and mud, respectively, with a marked increase of 15% of the gravel fraction in the northeast sector, due to its closer proximity to the hill. The mass percentage of organic matter was 0.5–2% (2–0.063 mm) and ~7% (<0.063 mm). These results are not proportional to the volume of the material present, since the density of the organic matter is less than that of the inorganic material.

The top and base contacts are well marked and parallels. This layer is extremely compact and dense, separating easily from the upper and lower layers. When drying, it fractures vertically (Fig. 14.2). The color is grayish brown (4/1 10 YR) in the lower part and dark brown (3/1 10 YR) in the upper section. The presence of diatoms in the entire layer indicates an aquatic sedimentation environment or at least high humidity.

The electromagnetic field radiation levels of the profile were measured with an EFM (EXTECH 480826) triaxial apparatus. It was observed that the electromagnetic radiation in layer LN-3 increases notably from 0.7 to 1.7 mG (milligauss) with respect to the upper and lower strata. Comparatively, the electromagnetic field on the site is 0.1 mG when considering the effect of high voltage cables located on the site.

LN-4: Black peat (2/3 2.5 YR) with at least 150 cm thickness in the northern part of the excavation. In the southern sector, it reaches 50 cm due to truncation at the top of the layer due to an anthropic activity. In general terms, this layer seems to outcrop subhorizontal over LN-3, given that the lower section is massive and does not exhibit internal structure. However, in the northern area of the site where the upper section is exposed, the layer presents lamination and seven different lithological levels of 2–40 cm thickness inclined 32° toward the NE (Figs. 14.3 and 14.4). The slope is opposite to the normal inclination of the hill, presenting a well develop unconformity with respect to layer LN-3.

The layer lower section is homogeneous. The analyses performed from three different sections revealed average gravel, sand and mud values of 21%, 36%, and 43%, respectively. This section presents a supported matrix fabric composed of sand, with abundant organic matter. Cobble clasts, ranging from 5 to 20 mm in diameter, are immersed in the matrix. They are mainly tuff and basalt and, to a lesser degree, volcanic scoria and granite. The clasts of tuffs are orange-yellow and are exclusive to this layer, coming from the adjacent northern hill where outcrops the San Pablo Sequence. Locally, these tuff intraclasts have centimetric to decimetric dimensions (Fig. 14.5). There are abundant remains of charcoal and partially burned wood fragments, of decimetric sizes, randomly concentrated at the site (Fig. 14.5). The lamination responds to variations in the number of clasts versus matrix. The lithological composition of the clasts is the same across all the lamina. According to the observed characteristics, it is not possible to determine a consistent interpretation and therefore the origin of this lamination is still under discussion.

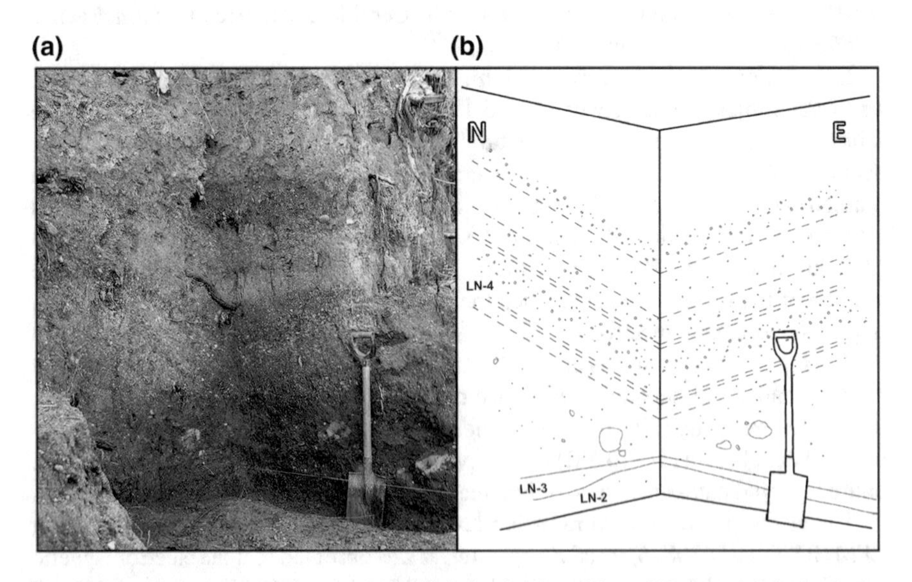

Fig. 14.3 **a** Photograph showing the slope of the lamination of layer LN-4 toward the northeast. **b** Schematic drawing of the profile

Fig. 14.4 Layer LN-3 showing lamination dipping to the NE (grids BJ5–BJ7, see Chap. 1, this volume)

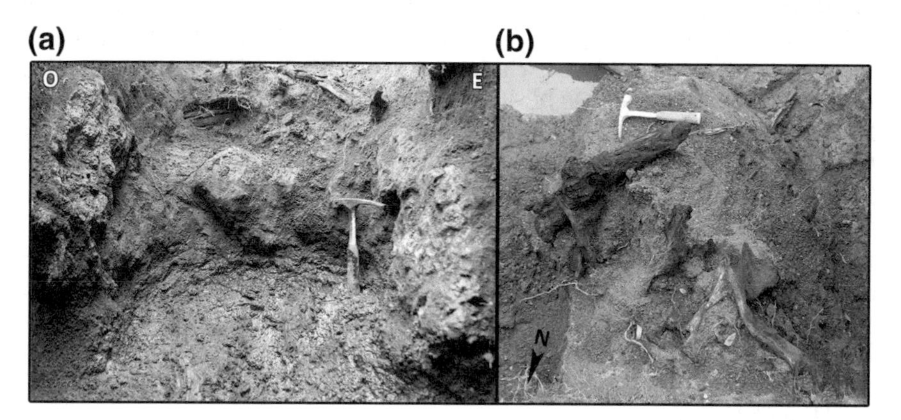

Fig. 14.5 Layer LN-4. **a** Fragment of yellow lapilli tuff with a minimum diameter of 90 cm. This rock was crushed to make the pit on each side of the wall. **b** Concentration of partially burned wood fragments in the northern area of the site (floor view)

14.4 Radiocarbon Dates and Correlations with the Pilauco Site

The base of LN-1 has an age of 12,449 ± 52 ^{14}C yr BP (14,155–14,860 cal. yr BP) in the southern sector. In the northern sector, an age of 13,585 ± 81 ^{14}C yr BP (16,048–16,614 cal. yr BP) was obtained in the Gomphotheriidae tusks.

Alternatively, layer PB-7 at Pilauco has a maximum age of 13,650 ± 70 ^{14}C yr BP (16,155–16,681 cal. yr BP), but in the eastern sector, the age of its base is 13,332 ± 72 ^{14}C yr BP (15,744–16,217 cal. yr BP). Thus, the layers LN-1 and PB-7 are not only sedimentologically analogous but also contemporaneous. Both layers lie discordantly over PB-1 and in some areas over PB-6; both yielded fossil remains of *Notiomastodon platensis* (described as *Stegomastodon platensis* by Recabarren 2007, Recabarren et al. 2014, Recabarren 2016). Likewise, layers LN-2 and PB-8 have striking sedimentological and compositional similarities, laying concordantly over strata LN-1 and PB-7 respectively. Also, both register fossil bones that belong from the Equidae family (Recabarren et al. 2011; Lira et al. 2016). At the base of layer LN-2, a Cervidae antler (Lira et al. 2016) was founded. The radiocarbon dating of which was performed more than once due to the low amount of keratin present in the piece, caused by bad preservation conditions. The dating produced an age of 10,860 ± 60 ^{14}C yr BP (12,660–12,806 cal. yr BP)—similar value to that obtained from the top of layer LN-2 despite the difference in stratigraphic position. Considering the difficulties encountered with radiocarbon dating of the fossil remains, it was decided not to use it to limit the layer age.

The top of layer LN-2 (base of LN-3) was dated from charcoal debris and yielded an age of 10,916 ± 67 ^{14}C yr BP (12,680–12,924 cal. yr BP), while the top of PB-8 has an age of 10,660 ± 30 ^{14}C yr BP (12,551–12,672 cal. yr BP). Thus, LN-1 and LN-2 layers chronostratigraphically correlated with PB-7 and PB-8. Furthermore, the contact at the top of LN-2 is chronologically related to the unconformity that exists between layers PB-8 and PB-9 at Pilauco. At Los Notros, the contact between layers LN-3 and LN-4 was dated twice, producing ages between 10,214 ± 37 ^{14}C yr BP (11,700–12,019 cal. yr BP) and 10,059 ± 36 ^{14}C yr BP (11,306–11,716 cal. yr BP), indicating that layer LN-3 represents the base of the Younger Dryas chronozone (see Chaps. 3 and 15, this volume). The northern hill area at Pilauco, adjacent to a steep slope, was likely affected by an erosional episode. In more distal portions of the hill—which exhibit less sloping compared to Los Notros site—a marshy environment developed where the material of layer LN-3 was deposited. This is the reason why the contact between PB-8 and PB-9 at Pilauco is represented by an erosional unconformity, whereas at Los Notros the contact between the layers LN-2 and LN-4 include the LN-3 layer.

Layers LN-4 and PB-9 display also similarities in their composition. Both are composed by a black peat with large tuff intraclasts in addition to a significant amount of wood and charcoal. LN-4 layer dip at 32° over LN-3. On the other hand, layer PB-9 lies in slight erosional discordance inclined 5–8° toward the south over layer PB-8. Layers LN-4 and PB-9 are chronostratigraphically correlated and constitute the top of the sequence at both sites (Fig. 14.6).

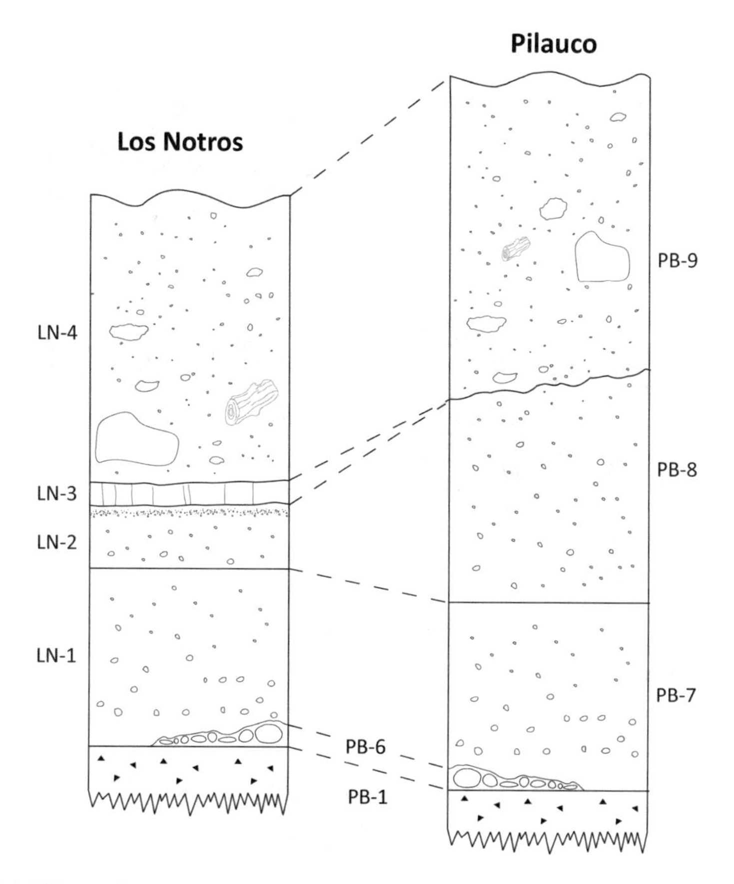

Fig. 14.6 Lithostratigraphic correlations between Los Notros and Pilauco columns

14.5 Megafauna Taxonomy

Table 14.1 shows a summary of fossil remains from layers LN-1 and LN-2 site Los Notros, Chile. The morphological characteristics of three isolated molars allow to identify MHMOP/LN/8 as a second molar, MHMOP/LN/9 as a first molar and MHMOP/LN/15 as an upper incisor, all dental pieces of *Notiomastodon platensis* (Figs. 14.7 and 14.8).

Class Mammalia Linnaeus (1758)
Order Proboscidea Illiger (1811)
Family Gomphotheriidae Cabrera (1929)
Subfamily Cuvieroniinae Cabrera (1929)
Notiomastodon Cabrera (1929)
Notiomastodon platensis Ameghino (1888) (Tables 14.2 and 14.3; Figs. 14.7 and 14.8).

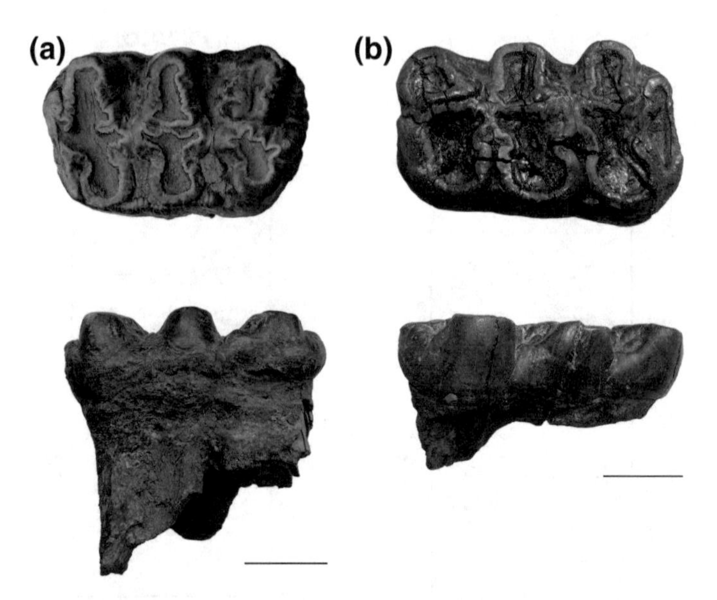

Fig. 14.7 Molars of *Notiomastodon platensis* (=*Stegomastodon platensis*) from site Los Notros, Chile. **a** MHMOP/LN/8 M2, occlusal and labial view; **b** MHMOP/LN/9 M1, occlusal and labial view. The scale bar = 5 cm

Fig. 14.8 Incisor of *Notiomastodon platensis* (=*Stegomastodon platensis*) from site Los Notros, Chile. Specimen MHMOP/LN/15. Scale bar = 10 cm

Specimens referenced. MHMOP/LN/8 s molar, MHMOP/LN/9 first molar, and MHMOP/LN/15 upper incisor.

Table 14.2 Size of molars (in mm) of *Notiomastodon platensis,* Los Notros site

Code	Specimen	Length max.	Width max.	Width Lopho 1	Width Lopho 2	Width Lopho 3	Enamel thickness
MHMOP/LN/8	M2	130.6	88.3	84.3	86.3	81.9	6.0
MHMOP/LN/9	M1	109.5	72.0	58.0	65.0	71.2	4.0

Table 14.3 Measurements (cm) of tusk MHMOP/LN/15 from the Los Notros site

Total length	Total curvature	Base diameter	Central diameter	Base length
109.3	136.4	45.3	45.6	15.3

Provenance. Layer LN-1 of Los Notros site. The layer is dated at its base as 13,585 ± 81 years ^{14}C B.P., which corresponds to a range of 16,048–16,614 years cal BP (See Table 14.1 for spatial details of provenance).
Description and comparison. The piece MHMOP/LN/8 (Fig. 14.7) is an almost complete left M2, with some root segments absent. It exhibits significant wear of the occlusal surface, generating trefoil figures on the pretrite side, very clearly in loph 1 and 2. In loph 3, there is wear of the main cone, the medial cone and in the anterior pretrite, meaning that no trefoil figures are recorded. Particularly in loph 1, the pretrite and posttrite sides are joint together due to advanced wear. The heel is without wear. Table 14.1 displays the measurements for this piece. For its part, the molar MHMOP/LN/9 (Fig. 14.7) is a left M1. The crown is complete, while the anterior root is absent and the posterior root is fractured. Trilophodon, exhibits considerable wear on the three lophs and on its talon, with the appearance of dentine. It has more complex lingual-facing wear patterns than labial, that is, forming a clear trefoil only on the pretrite side. On the posttrite side, there are incipient forms of trefoil only on loph 1 and loph 2. The posttrite and pretrite areas are defined in all the lophs, even though the main cones are attached to the mescones in all cases (pretrite and posttrite). Table 14.1 gives the measurements of this piece. Finally, piece MHMOP/LN/15 (Fig. 14.7) is a practically complete upper incisor or tusk. It is markedly curved, with significant wear at the apex. It has no twisting or enamel strip. Table 14.3 shows the measurements of this specimen.

Following Alberdi et al. (2002), Labarca et al. (2016) and Mothé et al. (2013, 2017), the main diagnostic features used to segregate the two forms of South American gomphotheres (Cuvieronius hyodon and Notiomastodon platensis) are located in the anterior part of the skull and in the morphology of the tusks. In relation to this, the presence of an incisor without enamel or torsion makes it possible to clearly assign the specimen to Notiomastodon platensis, a species that had already been described for the north Patagonia area (Labarca and Alberdi 2011; Recabarren et al. 2014). The molars, which judging by their disposition and wear probably correspond to the same individual, are similarly assigned to this taxon. Table 14.2 show the size of molars of *Notiomastodon platensis* from the Los Notros site.

In relation to the horses, the morphological characteristics of the MHMOP/LN/3 upper incisor, MHMOP/LN/1 first or second lower molar, and MHMOP/LN/5 first or second upper molar, identify the dental pieces as *Equus* sp.

Order Perissodactyla Owen (1848)
Family Equidae Gray (1821)
Subfamily Equinae Gray (1821)
Equus Linnaeus (1758)
Equus (Amerhippus) Hoffstetter (1950)
Equus sp. (Table 14.4 and 14.5; Fig. 14.9).

Specimens referenced. MHMOP/LN/3 upper incisor, MHMOP/LN/1 first or second lower molar, and MHMOP/LN/5, first or second upper molar.
Provenance. Layers LN-1 and LN-2 of the Los Notros site. The layers are found by 14C between 13,585 ± 81 years 14C B.P. (16,048–16,614 years cal B.P.) and 10,916

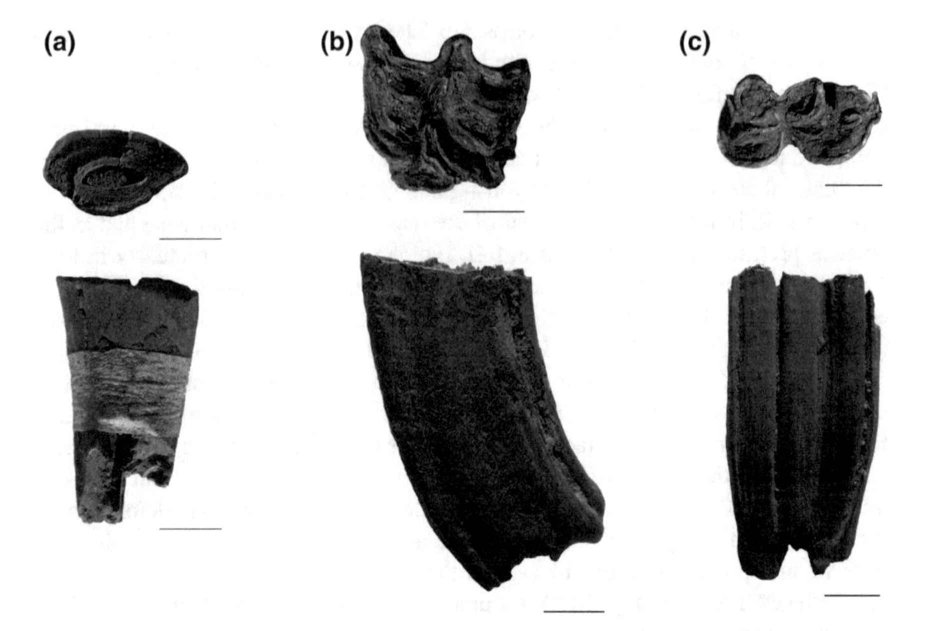

Fig. 14.9 Dental remains of *Equus (Amerhippus)* sp. from Los Notros. **a** MHMOP/LN/3 I1–2 occlusal and lingual view; **b** MHMOP/LN/5 M1–2 occlusal and lateral view; **c** MHMOP/LN/1 m1–2 occlusal and lingual view. The scale bar = 1 cm

± 67 years 14C B.P. (12,680–12,924 years cal B.P.), respectively (see Table 14.1 for spatial details of provenance).

Description and comparison. The MHMOP/LN/3 piece (Fig. 14.9) is the first or second upper incisor with medium occlusal wear. An occlusal view presents a sub-trapezoidal morphology, revealing an enamel line around the perimeter that encloses the dentin (partially exposed) and the infundibulum. The latter has a markedly ovoid shape and is delimited by an enamel ring, which gives way to the cup. The piece MHMOP/LN/1 (Fig. 14.9) is a practically complete left M1–2. In occlusal view, it is possible to observe a clearly rounded metaconid and a comparatively more angular

Table 14.4 Dimensions of molar m1-2 MHMOP/LN/1 from site Los Notros, Chile. Values according to Eisenmann et al. (1988). Measurements in mm

1	2	3	4	5	6
45	28.9	10	15	12	15.9

Table 14.5 Dimensions of M1-2 MHMOP/LN/5 of Los Notros. Values according to Eisenmann et al. (1988). Measurements in mm

1	2	3	4
59	29.3	11.9	27.8

metastilid. The ectoflexid is deep, but it does not contact the linguaflexid, due to the presence of an isthmus. The hypoconid is more extended anteriorly than the protoconid. A small hypoconulid, clearly separated from the paraconid, is observed posteriorly. The measurements of this piece, following Eisenmann et al. (1988) are given in Table 14.4. For its part, the MHMOP/LN/5 specimen (Fig. 14.9) is a practically complete left M1–2. On the occlusal surface, it is possible to see a well-developed protocone with a triangular morphology. Associated with the postprotoconal groove, it is possible to see a well-defined pli caballin. The prefossettes and postfossettes are irregular and present diverse folds in their figures. The measurements of this piece, in accordance with Eisenmann et al. (1988) are given in Table 14.5.
Discussion. Both the M1–2 and the m1–2 display attributes of the Equus genus: the presence of a plicaballin, the presence of folds in the fossettes and a triangular protocone in M1–2; and angular metastylid in m1–2. The metric comparison of Los Notros materials with other localities in Chile and South America (Fig. 14.10) suggests the presence of a "large" morphotype, similar to species such as E. neogeus and E. insulatus, whose measurements overlap each other. However, a recent review by Machado et al. (2017) suggests that the metric variability observed within true South American horses does not account for different species but are merely the phenotypic expression of different metapopulations. As shown above, the materials are assigned at the generic level. However, an important variation should be noted between the materials reviewed here and Equus molar measurements recorded among materials presented by Recabarren et al. (2011) from the western sector of Pilauco and pieces from the eastern sector of that same site. Equidae indet.

Reference specimen. MHMOP/LN/7, molariform.
Provenance. Layer LN-2 of Los Notros site (see Table 14.1 for spatial details of provenance).
Description. Fragment of equid molar, possibly superior, very deteriorated. It is not possible to describe its occlusal surface or obtain measurements.
Discussion. The quadrangular form of the occlusal surface allows us to suppose that it is an upper molar. Its high level of fragmentation prevents a more precise assignation (Fig. 14.10).

The specimen MHMOP/LN/2 belong to the Cervidae family.

Order Artiodactyla Owen (1848)
Suborder Ruminantia Scopoli (1777)
Family Cervidae Gray (1821)
Subfamily Capreolinae Brookes (1828)
Antifer Ameghino (1889)
Antifer ultra Ameghino (1889) (Fig. 14.11).

Specimen referenced. MHMOP/LN/2 practically complete antler
Provenance. Layer LN-2 of Los Notros site (see Table 14.1 for spatial details of provenance).

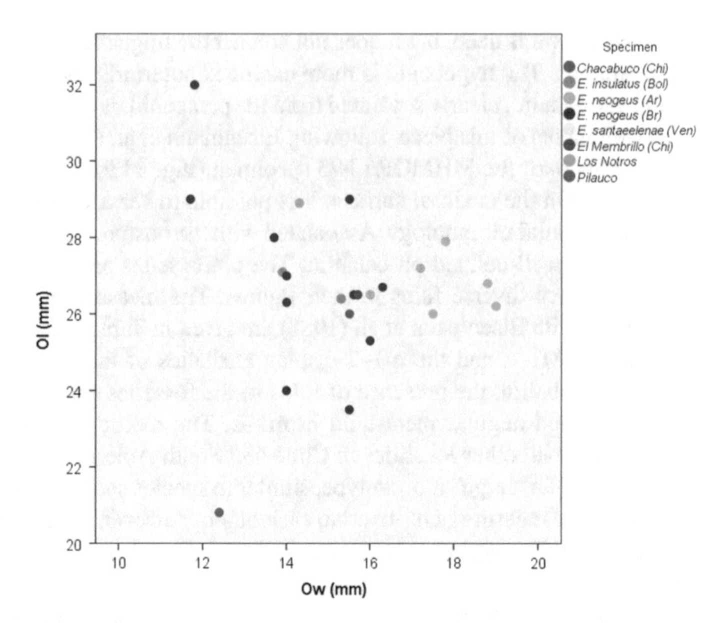

Fig. 14.10 Scatter diagram of surface length (Ol) and surface width (Ow) in molars m1–2 from Los Notros and Pilauco sites, and localities in South America (anatomical features according to Eisenmann et al. 1988). Data derived from Alberdi and Frassinetti (2000), Alberdi et al. (2003), López (2007), McFadden and Azzaroli (1987), Prado and Alberdi (2008), Prado et al. 2012), Rincón et al. (2006)

Fig. 14.11 *In situ* antler assigned to *Antifer ultra* (MHMOP/LN/2)

Description. The piece MHMOP/LN/2 is a practically complete antler, although in a poor condition of preservation. Part of the distal portion of the beam—along with the third and fourth tines—are absent. Despite its state of deterioration, it is possible to observe a relatively narrow, oblique base, situated immediately next to the burr. The pedicle is rounded, underdeveloped and its located very close to the basal tine, which in this case is forked. At the fork of the basal tine, it is possible to see a widened area. Due to its poor state of preservation, irregular marks and creases are observed. Considering the overall state of the piece, it was not possible to obtain measurements.
Discussion. The presence of a rounded pedicle, a compressed and forked basal tine, a well-developed stem, the existence of more than one tine and the presence of roughness and grooves along the entire antler are attributes of the *Antifer* genus (Alcaraz 2010). At present, two species are recognized within this genus: *A. ultra* and *A. ensendensis* Ameghino, which are differentiated by the morphology and proportions of the proximal section of the antlers (Labarca and Alcaraz 2011; Alcaraz 2010). Despite being incomplete, specimen MHMOP/LN/2 presents attributes of the *A. ultra* species, such as the little developed and relatively short pedicle, as well as a relatively oblique base.

Artiodacyla Indet (Fig 14.12).

Specimens referenced. MHMOP/LN/6 molar portion, MHMOP/LN/13 distal scapula.
Provenance. Layers LN-1 and LN-2 of the Los Notros site (see Table 14.1 for spatial details of provenance).
Description. The piece MHMOP/LN/13 (Fig. 14.12) is a distal part of the scapula, which retains a small distal segment of the blade, the neck, distal segment of the spine (missing acromion), glenoid cavity, and tubercle. It presents a generally robust appearance, with a thick spine on its lateral side, located very close to the cranial edge of the piece. It has a slight cranial orientation. On its medial side, the final

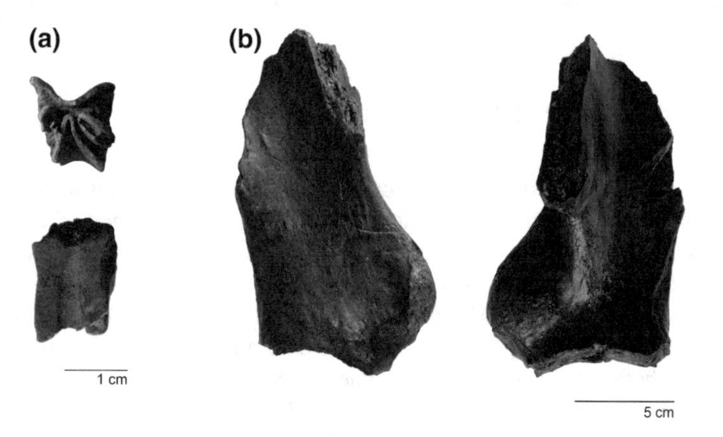

Fig. 14.12 Indeterminate Artiodactyla remains from site Los Notros. **a** Specimen MHMOP/LN/6 lower molar and **b** MHMOP/LN/13 distal scapula

portion of the blade is concave, varying toward a quite flat section up to the height of the neck. The acromion is absent, but it is possible to observe the base of it, which is distally projected. This feature is a characteristic of artiodactyls and is not present in perissodactyls, where the acromion is absent and the spine descends gradually up to the neck, generating a comparatively greater space between it and the beginning of the glenoid cavity. The final segment of the spine in this specimen is in line with the start of the supraglenoid tubercle. This is robust, large, rounded and well projected toward the cranium. The glenoid cavity has eroded to such an extent that it is not possible to discern its original shape; however, its concave morphology stands out. The tubercle is separated from the cavity by a narrow cavity, giving the impression of being practically joined, unlike equidae where both traits are separated by a comparatively larger space. For its part, the specimen MHMOP/LN/6 (Fig. 14.12) is a fragment of the lower molar. Here, it is possible to observe part of the cavity of the talonid and trigonid and segments of the hypoconid and protoconid.
Discussion. Specimen MHMOP/LN/13 has morphological features compatible with Order Artiodactyla, such as the presence of acromion and the proximity between the tubercle and the glenoid cavity. The absence of comparative material for Cervidae, particularly those of large extinct forms such as Antifer, does not allow a more precise generic assignation. However, the piece is markedly more robust than specimen MHMOP/PI/602, a scapula that was assigned to cf. Hemiauchenia paradoxa (Labarca et al. 2013). MHMOP/LN/6 is too fractured to discuss a more precise taxonomic assignment.

Mammlia indet.

Species referenced. MHMOP/LN/4, MHMOP/LN/10, MHMOP/LN/11, MHMOP/LN/12.
Provenance. Layer LN-1 of the Los Notros site (see Table 14.1 for spacial details of provenance).
Description. Unidentified mammalian bone fragments without a diagnostic value.

14.6 Discussion and Conclusions

The fossil assemblage from the Los Notros site, although small, has a unique value from a taxonomic and biogeographical perspective. The presence of a complete Notiomastodon platensis tusk stands out in that it confirms the presence of this species in the north Patagonia area and in coherence with previous evidence recorded in Nochaco, Mulpulmo and Monte Verde (Labarca and Alberdi 2011; Recabarren et al. 2014). Similarly, the presence of Antifer ultra on the site allows it to extend its known distribution by several hundred kilometers to the south in Chile, since this form had previous records in central Chile's (Taguatagua) and Quereo, (south of the semiarid region) (Labarca and Alcaraz 2011). Its presence is linked with other typically Pampean forms, such as *N. platensis* itself and cf. Hemiauchenia paradoxa (identified in the site of Pilauco), which are characteristic of the transandean Lujanense age. The record in Los Notros is complemented by equine molars, which have been assigned only at the generic level due to an uncertainty regarding the effective number of species that inhabited the South American Pleistocene (Machado et al. 2017). Excluding the presence of Antifer ultra, the Los Notros faunal assemblages are compatible with those registered at Pilauco, where Equus sp. (Labarca this volume) and Notiomastodon aff. *N. platensis*. Labarca et al. (2014, Recabarren this volume) have also been identified. A clear stratigraphical correlation has also been observed; the remains of gomphotheres come exclusively from the LN-1 layer, which is compatible with PB-7 from Pilauco, where all of the proboscid remains have been recorded. Equally, the main equidae record comes from LN-2, similarly to that observed in PB-8, where the greatest number of horse remains were discovered. Finally, it is important to point out from a paleoecological point of view the sympatry between two deer species that represent the two extremes of known sizes for the family in South America: Antifer ultra and cf. Pudu sp. In principle, this suggests similar environmental requirements for the two forms.

References

Alberdi MT, Frassinetti D (2000) Presencia de Hippidion y Equus (Amerhippus) (Mammalia, Perissodactyla) y su distribución en el Pleistoceno superior de Chile. Estud Geol (Madr) 56(5–6):279-290

Alberdi MT, Cartelle C, Prado JL (2002) El registro de Stegomastodon (Mammalia, Gomphotheriidae) en el Pleistoceno superior de Brasil. Rev Esp Paleontol 17(2):217–235

Alberdi MT, Cartelle C, Prado JL (2003) El registro Pleistoceno de Equus (Amerhippus) e Hippidion (Mammalia, Perissodactyla) de Brasil. Consideraciones paleoecológicas y biogeográficas. Ameghiniana 40:173–196

Alcaraz MA (2010) Sistemática de los cérvidos [Mammalia, Artiodactyla] del Pleistoceno de las áreas extraandinas de Argentina. Dissertation Universidad Nacional de La Plata

Eisenmann V, Alberdi MT, De Giuli C, Staesche U (1988) Methodology. Studying fossil horses 1: 1–71.

Frassinetti D, Alberdi MT (2000) Revisión y estudio de los restos fósiles de Mastodontes de Chile (Gomphotheriidae): Cuvieronius hyodon, Pleistoceno superior. Estudios Geológicos 56(3–4): 197–208.

Hogg AG, Hua Q, Blackwell PG, Niu M, Buck CE, Guilderson TP, Heaton TJ, Palmer JG, Reimer PJ, Reimer RW, Turney CSM, Zimmerman SRH (2003) ShCal13 Southern Hemisphere calibration, 0–50,000 years cal BP. Radiocarbon 55(4):1889–1903

Labarca R, Alberdi MT (2011) An updated taxonomic view on the family Gomphotheriidae (Proboscidea) in the final Pleistocene of south-central Chile. Neues Jahrb Geol Paläontol, Abhandlungen 262(1):43–57

Labarca R, Alcaraz MA (2011) Presencia de Antifer ultra Ameghino (=Antifer niemeyeri Casamiquela) (Artiodactyla, Cervidae) en el Pleistoceno tardío-Holoceno temprano de Chile central (30–35° S). Andean Geol. 38(1):156–170

Labarca R, Pino M, Recabarren O (2013) Los Lamini (Cetartiodactyla: Camelidae) extintos del yacimiento de Pilauco (Norpatagonia chilena): aspectos taxonómicos y tafonómicos preliminares. Estud Geol (Madr) 69(2):255–269

Labarca R, Recabarren O, Canales-Brellenthin P, Pino M (2014) The Gomphotheres (Proboscidea: Gomphotheriidae) from Pilauco site: scavenging evidence in the late Pleistocene of the chilean Patagonia. Quat Int 352:75–84

Labarca R, Alberdi MT, Prado JL, Mansilla P, Mourgues FA (2016) Nuevas evidencias acerca de la presencia de Stegomastodon platensis, Ameghino, 1988, Probosicdea Gomphotheriidae en el Pleistoceno Tardío de Chile central. Estud Geol (Madr) 76(1):e046

Lira MP, Labarca R, Navarro X, Fritte D, Oyarzo H, Pino M (2016) El sitio los Notros, Pleistoceno tardío, Osorno Norpatagonia Occidental de Chile, In: Abstracts of V Simposio de Paleontología en Chile, 7–11 November 2006

Machado H, Grillo O, Scott E, Avilla L (2017) Following the footsteps of South American Equus: are the autopodia taxonomically informative? J Mamm Evol 25(3):397–405

McFadden BJ, Azzaroli A (1987) Cranium of Equus insulatus (Mammalia, Equidae) from the middle Pleistocene of Tarija, Bolivia. J Vert Paleontol 7:325–334

Mothé D, Avilla LS, Cozzuol MA (2013) The South American gomphotheres (Mammalia, Proboscidea, Gomphotheriidae): taxonomy, phylogeny and biogeography. J Mamm Evol 20:23–32

Mothé D, Avilla LS, Asevedo L, Borges-Silva L, Rosas M, Labarca-Encina R, Souberlich R, Soibelzon E, Roman-Carrion JL, Ríos SD, Rincón AD, Oliveira GC, Lopes RP (2017) Sixty years after 'the mastodonts of Brazil': the state of the art of South American proboscideans (Proboscidea, Gomphotheriidae). Quat Int 443:52–64

Pino M, Miralles C (2008) La geología cuaternaria de Pilauco. In: Pino M (ed) Pilauco, un sitio complejo del Pleistoceno tardío. Osorno, Norpatagonia Chilena. Universidad Austral de Chile, Imprenta América, Valdivia, Chile, pp 37–42

Pino M, Chavez M, Navarro X, Labarca R (2013) The late Pleistocene Pilauco site, south central Chile. Quat Int 299:3–12

Pino M, Martel-Cea A, Vega R, Fritte D, Soto-Bollmann K (2016) Geología y geomorfología del sitio Pilauco. In: Pino, M (ed) El sitio Pilauco Osorno, Patagonia Noroccidental de Chile. Universidad Austral de Chile, Imprenta América, Valdivia, Chile, pp 12–46

Pino M, Martel-Cea A, Astorga G, Abarzúa AM, Cossio N, Navarro X, Lira MP, Labarca R, Lecompte MA, Adedeji V, Moore CR, Bunch TE, Mooney C, Wolbach WS, West A, Kennett JP (2019) Sedimentary record from Patagonia, Southern Chile supports cosmic-impact triggering of biomass burning, climate change, and megafaunal extinctions at 12.8 ka. Sci Rep 9(1):4413

Prado JL, Alberdi MT (2008) Restos de Hippidion y Equus (Amerhippus) procedentes de las barrancas de San Lorenzo, Pleistoceno tardío (provincia de Santa Fe, Argentina). Rev Esp Paleontol 2:225–236

Prado JL, Alberdi MT, Di Martino VJ (2012). Équidos y Gonfoterios del Pleistoceno tardío del Sudeste de la Provincia de Buenos Aires. Ameghiniana 49(4):623–642

Recabarren O (2007) Análisis de restos óseos de gonfoterios del área comprendida entre los 39° 39′ y 42° 49′ S, centro-sur de Chile. Undergraduate Dissertation, Universidad Austral de Chile

Recabarren O (2016) Los proboscídeos gonfotéridos (Mammalia, Gomphotheriidae) del extremo sur de América. In: Pino, M (ed) El sitio Pilauco Osorno, Patagonia Noroccidental de Chile. Universidad Austral de Chile, Imprenta América, Valdivia, Chile, pp 75–95

Recabarren O, Pino M, Cid I (2011) A new record of Equus (Mammalia: Equidae) from the late Pleistocene of central-south Chile. Rev Chil Hist Nat 84:535–542

Recabarren O, Pino M, Alberdi MT (2014) La Familia Gomphotheriidae en América del Sur: evidencia de molares al norte de la Patagonia chilena. Estud Geol (Madr) 70(1)

Rincón AD, Alberdi MT, Prado JL (2006) Nuevo registro de Equus (Amerhippus) santaeelenae (Mammalia, Perissodactyla) del pozo de asfalto de Inciarte (Pleistoceno Superior), estado Zulia, Venezuela. Ameghiniana 43:529–538

Smutz MMS, Bezuidenhout AJ (1987) Anatomy of the dromedary. Clarendon Press

Stuiver M, Reimer PJ (1993) Extended 14C database and revised CALIB radiocarbon calibration program. Radiocarbon 35:215–230

Chapter 15
Evidence from Pilauco, Chile Suggests a Catastrophic Cosmic Impact Occurred Near the Site ~12,800 Years Ago

Allen West, Ted Bunch, Malcolm A. Lecompte, Víctor Adedeji, Christopher R. Moore and Wendy S. Wolbach

Abstract The Younger Dryas (YD) impact hypothesis proposes that fragments of a large, disintegrating asteroid/comet struck the Earth ~12,800 years ago. This event simultaneously deposited high concentrations of platinum, high-temperature spherules, melt glass and nanodiamonds into the YD boundary layer (YDB) at >50 sites worldwide. Here, we report on a ~12,800-year-old sequence at Pilauco that exhibits peak YD boundary concentrations of platinum, gold, high-temperature iron- and chromium-rich spherules and native iron particles uncommonly found together in sedimentary deposits. In addition, an erosional unconformity is associated with a large abundance peak in charcoal, representing an intense biomass-burning episode correlated with intense changes in vegetation. At Pilauco, the disappearance of megafaunal remains and dung fungi in the YDB lamina correlates well with megafaunal extinctions across the Americas. The Pilauco record is consistent with YDB impact evidence found at multiple sites on four continents.

Keywords Younger Dryas · Asteroid · Platinum · High-temperature spherules

A. West (✉)
Comet Research Group, Prescott, AZ 86301, USA
e-mail: cometresearchgroup@gmail.com

T. Bunch
Geology Program, School of Earth Science and Environmental Sustainability, Northern Arizona University, Flagstaff, AZ 86011, USA

M. A. Lecompte
Center of Excellence in Remote Sensing Education and Research, Elizabeth City State University, Elizabeth City, NC 27909, USA

V. Adedeji
Department of Natural Sciences, Elizabeth City State University, Elizabeth City, NC 27909, USA

C. R. Moore
Savannah River Archaeological Research Program, South Carolina Institute of Archaeology and Anthropology, University of South Carolina, Columbia, SC 29208, USA

W. S. Wolbach
Department of Chemistry, DePaul University, Chicago, IL 60614, USA

M. Pino and G. A. Astorga (eds.), *Pilauco: A Late Pleistocene Archaeo-paleontological Site*, The Latin American Studies Book Series,
https://doi.org/10.1007/978-3-030-23918-3_15

15.1 Background

In 1986, a construction company that was excavating in Pilauco, a suburb of the city of Osorno, Chile, discovered megafaunal bones of mastodons and horses, and in November 2007, Mario Pino organized a team of scientists to explore the site. Their investigation determined that from the Late Pleistocene into the Early Holocene, the site was a wetland that was occupied by megafauna and humans, spanning the onset of the Younger Dryas (YD), a severe climate episode that began ~12,800 years ago (see Chap. 3, this volume). At Pilauco, these researchers noticed a distinct, sedimentological change, dating to the YD onset and followed by the deposition of dark, organic-rich sediments. Aware of similar sediments in North America called 'black mats' that are related to a cosmic impact event, Mario Pino contacted the co-authors of this chapter in 2015 to determine whether Pilauco's stratigraphic profile contains cosmic impact material. This chapter presents the results of that collaboration and studies.

In 2007, Firestone et al. proposed that a large cosmic object, either an asteroid or a comet, collided with Earth approximately 12,800 years ago (cal BP), triggering a cascade of catastrophic climatic, environmental, and biotic changes. Firestone et al. (2007) and Napier (2010) proposed that these widespread effects resulted from a cosmic airburst/impact, a term referring to atmospheric collisions with extraterrestrial bodies, typically producing aerial explosions and sometimes, small crater-forming ground impacts. Such events rarely leave large visible craters, but instead produce a wide array of high-temperature, melted materials called impact proxies. This cosmic event is described in the Younger Dryas Impact Hypothesis, and now after more than a decade of scientific research, it is supported by an increasing body of evidence across at least four continents, North America, South America, Europe and Asia. This impact is proposed to have triggered the abrupt cooling of the Younger Dryas climate episode and caused major biomass burning, destroying nearly 10% of trees and plants within a few days across most of the Northern Hemisphere and parts of the Southern Hemisphere (Wolbach et al. 2018a, b). These catastrophic changes contributed to the end-Pleistocene megafaunal extinctions, during which South America lost more genera (~52) and more species (~66) than North America, Europe or Asia. In addition, the proposed impact event is synchronous with cultural changes and human population declines, amounting to losses of >50% across the Northern Hemisphere (Firestone et al. 2007; Kennett et al. 2008, 2015; Anderson et al. 2011; Wittke et al. 2013).

The Younger Dryas impact theory evolved based on the discovery of abundance peaks in a variable assemblage of more than a dozen high-temperature, impact-related proxies. These are widely found in a unique, multi-continental, sedimentary lamina called the Younger Dryas boundary layer (YDB), which is typically less than a few centimetres in thickness (Firestone et al. 2007). Table 15.1 provides is a brief list of YDB impact-related proxies and of selected published papers, describing both positive and negative findings.

Table 15.1 YDB proxies and a bibliography of selected references

PROXY	Proponents	Independent workers	Critics
Cosmic impact spherules	Firestone et al. (2007, 2009, 2010), Bunch et al. (2012), Israde-Alcántara et al. (2012), Wittke et al. (2013), LeCompte et al. (2017), Pino et al. (2019)	Mahaney et al. (2008, 2010a, b, 2011a, b), Mahaney and Keiser (2012), Redmond and Tankersley (2012), Fayek et al. (2012), LeCompte et al. (2012), Andronikov and Andronkova (2016)	Surovell et al. (2009), Haynes et al. (2010), Pinter et al. (2011), Pigati et al. (2012), Boslough et al. (2012), Holliday et al. (2016)
Melt glass (scoria-like objects)	Bunch et al. (2012)	Mahaney et al. (2008, 2010a, b, 2011a, b), Mahaney and Keiser (2012), Fayek et al. (2012)	–
Carbon spherules, glass-like carbon, aciniform carbon, PAHs, fullerenes	Firestone et al. (2007, 2009, 2010), Israde-Alcántara et al. (2012), Maiorana-Boutilier et al. (2016)	Baker et al. (2008), Redmond and Tankersley (2012), Mahaney et al. (2010a, b, 2011a, b), Mahaney and Keiser (2012)	Scott et al. (2010), Pinter et al. (2011), Van Hoesel et al. (2012, 2014), Boslough et al. (2012)
Nanodiamonds	Firestone et al. (2007), Kennett et al. (2009a, b), Kurbatov et al. (2011), Israde-Alcántara et al. (2012), Kinzie et al. (2014)	Baker et al. (2008), Tian et al. (2011), Redmond and Tankersley (2012), Bement et al. (2014)	Daulton et al. (2010), Pinter et al. (2011), Van Hoesel et al. (2012, 2014), Boslough et al. (2012)

(continued)

Table 15.1 (continued)

PROXY	Proponents	Independent workers	Critics
Iridium	Firestone et al. (2007, 2009, 2010)	Andronikov et al. (2011), Marshall et al. (2011), Andronikov et al. (2016a)	Paquay et al. (2009), Haynes et al. (2010), Pinter et al. (2011), Pigati et al. (2012), Boslough et al. (2012)
Platinum	Mahaney et al. (2016), Moore et al. (2017), LeCompte et al. (2017), Pino et al. (2019)	Petaev et al. (2013)	–
Osmium	–	Beets et al. (2008), Sharma et al. (2009), Wu et al. (2013)	Paquay et al. (2009)
Nickel, cobalt, chromium, thorium, 14C, 10Be, 26Al, REEs	Firestone et al. (2007, 2009, 2010)	Melott et al. (2010), Andronikov et al. (2014, 2015, 2016a, b)	–
Impact-related biomass burning	Firestone et al. (2007), Wolbach et al. (2018a, b), Pino et al. (2019)	–	–
Extinctions, human population declines	Firestone et al. (2007, 2009, 2010), Anderson et al. (2011), Pino et al. (2019)	–	Marlon et al. (2009)

Most recent research has attributed the YDB proxies to a cosmic impact event, although some studies offer alternate explanations. However, the negative studies suffered from one or more flaws. Frequently, (1) they did not follow the requisite analytical protocol, for example, by not using scanning electron microscopy (SEM) and electron diffraction spectroscopy (EDS); (2) they did not use rigorous dating methods, for example, by not using Bayesian analyses; and/or (3) they propose nonimpact hypotheses for a specific proxy that are not viable alternative explanations for the synchronous deposition of the entire suite of proxies.

Previously, the YDB layer has been identified in ~40 stratigraphic sections in 12 countries on four continents and has a modelled age range of 12,835–12,735 cal yr BP at 95% probability (Kennett et al. 2015). The widespread distribution of this well-dated, synchronously deposited layer makes it a highly valuable datum layer for stratigraphic correlation across wide areas. Cosmic impact-related proxies in the YDB layer are generally present in small quantities and/or as small particles, ranging in size from nanometers (for nanodiamonds) to several cms (for melt glass). Deposition of these proxies is typically synchronous with the upper biostratigraphic limits of many extinct taxa of Late Pleistocene megafauna.

The primary purpose of this chapter is to summarize the general characteristics, origin, and distribution of the various impact-related proxies that exhibit abundance peaks in a ~12,800-year-old stratum at Pilauco and to compare them with other YDB sites.

This investigation included five groups of material: Group 1 high-temperature, Fe-rich spherules; Group 2 high-temperature, Cr-rich spherules; Group 3 basaltic volcanic spherules; Group 4 authigenic framboidal spherules, resulting from impact-related environmental degradation; and Group 5 platinum-group elements (platinum, palladium, iridium, and osmium) and gold. Stratigraphy, charcoal, pollen, lithic artefacts and megafaunal remains are covered in Chaps. 3, 9 and 16 of this volume and in Pino et al. (2019).

15.2 Age–Height Model

Sixteen radiocarbon dates were calibrated within the OxCal program 4.3.3 (Bronk Ramsey 2008, 2009), using the Southern Hemispheric calibration curve, SHCal13 (Hogg et al. 2013), and from those, a Bayesian age–height model was generated (Pino et al. 2019). The YDB layer was identified in grid 8AD at an elevation of ~550 cm above the site datum (height, not depth) with a Bayesian-derived age of ~12,770 ± 160 cal yr BP (Fig. 15.1).

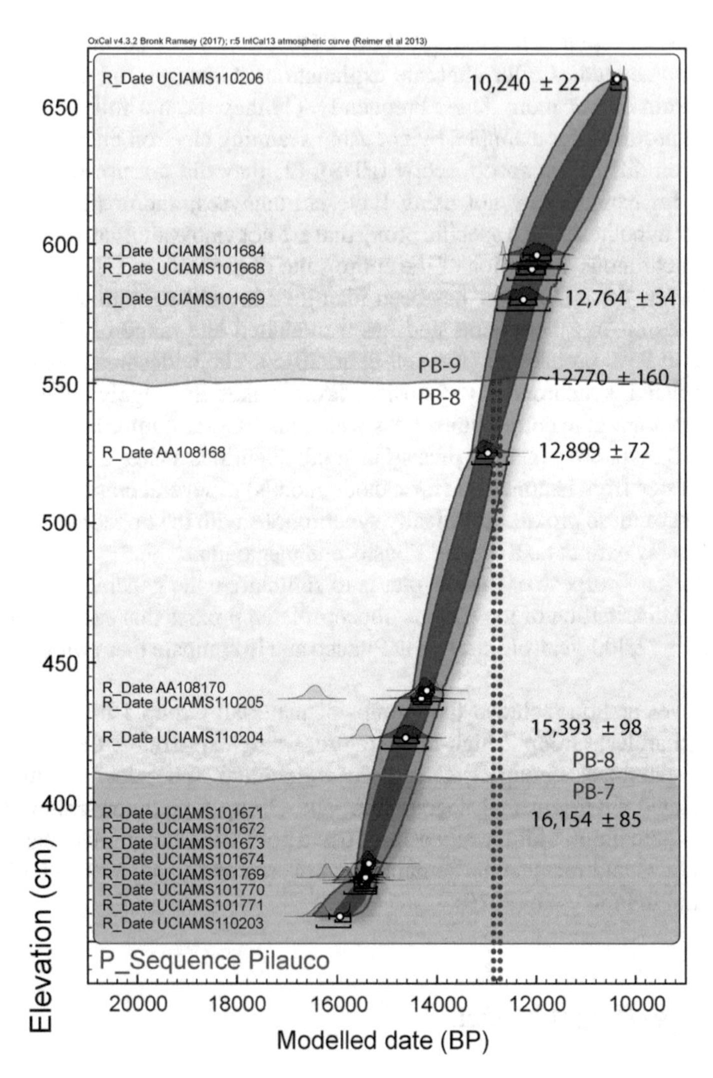

Fig. 15.1 Bayesian age–height model based on 16 radiocarbon dates (Pino et al. 2019). Green lines represent the stratigraphic transitions. Vertical red dotted lines represent age range of 12,835–12,735 cal yr BP, as previously reported for the YDB impact event (Kennett et al. 2015). This age range overlaps the PB-8/PB-9 transition, with an age of 12,770 ± 160 cal yr BP, consistent with that being the YDB layer. Age model produced with OxCal v. 4.3.3, and the SHCal13 calibration curve (Pino et al. 2019)

15.3 Iron-Rich Spherules (Group 1)

The diameters of Fe-rich spherules at other YDB sites across four continents range from 5 μm to 5.5 mm, averaging 135 μm, with a median diameter of 30 μm. Most spherules (>95%) are rounded, and the remainder appear as ovoids, aerodynamically shaped teardrops or fused clusters of one or more spherules. Spherule concentrations in the YDB layer average 955/kg with a median of 388/kg (Wittke et al. 2013). Found distributed across wide areas of four continents, YDB spherules are mostly black or brown in colour (Fig. 15.2), but may also be red, blue, green, grey, tan, or white, ranging in clarity from opaque to transparent (Wittke et al. 2013).

Based on all available evidence, YDB spherules are inferred to have formed when high-temperature, hypervelocity jets descended to the ground from extraterrestrial explosions in the atmosphere. The descending plumes melted terrestrial sediment, whether located on land or in glacial ice as detritus. After the spherules formed, the rising impact plumes dispersed then into the atmosphere and distributed them widely across multiple continents.

For YDB spherules at the Pilauco site, the average size detected is ~45 μm (range: ~95 to ~10 μm). They were observed only in samples at local elevations of 550–558 cm and were not detected in any other samples (Fig. 15.3).

YDB spherules at ~40 other sites have elemental compositions ranging from iron (FeO) to silica (SiO_2). FeO averages 44.9 wt%; SiO_2 averages 30.9 wt%; and Al_2O_3 averages 12.2 wt%. A few spherules contain platinum, iridium, osmium and rare-earth elements (lanthanum, cerium), ranging from <1 to ~40 wt%. (LeCompte et al. 2012; Wu et al. 2013; Andronikov et al. 2016b).

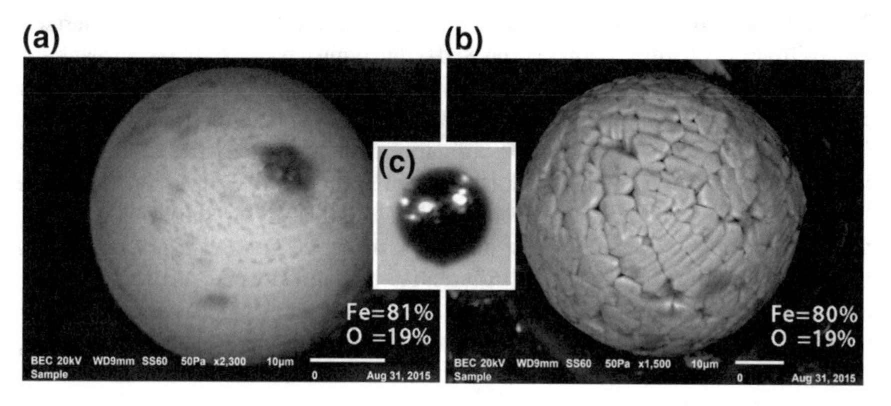

Fig. 15.2 Fe-rich YDB spherules. **a** and **b** SEM images of YDB spherules displaying distinctive dendritic texturing indicative of melting and rapid quenching at high temperatures above ~1450 °C, the melting point of iron. **c** Photomicrograph of YDB spherule from Pino et al. (2019). The percentages listed (in yellow) indicate that the composition is FeO (wustite), an oxygen-deficient mineral that is very rare under natural terrestrial conditions, but common in meteorites and impact events

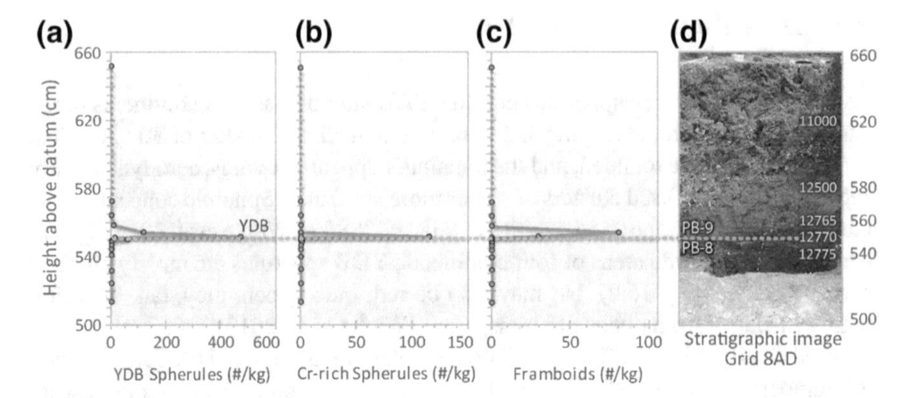

Fig. 15.3 Peak concentrations of spherules and framboids from Pilauco, Chile. **a** Fe-rich YDB spherules reach abundance peak of ~520/kg at ~552 cm; **b** Cr-rich spherules peak at ~115/kg at ~552 cm; **c** Authigenic framboids peak at ~80/kg at ~554 cm; **d** Stratigraphic profile of grid 8AD, showing where spherules were found. Age of the unconformity PB-8/PB-9 is 12,770 ± 160 cal yr BP. Adapted from Pino et al. (2019)

For Pilauco, 11 analyses by scanning electron microscopy–energy dispersive spectroscopy (SEM-EDS, Fig. 15.4) show that the site's Fe-rich spherules contain just two oxides, FeO (average: 93.7 wt%; range 100 to 86.6 wt%) and SiO_2 (average: 6.3 wt%; range 13.4 to 0.0 wt%), with no other detectable oxides (Pino et al. 2019). High formation temperatures for these spherules are indicated by the presence of FeO (melting point: ~1450 °C) and SiO_2 (melting point: ~1750 °C).

Pilauco spherules are highly unusual, showing a considerable variation, ranging from dendritic, oxygen-deficient, Fe-rich spherules to oxidized iron oxide spherules. Some Group 1 spherules are not composed of the typical terrestrial Fe_2O_3 (hematite; Fe/O ratio of ~70:30) and/or Fe_3O_4 (magnetite; Fe/O ratio of ~72:28) but rather are composed of FeO (wustite; Fe/O ratio of ~78:22 to ~81:19) (Pino et al. 2019). Wustite almost never occurs under ordinary conditions on Earth but is common in meteorites and impact melts that have much lower oxygen percentages than are typical for terrestrial crustal environments. SEM-EDS analyses show that Pilauco's volcanic spherules, framboids, and non-YDB magnetic grains are not reduced in oxygen, providing a simple test for differentiation. Such low percentages of oxygen are the expected result of highly variable temperatures and chemical heterogeneities within a cosmic impact plume but are inconsistent with other naturally occurring geological and anthropogenic mechanisms.

15.4 Chromium-Rich Spherules (Group 2)

The second group of spherules was detected only in the YDB layer at an elevation of 552 cm (Fig. 15.3). These have a concentration of ~115 spherules/kg with average diameters of ~43 μm, ranging from ~45 to ~42 μm (Fig. 15.3a, b). Group 2's

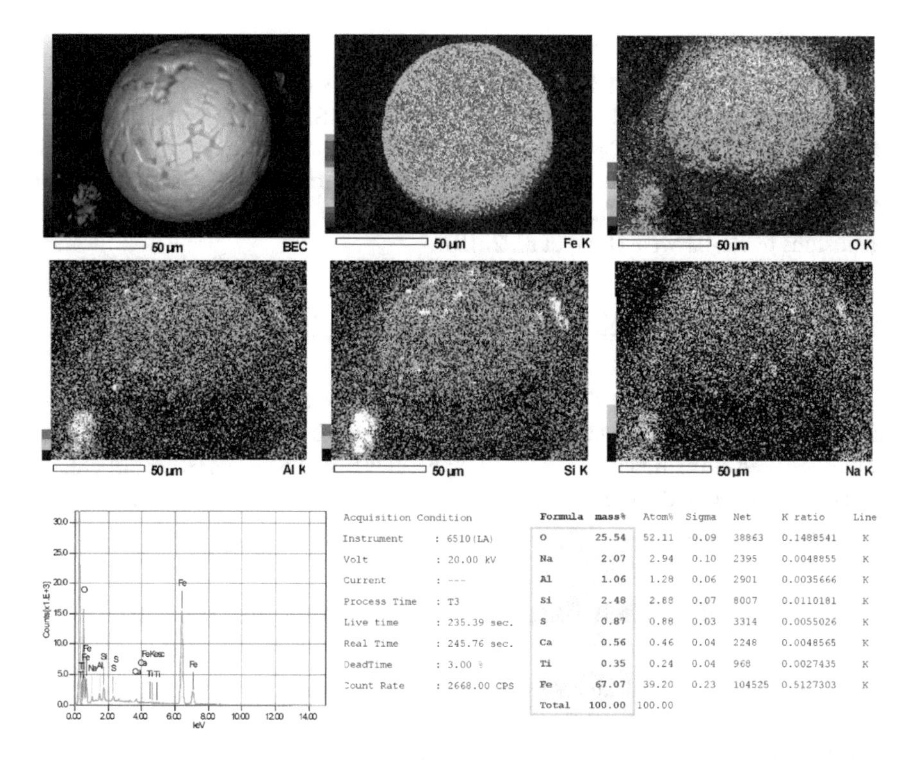

Acquisition Condition	
Instrument	: 6510(LA)
Volt	: 20.00 kV
Current	: ---
Process Time	: T3
Live time	: 235.39 sec.
Real Time	: 245.76 sec.
DeadTime	: 3.00 %
Count Rate	: 2668.00 CPS

Formula	mass%	Atom%	Sigma	Net	K ratio	Line
O	25.54	52.11	0.09	38863	0.1488541	K
Na	2.07	2.94	0.10	2395	0.0048855	K
Al	1.06	1.28	0.06	2901	0.0035666	K
Si	2.48	2.88	0.07	8007	0.0110181	K
S	0.87	0.88	0.03	3314	0.0055026	K
Ca	0.56	0.46	0.04	2248	0.0048565	K
Ti	0.35	0.24	0.04	968	0.0027435	K
Fe	67.07	39.20	0.23	104525	0.5127303	K
Total	100.00	100.00				

Fig. 15.4 SEM-EDS elemental map of YDB spherule. SEM image of YDB spherule at upper left. Elemental maps show that spherule is composed of nearly all ironed and oxygen (Pino et al. 2019). It contains minor amounts of other elements, most of which are probable surface contamination, as indicated by white areas in lower row of images above. From Pino et al. (2019)

Cr-rich spherules are differentiated from Group 1's Fe-rich spherules by the presence of three major oxides: chromium oxide, Cr_2O_3, averaging 6.2 wt%; SiO_2, averaging 11.3 wt%; and FeO, averaging 71.5 wt% (Pino et al. 2019). Approximately half of these Cr-rich spherules also contain TiO_2, averaging 6.1 wt%. No other Pilauco spherules observed outside of the YDB layer contain any detectable Cr. Unlike the Fe-rich YDB spherules, this group of spherules is not depleted in oxygen.

The origin of these Cr-rich spherules is suggested by the presence in all of Pilauco's 23 sedimentary samples of variable amounts of Cr-rich vesicular basalt (Pino et al. 2019). These typically range in diameter from a few millimetres to ~1 cm and are always composed of aluminosilicate glass (Al: 2.3 wt%; Si: 19.1 wt%; Mg: 8.1 wt%; and Ca: 8.6 wt%). The aluminosilicate matrix is not enriched in Cr, but the matrix of each piece typically contains numerous chromium–magnetite inclusions (Cr: 10.1 wt%; Fe: 74.8 wt%; Si: 5.5 wt%; and Al: 2.6 wt%) that crystallized from the molten magma as it cooled or formed from supergene or hypogene processes (deposition or enrichment of mineral deposits by solutions moving downward through magmatic rocks). These crystalline inclusions contain nearly identical relative concentrations of Cr and Fe as found in the Cr-rich spherules of Group 2 (Cr: 10.1 wt%; Fe: 74.8 wt%; Si: 5.5 wt%; and Al: 2.6 wt%).

The comparison of oxide abundances performed by SEM-EDS shows that Cr-rich spherules contain 4–14 wt% Cr_2O_3, values that overlap the range of 10–13 wt% for Cr-rich inclusions in vesicular basalt at Pilauco. Similarly, Cr-rich spherules contain 63–79 wt% FeO, concentrations that fall within the range of Pilauco's Cr-rich inclusions at 70–81 wt% (Pino et al. 2019).

15.5 Ca- and Si-Rich Basaltic Volcanic Spherules (Group 3)

Group 3 is comprised of Ca- and Si-rich spherules that were found randomly distributed at low concentrations (0 to ~10 spherules/kg) in five sedimentary samples tested at Pilauco (Fig. 15.8a, b). Spherules from this group were not present in the YDB layer and were not depleted in oxygen, as are most spherules found in the YDB layer (Fig. 15.5).

15.6 Ca- and Si-Rich Basaltic Volcanic Spherules (Group 3)

Pilauco is adjacent to Andean volcanoes, and therefore, we investigated whether these spherules resulted from local eruptions. Volcanic spherules are produced during the low-intensity eruptions of low-viscosity magmas, during which shards of molten tephra coalesce into Si-rich droplets with the same chemical composition as the erupting Si-rich magma (Heiken 1972). Volcanic spherules that form during such low-energy eruptions typically fall only onto the slopes of the volcano and not beyond. On the other hand, volcanic eruptions capable of distributing tephra across wide areas are violent, high-energy events driven by high gas pressures (Smit 1990). Such highly energetic eruptions produce angular tephra, rather than spherules like those found at Pilauco. For example, the largest known eruption in the last 5 million years, the highly explosive, 75,000-year-old Toba-lake eruption, distributed angular tephra as much as 2500 km away from the source volcano, but did not produce any detectable spherules (Smit 1990). Thus, it is highly unlikely that volcanism accounts for the wide distribution of YDB spherules across multiple continents, representing ~10% of the planet (Bunch et al. 2012; Wittke et al. 2013).

Fe content is another important difference between the Fe-rich spherules in Group 1 and the Si-rich volcanic spherules in Group 3. Typical magma can contain up to ~13 wt% Fe by volume and can also contain non-spherulitic, high-Fe inclusions that exsolve slowly from the molten magma as it cools. However, spherules containing 85–100 wt% FeO cannot form from erupting magma composed of no more than ~13 wt% iron. This is confirmed by the research of Wright and Hodge (1965), who reported finding ~150 volcanic spherules with Fe compositions that ranged from 4 wt% to a maximum of 50 wt% and averaging ~20 wt%. No spherules were found with compositions approaching the minimum iron concentration of >85 wt% found in Pilauco spherules.

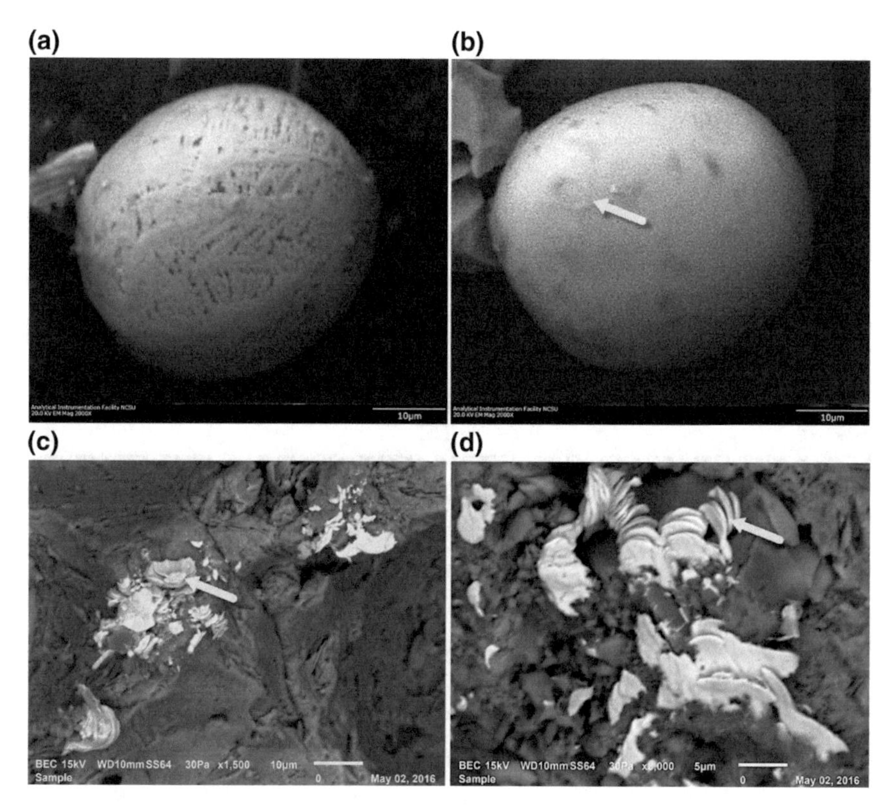

Fig. 15.5 Cr-rich YDB spherules from Pilauco (Group 2). **a** and **b** SEM images of Cr-rich ovoid spherules with SEM-EDS compositions corresponding to chromium–magnetite inclusions in local volcanic basalt. **a** displays distinctive dendritic texturing indicative of melting and rapid quenching at temperatures above ~1670 °C, the melting point of chromium–magnetite. **b** displays a shallow crater (yellow arrow), indicating high-velocity collision with smaller object while molten. **c** and **d** Typical chromium–magnetite inclusions in basaltic melt (yellow arrows). Scalloped stack of individual Cr-rich crystals indicates exsolution from molten basalt. From Pino et al. (2019)

Cr content is another important difference between Pilauco's YDB spherules and volcanic spherules. Wright and Hodge (1965) reported analyses of ~380 basaltic spherules from volcanoes, including two volcanoes in Peru, Mount Ubinas and Mount Huainaputina, along with Kilauea (Hawaii), Irazu (Costa Rica), and Surtsey (Iceland). The volcanic spherules they analysed contained an average of <0.1 wt% Cr_2O_3 with a range of 0.00–0.73 wt%, far below the concentrations of 4–14 wt% Cr found in Group 2 spherules from Pilauco. These results further confirm that Pilauco's Cr-rich spherules are nonvolcanic in origin.

On the other hand, the compositions of the known volcanic spherules are consistent with the compositions of Pilauco's Group 3 spherules, which closely match the SEM-EDS analyses of the vesicular basalt from Pilauco, thus confirming their identification as likely volcanic spherules (Fig. 15.6).

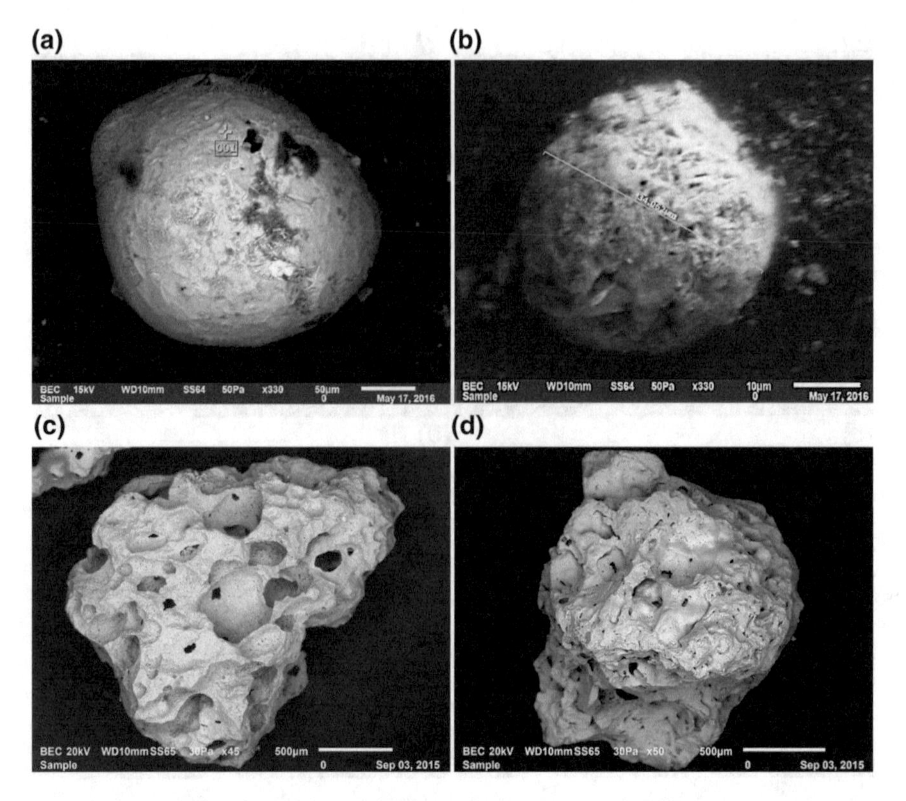

Fig. 15.6 Ca- and Si-rich basaltic volcanic spherules. **a** and **b** Group 3: Ca- and Si-rich volcanic spherules that contain no detectable chromium. **c** and **d** Examples of Ca- and Si-rich volcanic glass commonly found at Pilauco. Pilauco's vesicular basalt is geochemically similar to spherules in panels (**a**) and (**b**). From Pino et al. (2019)

15.7 Authigenic Framboidal Spherules (Group 4)

At Pilauco, authigenic spherules, called framboids, reach a peak concentration of ~80/kg at an elevation of 554 cm (Fig. 15.3c) and are closely associated with Group 1 Fe-rich spherules and Group 2 Cr-rich spherules found approximately 2 cm below at a height of 552 cm. These spherules are typically composed of pyrite, FeS_2, formed into unmelted, cube-like crystals (Fig. 15.7). Framboids usually form slowly under hypoxic conditions, rather than instantaneously, as expected for impact-related spherules (Wittke et al. 2013).

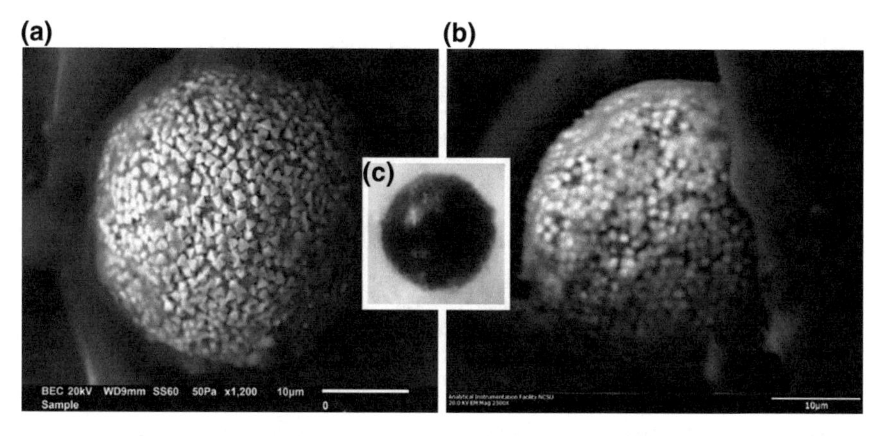

Fig. 15.7 Authigenic pyritic framboids. **a** and **b** SEM-EDS images of framboidal spherules composed of distinctive cube-like crystals that formed slowly over time. Composition mostly iron, sulphur and oxygen; **c** Photomicrograph of framboid, demonstrating that using only a light microscope make impossible to distinguish framboids from the YDB spherule shown in the photomicrograph in Fig. 15.2. Adapted from Pino et al. (2019)

15.8 Anomalous Concentrations of Platinum, Palladium and Gold

Platinum, iridium and osmium (platinum-group elements) are common constituents in meteoritic and cometary material, and so, are commonly used as indicators of extraterrestrial impact events. For the YDB impact event, Petaev et al. (2013) were first to identify a large peak in platinum concentration that dates to the onset of Younger Dryas climate change in the Greenland ice core (Greenland Ice Sheet Project 2). These authors attributed this platinum deposition to a cosmic impact event. Several subsequent investigations have found additional high concentrations of platinum and other platinum-group elements in YD-age sediments at more than 20 widely separated sites across North America, Europe and Asia (LeCompte et al. 2012; Andronikov et al. 2014, 2015, 2016a, b; Moore et al. 2017). Prior to this study, YDB-age concentrations of platinum have not been investigated for any site in South America.

At Pilauco, we measured platinum, palladium, and gold concentrations in 17 samples, using fire assay and inductively coupled plasma mass spectrometry (ICP-MS) (Pino et al. 2019). The results show one significant platinum peak that begins to rise at 550 cm in grid 8AD, reaching 9.9 ppb at 551 cm (Fig. 15.8a). After normalizing platinum (Pt) to palladium (Pd), we found Pt/Pd ratios 2 × higher in the YDB layer than background ratios (Fig. 15.8b). Similarly, Au/Pt ratios are 5 × higher than background ratios (Fig. 15.8c). Relative enrichments: Au > Pt > Pd.

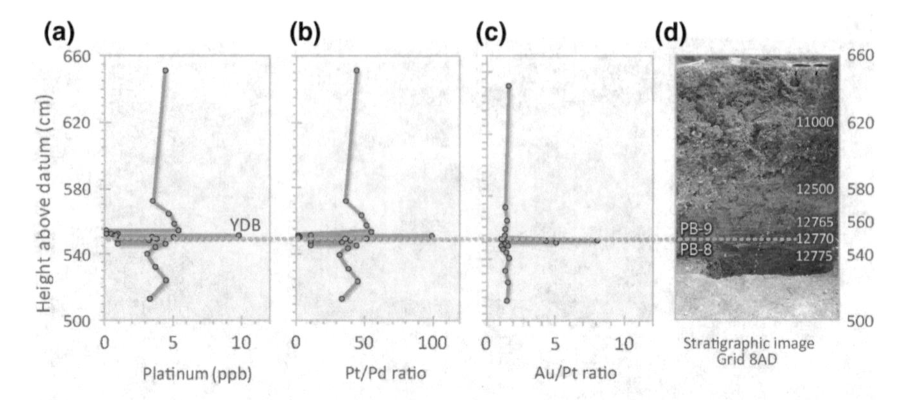

Fig. 15.8 Abnormal YDB concentrations of platinum, palladium, and gold. **a** Platinum (Pt) abundance peak of 9.9 ppb in the YDB layer at 551 cm. **b** Anomalous Pt/Pd ratio occurs only in the YDB layer, indicating the influx of nonlocal platinum at the YDB. **c** Anomalous Au/Pt ratio occurs only in the YDB layer, indicating the influx of nonlocal gold (Au) in higher concentrations than the nonlocal platinum. **d** Stratigraphic profile of grid 8AD, showing the location of the YDB layer (orange dashed line) where anomalies were found. Age of transition PB-8/PB-9 is 12,770 ± 160 cal yr BP Adapted from Pino et al. (2019)

15.9 Charcoal and Biomass Burning

Biomass burning is another crucial aspect of the YDB theory. Wolbach et al. (2018a, b) reported that a multi-continental episode of biomass burning at the YD onset is the largest in >100,000 years.

From ~16,000 to 13,000 cal yr BP, charcoal was almost absent from the sedimentary record at Pilauco, indicating low fire activity that was associated with cold and humid conditions, as inferred from the pollen assemblage (see Chap. 9, this volume). In the YDB layer just above that, charcoal concentrations began their rise to the highest concentration in the entire record at the same time that high concentrations of platinum and spherules were being deposited. This peak in charcoal concentrations is the highest in the entire 2500-year span from 15,000 to 12,500 cal yr BP, and it suggests a highly unusual episode of wildfires after ~12,800 cal yr BP. Investigations have not yet been undertaken for other key carbon-rich YDB proxies, including nanodiamonds, carbon spherules and glass-like carbon. Aciniform carbon was not found in detectable concentrations (Pino et al. 2019).

15.10 Discussion and Conclusions

The age of the YDB of ~12,770 ± 160 cal yr BP at Pilauco is even better defined with less standard deviation at the neighbouring site Los Notros (see Chap. 14, this volume).

The shapes, composition and surface textures of most YDB spherules are similar to those formed in the Tunguska airburst in 1908, the Australasian Tektite Field at ~680 ka, Meteor Crater at ~50 ka, the Chesapeake Bay impact at ~35 Ma and the K-Pg impact ~65 Ma (Wittke et al. 2013). After eliminating all other known possibilities, a cosmic impact is the only remaining plausible explanation for high-temperature, rapidly quenched YDB spherules. The spherule maximum of ~520/kg is higher than the average of 389 spherules/kg reported for other YDB sites, including some of the best-known, previously investigated YDB sites, such as Murray Springs, Arizona; Blackwater Draw, New Mexico; and the Topper site, South Carolina, the closest of which is ~9000 km north.

All Fe-rich spherules display the distinctive fine-grained, dendritic texturing of FeO crystals (Fig. 15.2) that formed when molten material quenched rapidly from high temperatures down to 1450 °C, iron's solidus temperature (LeCompte et al. 2012; Israde-Alcántara et al. 2012; Bunch et al. 2012; Wittke et al. 2013; LeCompte et al. 2017). The presence of reduced magnetite, wüstite and native iron strongly suggests low-oxygen conditions over very short distances, together with rapid cooling rates, as exemplified by quench textures on spherule surfaces. Similar objects were observed in melt glass and spherules from Blackville, South Carolina; Melrose, Pennsylvania; Abu Hureyra, Syria; and the Trinity atomic bomb test, Alamogordo, New Mexico (Bunch et al. 2012).

Cr-rich spherules from Pilauco are unlike any known volcanic spherules, including those from two Peruvian volcanoes to the north (Huainaputina and Ubinas), whose spherules contain an average of only 0.1 wt% Cr_2O_3 (Wright and Hodge 1965), far below the average of 6.1 wt% Cr_2O_3 (max: 8.6 wt%) at Pilauco. Similarly, there are no known volcanic spherules reported in the literature that contain more than ~1.1 wt% Cr_2O_3 (Wright and Hodge 1965), compared to a minimum of 3.9 wt% in the Cr-rich spherules from Pilauco. Group 2 spherules also contain high concentrations of Fe that are unlike those found in other known volcanic spherules, including those from volcanoes in Hawaii, Iceland, Costa Rica, Peru and Japan. Those volcanic spherules are mainly composed of SiO_2, averaging 50.9 wt%; Al_2O_3, averaging 18.5 wt%; FeO, averaging 12.2 wt%; and CaO, averaging 10.0 wt%. The maximum amount of Fe present in volcanic spherules found in the above referenced five countries is 35.5 wt%, compared to the minimum amount of 74.3 wt% in Cr-rich spherules from Pilauco. These results suggest that even though Group 2 spherules did not result directly from volcanic eruptions but rather, are secondarily derived from volcanic material. The geochemical evidence indicates that Group 2 spherules are very similar to Cr-rich inclusions in volcanic material from volcanoes close to Pilauco, suggesting that those volcanoes the parent material of the Cr-rich spherules in the YDB layer.

The question arises of how Pilauco's Cr-rich spherules formed. After a volcanic eruption, chromium–magnetite can crystallize from molten magma to form Cr-rich inclusions at temperatures of less than ~1300 °C, the typical temperature of molten basalt (Rutherford and Devine 2008). However, much higher temperatures of ~1670–2160 °C are required to remelt these Cr-rich inclusions once they have formed (McEwan et al. 2011; Kracek 1942). Because these temperatures are higher than those of erupting magma, the Cr-rich spherules at Pilauco could not have been produced by an eruption, but rather formed through some nonvolcanic process. The high temperatures required to produce the Cr-rich spherules are limited to only a few processes, such as anthropogenesis, lightning and cosmic impact (Bunch et al. 2012). Of those, anthropogenesis is highly unlikely because we found no similar Cr-rich spherules at the surface, where anthropogenic spherules have been found at Pilauco (Pino et al. 2019). Lightning strikes are also a highly unlikely source because lightning activity should be a common occurrence in all strata. Instead, these spherules are found only in the 12,800-year-old YDB layer, suggesting that they were not generated by lightning.

Even though pieces of Cr-rich basalt are found in every sedimentary sample at Pilauco, Cr-rich spherules occur only in association with ~12,800-year-old Cr-rich and Fe-rich YDB-like spherules (Pino et al. 2019). This suggests that Pilauco's Fe-rich spherules were produced during the multi-continental YDB impact event. However, Cr-rich spherules have never been reported previously in the YDB layer at any other site, suggesting that unique conditions occurred at or near Pilauco that did not occur at other YDB sites. One possible explanation for the unique presence of Cr-rich spherules is that one or more local YDB-age impacts occurred in basaltic terrain near Andean volcanoes and remelted the Cr-rich inclusions commonly found in local basalt, thus producing the Cr-rich spherules in the YDB layer. More research would be necessary to confirm this possibility.

Pilauco is adjacent to Andean volcanoes, and therefore, we investigated whether Ca- and Si-rich basaltic volcanic spherules resulted from local eruptions. Volcanic spherules are produced during the low-intensity eruptions of low-viscosity magmas, during which shards of molten tephra coalesce into Si-rich droplets with the same chemical composition as the erupting Si-rich magma (Heiken 1972). Volcanic spherules that form during such low-energy eruptions typically fall only onto the slopes of the volcano and not beyond. On the other hand, volcanic eruptions capable of distributing tephra across wide areas are violent, high-energy events driven by high gas pressures (Smit 1990). Such highly energetic eruptions produce angular tephra, rather than spherules like those found at Pilauco. For example, the largest known eruption in the last 5 million years, the highly explosive, 75,000-year-old Toba-lake eruption, distributed angular tephra as much as 2500 km away from the source volcano but did not produce any detectable spherules (Smit 1990). Thus, it is highly unlikely that volcanism accounts for the wide distribution of YDB spherules across multiple continents, representing ~10% of the planet (Bunch et al. 2012; Wittke et al. 2013).

Iron content is another important difference between the Fe-rich spherules in Group 1 and the Si-rich volcanic spherules in Group 3. Typical magma can contain up to ~13 wt% Fe by volume and can also contain non-spherulitic, high-Fe inclusions that exsolve slowly from the molten magma as it cools. However, spherules containing 85–100 wt% FeO cannot form from erupting magma composed of less than ~13 wt% iron. This is confirmed by the research of Wright and Hodge (1965), who reported finding ~150 volcanic spherules with Fe compositions that ranged from 4 wt% to a maximum of 50 wt% and averaging ~20 wt%. No spherules were found with compositions approaching the minimum iron concentration of >85 wt% found in Pilauco spherules.

Chromium content is another important difference between Pilauco's YDB spherules and volcanic spherules. Wright and Hodge (1965) reported analyses of ~380 basaltic spherules from volcanoes, including two volcanoes in Peru, Mount Ubinas and Mount Huainaputina, along with Kilauea (Hawaii), Irazu (Costa Rica) and Surtsey (Iceland). The volcanic spherules they analysed contained an average of <0.1 wt% Cr_2O_3 with a range of 0.00–0.73 wt%, far below the concentrations of 4–14 wt% Cr found in Group 2 spherules from Pilauco. These results further confirm that Pilauco's Cr-rich spherules are nonvolcanic in origin.

On the other hand, the compositions of the known volcanic spherules are consistent with the compositions of Pilauco's Group 3 spherules, which closely match the SEM-EDS analyses of the vesicular basalt from Pilauco, thus confirming their identification as likely volcanic spherules.

Framboids often are associated with high-temperature spherules in the YDB layer at other sites (LeCompte et al. 2012; Bunch et al. 2012; Wittke et al. 2013; LeCompte et al. 2017), but because they grew slowly, they cannot be the direct result of the proposed impact event. Instead, they are most likely the result of impact-related climate change, during which environmental degradation sometimes triggered hypoxic conditions that promoted the growth of these framboids (Wittke et al. 2013).

The Pt concentration of 9.9 ppb is more than 3 × higher than average Pilauco background values of 2.7 ppb (range: 0.1–5.4 ppb), which are higher than background at most other platinum-rich YDB sites (Moore et al. 2017). That would possibly because Pilauco is close to Andean volcanoes and platinum is a common component of volcanic material. Palladium and gold also reach large abundance peaks at 551 cm. We considered three alternate explanations for the high concentrations of platinum, palladium and gold in the YDB layer at Pilauco: (1) The enrichments are part of a lag deposit, which often is produced when water or wind action concentrate heavy metals, through an action that is much like panning for gold. However, the conditions that produce lag deposits are expected to be common through most stratigraphic profiles, and so, high platinum concentrations should be found outside of the YDB layer, as well. Instead, such enrichments only occur in the YDB layer at Pilauco and at nearly two dozen other YDB sites on four continents, making this explanation unlikely. In addition, lag deposits would have concentrated local platinum, palladium, and gold, and therefore, the ratios among those elements should be the same. It is the same above and below the YDB, but because the YDB has a distinctly different ratio, it is unlikely that the anomaly resulted from local minerals. (2) Alternatively, their co-occurrence may simply be coincidental. If so, peaks in platinum and magnetic spherules should co-occur at random in non-YDB layers. However, they do not. No such synchronous, co-occurrence has ever been found outside of the YDB layer at any site worldwide that is not known to result from a previous impact event. In addition, if the layer coincidentally formed, it would have done so from local minerals with the same elemental ratios, which is not the case. (3) The platinum enrichment is extraterrestrial and/or impact-related. If so, the anomalous ratios resulted from an influx of nonlocal, impact-related material that is enriched above local background in platinum, palladium and gold. Of the three options, the last is the only plausible one.

The presence of ~12,800-year-old peaks at Pilauco in magnetic spherules, platinum, palladium, gold and charcoal is consistent with impact-related evidence at more than 50 YDB sites across North America, northern South America, Europe and western Asia, suggesting that the YDB layer at Pilauco is directly related to the multi-continental YDB impact event.

The suggest that impacts occurred 12,800 years ago in basaltic terrain near Pilauco. This supports an expansion of the original YDB impact theory to include multiple airburst/impacts that were more widely distributed across the Earth's surface, implying possible global effects. The YDB impact proxy field now extends ~6000 km farther south of the nearest YDB site in Venezuela and 12,000 km south of the northernmost YDB site in Canada, a distance equaling 30% of Earth's circumference.

Chapters 3 and 9 (this volume) present the Pilauco charcoal record. In relation to biomass burning, other area research indicates that regional fires occurred on a regional scale, not just the local one (Pino et al. 2019). The evidence shows that the largest episode of biomass burning in the 2500-year record occurred at the YDB layer, coeval with the widespread YDB biomass-burning event recorded in >100 ice, lake, and terrestrial records on four continents. The beginning of the charcoal peak is synchronous with the deposition of high concentrations of YDB spherules, platinum, palladium and gold.

In summary, all evidence observed in this study of Pilauco is consistent with the proposed effects of the YDB impact event, and none contradicts the impact theory.

References

Anderson DG, Goodyear AC, Kennett J, West A (2011) Multiple lines of evidence for possible human population decline/settlement reorganization during the early Younger Dryas. Quat Int 242:570–583

Andronikov AV, Andronikova IE (2016) Sediments from around the lower Younger Dryas Boundary (SE Arizona, USA): implications from LA-ICP-MS multi-element analysis. Geogr Ann A: Series A, Phys Geogr 98:221–236

Andronikov AV, Lauretta DS, Andronikva IE, Maxwell RJ (2011) On the possibility of a late Pleistocene, extraterrestrial impact: LA-ICP-MS analysis of the Black Mat and Usselo horizon samples. In: Abstract 74th meteoritic society meeting, London UK

Andronikov AV, Subetto DA, Lauretta DS, Andronikova IE, Drosenko DA, Kuznetsov DD, Sapelko T, Syrykh LS (2014) In search for fingerprints of an extraterrestrial event: trace element characteristics of sediments from the lake Medvedevskoye (Karelian Isthmus, Russia). Dokl Earth Sci 457(1):819–823

Andronikov AV, Rudnickaitė E, Lauretta DS, Andronikova IE, Kaminskas D, Šinkūnas P, Melešytė M (2015) Geochemical evidence of the presence of volcanic and meteoritic materials in Late Pleistocene lake sediments of Lithuania. Quat Int 386:18–29

Andronikov AV, Van Hoesel A, Andronikova I E, Hoek WZ (2016a) Trace element distribution and implications in sediments across the Allerød–Younger Dryas boundary in the Netherlands and Belgium. Geogr Ann A: Ser A, Phys Geogr 98:325–345

Andronikov, AV Andronikova IE, Loehn CW, Lafuente B, Ballenger JA, Crawford GT, Lauretta DS (2016b) Implications from chemical, structural and mineralogical studies of magnetic microspherules from around the lower Younger Dryas boundary (New Mexico, USA). Geogr Ann A: Ser A, Phys Geogr 98:39–59

Baker DW, Miranda PJ, Gibbs KE (2008) Montana evidence for extra-terrestrial impact event that caused ice-age mammal die-off. In: Abstract of American geophysical union, spring meeting 2008

Beets C, Sharma M, Kasse K, Bohncke S (2008) Search for extraterrestrial osmium at the Allerod-Younger Dryas boundary. In: Abstracts AGU fall meeting, 15–19 December 2008

Bement LC, Madden AS, Carter BJ, Simms AR, Swindle AL, Alexander BJ, Fine S, Benamara M (2014) Quantifying the distribution of nanodiamonds in pre-Younger Dryas to recent age deposits along Bull Creek, Oklahoma Panhandle, USA. Proc Natl Acad Sci USA 111(5):1726–1731

Boslough M, Nicol K, Holliday V, Daulton TL, Meltzer D, Pinter N, Scott AC, Surovell T, Claeys P, Gill P, Paquay F, Marlon J, Bartlein P, Whitlock C, Grayson D, Jull AJT (2012) Arguments and evidence against a Younger Dryas impact event. In: Giosan L, Fuller DQ, Nicoll K, Flad RK, Clift PD (eds) Climates, landscapes, and civilizations. Geophysical monograph series, vol 198. American Geophysical Union, Washington, DC, pp 13–26

Bronk Ramsey C (2008) Deposition models for chronological records. Quat Sci Rev 27:42–60

Bronk Ramsey C (2009) Bayesian analysis of radiocarbon dates. Radiocarbon 51:337–360

Bunch TE, Hermes RE, Moore AMT, Kennett DJ, Weaver JC, Wittke JH, DeCarli PS, Bischoff JL, Hillman GC, Howard GA, Kimbel DR, Kletetschka G, Lipo CP, Sakai S, Revay Z, West A, Firestone RB, Kennett JP (2012) Very high-temperature impact melt products as evidence for cosmic airbursts and impacts 12,900 years ago. Proc Natl Acad Sci USA 109(28):11066–11067

Daulton T, Pinter N, Scott N (2010) No evidence of nanodiamonds in Younger-Dryas sediments to support an impact event. Proc Natl Acad Sci USA 107(37):16043–16047

Fayek M, Anovitz LM, Allard LF, Hull S (2012) Framboidal iron oxide: chondrite-like material from the Black Mat, Murray Springs, Arizona. Earth Planet Sci Lett 319:251–258

Firestone RB (2009) The case for the Younger Dryas extraterrestrial impact event: mammoth, megafauna, and Clovis extinction, 12,900 years ago. J Cosmol 2:256–285

Firestone RB, West A, Kennett JP, Becker L, Bunch TE, Revay ZS, Schultz PH, Belgya T, Kennett DJ, Erlandson JM, Dickenson OJ, Goodyear AC, Harris RS, Howard GA, Kloosterman JB, Lechler P, Mayewski PA, Montgomery J, Poreda R, Darrah T, Que Hee SS, Smith AR, Stich A, Topping W, Wittke JH, Wolbach WS (2007) Evidence for an extraterrestrial impact 12,900 years ago that contributed to the megafaunal extinctions and the Younger Dryas cooling. Proc Natl Acad Sci USA 104(41):16016–16021

Firestone RB, West A, Revay Z, Hagstrum JT, Belgya T, Smith AR, Que Hee (2010) Analysis of the Younger Dryas impact layer. J Sib Fed Univ Eng Technol 1(3):30–62

Haynes CV Jr, Boerner J, Domanik K, Lauretta D, Ballenger J, Goreva J (2010) The Murray Springs Clovis site, Pleistocene extinction, and the question of extraterrestrial impact. Proc Natl Acad Sci USA 107(9):4010–4015

Heiken G (1972) Morphology and petrography of volcanic ashes. Geol Soc Am Bull 83:1961–1988

Holliday V, Surovell T, Johnson E (2016) A blind test of the Younger Dryas impact hypothesis. PloS one 11:e0155470

Israde-Alcántara I, Bischoff JL, Domínguez-Vázquez G, Li H-C, DeCarli PS, Bunch TE, Wittke JH, Weaver JC, Firestone RB, West A, Kennett JP, Mercer C, Xie S, Richman EK, Kinzie CR, Wolbach WS (2012) Evidence from Central Mexico supporting the Younger Dryas extraterrestrial impact hypothesis. Proc Natl Acad Sci USA 109(13):E738–E747

Kinzie CR, Que Hee SS, Stich A, Tague KA, Mercer C, Razink JJ, Kennett DJ, DeCarli PS, Bunch TE, Wittke JH, Israde-Alcántara I, Bischoff JL, Goodyear AC, Tankersley KB, Kimbel DR, Culleton J, Erlandson JM, Stafford TW, Kloosterman JB, Moore AMT, Firestone RB, Aura Tortosa JE, Jordá Pardo JF, A, Kennet JP, Wolbach WS (2014) Nanodiamond-rich layer across three continents consistent with major cosmic impact at 12,800 cal BP. J Geol 122(5):475–506

Kennett DJ, Kennett JP West GJ, Erlandson JM, Johnson JR, Hendy IL, West A, Culleton BJ, Jones TL, Stafford TW Jr (2008) Wildfire and abrupt ecosystem disruption on California's northern Channel Islands at the Ållerød–Younger Dryas boundary (13.0–12.9 ka). Quat Sci Rev 27(27–28):2530–2545

Kennett DJ, Kennett JP, West A, Mercer C, Que Hee SS, Bement L (2009a) Nanodiamonds in the Younger Dryas boundary sediment layer. Science 323(5910):94

Kennett DJ, Kennett JP, West A, West GJ, Bunch TE, Culleton BJ, Erlandson JM, Que Hee SS, Johnson JR, Mercer C, Shen F, Sellers M, Stafford TW, Stich A, Weaver JC, Wittke JH, Wolbach WS (2009b) Shock-synthesized hexagonal diamonds in Younger Dryas boundary sediments. Proc Natl Acad Sci USA 106(31):12623–12628

Kennett JP, Kennett DJ, Culleton BJ, Aura Tortosa JE Bischoff JL, Bunch TE, Daniel IR, Erlandson JM, Ferraro D, Firestone RB, Goodyear AC, Israde-Alcántara I, Johnson JR, Jordá Pardo JF, Kimbe DR l, LeCompte MA, Lopinot NH, Mahaney WC, Moore AMT, Moore CR, Ray JH, Stafford TW, Tankersley KB, Wittke JH, Wolbach WS, West A (2015) Bayesian chronological analyses consistent with synchronous age of 12,835–12,735 Cal B.P. for Younger Dryas boundary on four continents. Proc Natl Acad Sci USA 112(32): E4344–E4353

Kracek FC (1942) Melting and transformation temperatures of mineral and allied substances. Geol Soc Am Spec Pap 36:139–174

Kurbatov AV, Mayewski PA, Steffensen JP, West A, Kennett DJ, Kennett JP, Bunch TE, Handley M, Introne DS, Que Hee DS, Mercer C, Sellers M, Shen F, Sneed SB, Weaver JC, JH, Stafford TW Jr, Donovan JJ, Xie S, Razink JJ, Stich A, Kinzie CR, Wolbach WS (2011) Discovery of a nanodiamond-rich layer in the Greenland ice sheet. J Glaciol 56:749–759

LeCompte MA, Goodyear SC, Demitroff MN, Batchelor D, Vogel RK, Mooney C, Rock BN, Seidel AW (2012) Independent evaluation of conflicting microspherule results from different investigations of the Younger Dryas impact hypothesis. Proc Natl Acad Sci USA 109(44):E2960–E2969

LeCompte MA, West A, Adededji AV, Demitroff M, Witwer T, Langenburg RA (2017) The Bowser Road Mastodon and the Younger Dryas impact hypothesis, Appendix, 3 of the archaeological recovery of the Bowser Road Mastodon. RM Gramly, Persimmon Press, Orange County, NY

Mahaney WC, Keiser L (2012) Weathering rinds—unlikely host clasts for an impact-induced event. Geomorphology 184:74–83

Mahaney WC, Milner MW, Kalm V, Dirszowsky RW, Hancock RGV, Beukens RP (2008) Evidence for a Younger Dryas glacial advance in the Andes of northwestern Venezuela. Geomorphology 96(1):199–211

Mahaney WC, Krinsley D, Kalm V (2010a) Evidence for a cosmogenic origin of fired glaciofluvial beds in the northwestern Andes: correlation with experimentally heated quartz and feldspar. Sediment Geol 231(1–2):31–40

Mahaney WC, Kalm V, Krinsley DH, Tricart P, Schwartz S, Dohm J, Kim KJ et al (2010b) Evidence from the northwestern Venezuelan Andes for extraterrestrial impact: the black mat enigma. Geomorphology 116(1–2):48–57

Mahaney WC, Krinsley DH, Kalm V, Kurt L, Ditto J (2011a) Notes on the black mat sediment, Mucuñuque catchment, northern Mérida Andes, Venezuela. J Adv Microscop Res 6(3):177–185

Mahaney WC, Krinsley D, Langworthy K, Kalm V, Havics T, Hart KM, Kelleher BP (2011b) Fired glaciofluvial sediment in the northwestern Andes: biotic aspects of the Black Mat. Sediment Geol 237(1–2):73–83

Maiorana-Boutilier A, Mitra S, Norwood M, Louchouarn P, Bischoff J, Silva S, West A, Kennett J (2016) Organic composition of Younger Dryas Black Mat. In: Abstract of GSA conference, Southeastern Section, 65th annual meeting, 31 March–1 April 2016

Marlon JR, Bartlein PJ, Walsh MK, Harrison SP, Brown KJ, Edwards ME, Higuera PE, Power MJ, Anderson RS, Briles C, Brunelle A, Carcaillet C (2009) Wildfire responses to abrupt climate change in North America. Proc Natl Acad Sci USA 106:2519–2524

Marshall W, Head K, Clough R, Fisher A (2011) Exceptional iridium concentrations found at the Allerød-Younger Dryas transition in sediments from Bodmin Moor in southwest England. In: Abstract of XVIII INQUA-congress, Bern, Switzerland, 21–27 July 2011

McEwan N, Courtney T, Parry R, Knupfer P (2011) Chromite—a cost-effective refractory raw material for refractories in various metallurgical applications. In: Southern African pyrometallurgy, pp 359–372
Melott AL, Thomas BC, Dreschhoff G, Johnson CK (2010) Cometary airbursts and atmospheric chemistry: Tunguska and a candidate Younger Dryas event. Geology 38:355–358
Moore CR, West A, LeCompte MA, Brooks MJ, Daniel IR Jr, Goodyear AC, Ferguson TA, Ivester AH, Feathers JK, Kennett JP, Tankersley KB, Adedeji AV, Bunch TE (2017) Widespread platinum anomaly documented at the Younger Dryas onset in North American sedimentary sequences. Sci Rep 7:44031
Napier WM (2010) Palaeolithic extinctions and the Taurid complex. Mon Not R Astron Soc 405:1901–1906
Paquay FS, Goderis S, Ravizza G, Vanhaeck F, Boyd M, Surovell TA, Holliday VT, Haynes CV Jr, Claeys P (2009) Absence of geochemical evidence for an impact event at the Bølling–Allerød/Younger Dryas transition. Proc Natl Acad Sci USA 106(51):21505–21510
Petaev MI, Huang S, Jacobsen SB, Zindler A (2013) Large Pt anomaly in the GISP2 ice core points to a cataclysm at the onset of Younger Dryas. Proc Natl Acad Sci USA 110(32):12917–12920
Pigati JS, Latorre C, Rech JA, Betancourt JL, Martinez KE, Budahn JR (2012) Accumulation of impact markers in desert wetlands and implications for the Younger Dryas impact hypothesis. Proc Natl Acad Sci USA 109(19):7208–7212
Pino M, Martel-Cea A, Astorga G, Abarzúa AM, Cossio N, Navarro X, Lira MP, Labarca R, Lecompte MA, Adedeji V, Moore CR, Bunch TE, Mooney C, Wolbach WS, West A, Kennett JP (2019) Sedimentary record from Patagonia, southern Chile supports cosmic-impact triggering of biomass burning, climate change, and megafaunal extinctions at 128 ka. Sci Rep 9(1):4413
Pinter N, Scott AC, Daulton TL, Podoll A, Koeberl C, Anderson RS, Ishman SE (2011). The Younger Dryas impact hypothesis: a requiem. Earth-Sci Rev 106:247–264
Redmond BG, Tankersley KB (2012) Species response to the theorized Clovis Comet impact at Sheriden Cave, Ohio. Curr Res Pleistocene 28:141–143
Rutherford MJ, Devine JD (2008) Magmatic conditions and processes in the storage zone of the 2004–2006 Mount St. Helens dacite. A volcano rekindled: the renewed eruption of Mount St. Helens 2006. US Geological Survey No. 1750–31, pp 703–725
Scott AC, Pinter N, Collinson ME, Hardiman M, Anderson RS, Brain APR, Smith SY, Marone F, Stampanoni M (2010) Fungus, not comet or catastrophe, accounts for carbonaceous spherules in the Younger Dryas "impact layer". Geophys Res Lett 37(14):L14302
Sharma M, Chen C, Jackson BP, Abouchami W (2009) High resolution Osmium isotopes in deep-sea ferromanganese crusts reveal a large meteorite impact in the Central Pacific at 12 $\pm$ 4 ka. In: Abstract of American geophysical union, fall meeting, 14–18 December 2009
Smit J (1990) Meteorite impact, extinctions and the Cretaceous-Tertiary boundary. Geol Mijnbouw 69:187–204
Surovell T, Holliday VT, Gingerich JM, Ketron C, Haynes CV, Hilman I, Wagner DP, Johnson E, Claeys P (2009) An independent evaluation of the Younger Dryas extraterrestrial impact hypothesis. Proc Natl Acad Sci USA 104:18155–18158
Tian H, Schryvers D, Claeys P (2011) Nanodiamonds do not provide unique evidence for a Younger Dryas impact. Proc Natl Acad Sci USA 108(1):40–44
van Hoesel A, Hoek WZ, Braadbaart F, van der Plicht J, Pennock GM, Drury MR (2012) Nanodiamonds and wildfire evidence in the Usselo horizon postdate the Allerød-Younger Dryas boundary. Proc Natl Acad Sci USA 109(2):7648–7653
Van Hoesel A, Hoek WZ, Pennock GM, Drury MR (2014) The Younger Dryas impact hypothesis: a critical review. Quat Sci Rev 83:95–114

Wittke JH, Weaver JC, Bunch TE, Kennett JP, Kennett DJ, Moore AMT, Hillman GC, Tankersley KB, Goodyear AC, Moore CR, Daniel IR Jr, Ray JH, Lopinot NH, Ferraro D, Israde-Alcántara I, Bischoff JL, DeCarli PS, Hermes RE, Kloosterman JB, Revay Z, Howard GA, Kimbel DR, Kletetschka G, Nabelek L, Lipo CP, Sakai S, West A Firestone RB (2013) Evidence for deposition of 10 million tonnes of cosmic impact spherules across four continents 12,800 years ago. Proc Natl Acad Sci USA 110 (23):E2088–E2097

Wolbach WS, Ballard JP, Mayewski PA, Adedeji V, Bunch TE, Firestone RB, Timothy A. French TA, Howard GA, Israde-Alcántara I, Johnson JR, Kimbel D, Kinzie Chr, Kurbatov A, Kletetschka G, LeCompte MA, Mahaney WC, Melott AL, Mitra S, Maiorana-Boutilier A, Moore CR, Napier WM, Parlier J, Tankersley KB, Thomas BC, Wittke JH, West A, Kennett JP (2018a) Extraordinary biomass-burning episode and impact winter triggered by the Younger Dryas cosmic impact ~12,800 years ago; Part 1: ice cores and glaciers. J Geol 126 (2):165–184

Wolbach WS, Ballard JP, Mayewski PA, Parnell AC, CNahill, Adedeji V, Bunch TE, Domínguez-Vázquez G, Erlandson JM, Firestone RB, French TA, Howard GA, Israde-Alcántara I, Johnson JR, Kimbel D, Kinzie Chr, Kurbatov A, Kletetschka G, LeCompte MA, Mahaney WC, Melott AL, Mitra S, Maiorana-Boutilier A, Moore CR, Napier WM, Parlier J, Tankersley KB, Thomas BC, Wittke JH, West A, Kennett JP (2018b) Extraordinary biomass-burning episode and impact winter triggered by the Younger Dryas cosmic impact ~12,800 years ago; Part 2: lake, marine, and terrestrial sediments. J Geol 126(2):185–205

Wright FW, Hodge PW (1965) Studies of particles for extraterrestrial origin: 4 Microscopic spherules from recent volcanic eruptions. J Geophys Res 70:3889–3898

Wu Y, Sharma M, LeCompte MA, Demitroff M, Landis JD (2013) Origin and provenance of spherules and magnetic grains at the Younger Dryas boundary. Proc Natl Acad Sci USA 110(38):E3557–E3566110

Chapter 16
The Cultural Materials from Pilauco and Los Notros Sites

Ximena Navarro-Harris, Mario Pino and Pedro Guzman-Marín

Abstract Investigations of the first South American settlements have revealed great environmental diversity, to which the first humans adapted as they colonized the continent. In this chapter we focus mainly on the description and interpretation of a set of expeditive lithics, from the sites Pilauco and Los Notros, knapped in local and exotic rocks ascribed to the unifacial edge-trimmed tradition. The lithic assemblage is composed of 140 knapped and 7 non-knapped stones that were classified as cores, knapped pebbles, and other pieces. To determine the provenance of the raw materials used for the production of archaeological pieces we integrated geological and geochemical analyses of archaeological and non-archaeological samples. Additionally, we describe and interpret other cultural materials present at the sites. The most likely source of the rhyodacitic and dacitic glass artifacts is the Puyehue-Cordón Caulle Volcanic Complex. The sites show clear evidence of a chain of production tasks, suggesting that most of the work process was executed and maintained in the same place. Thus, the early cultural context of Pilauco and Los Notros sites is supported by the presence of an operative chain production, but also by a clear stratigraphic record and precise age dating.

Keywords Archaeology · Dacitic glass · Pilauco · Los Notros

X. Navarro-Harris (✉)
Departamento de Antropología, Universidad Católica de Temuco, Temuco, Chile
e-mail: ximenavaharris@gmail.com

X. Navarro-Harris · M. Pino
TAQUACH, Transdisciplinary Center for Quaternary Research in the South of Chile, Universidad Austral de Chile, Valdivia, Chile

M. Pino · P. Guzman-Marín
Instituto de Ciencias de la Tierra, Universidad Austral de Chile, Valdivia, Chile

M. Pino and G. A. Astorga (eds.), *Pilauco: A Late Pleistocene Archaeo-paleontological Site*, The Latin American Studies Book Series,
https://doi.org/10.1007/978-3-030-23918-3_16

16.1 Background

The paleo-archaeological site of Pilauco is located in the Central Depression of the Lake Region in Chile (40°34′S, 73°06′W, 30 m a.s.l.) and can be geographically assigned to the northern Patagonian region. The altitude of the northern limit of the site offers a wide view of the landscape, which is dominated by the volcanic chain of the Southern Andes towards the east, the Central Depression around the city of Osorno, and the Osorno Volcano towards the southwest (Fig. 16.1).

The Lake Region of southern Chile possesses two sites exhibiting the oldest records of the association humans and extinct megafauna of late Pleistocene age in the country. The Monte Verde II site (41°31′S, 73°17′W) is located 21 km WSW from the city of Puerto Montt, and the cultural and fossil findings at the site provide age constrains between 12,310 ± 40 and 12,290 ± 60 ^{14}C yr BP (c. 14,600 cal yr BP) (Dillehay 1989, 1997, 2008, 2009). The Pilauco site is located 100 km to the north, in the city of Osorno and reveals fossil materials dating between 10,660 ± 30 and 13,650 ± 70 ^{14}C yr BP, 16,400–12,600 cal yr BP, respectively. Other smaller fossil-bearing sites that present remains of extinct megafauna in the surroundings of Osorno include Fundo Lomas Blancas, Fundo Los Pinos, Nochaco, and Mulpulmo (Weischet 1958, 1959; Heusser 1966; Seguel and Campana 1975; Recabarren et al. 2014). None of these last four sites has been systematically studied, although there are some radiocarbon age data on gomphothere tusks from Nochaco and Mulpulmo with ages of 16,150 ± 750 and 18,700 ± 900 ^{14}C yr BP (I-1249 and I-125, respectively; Heusser 1966). To the best of our knowledge, these are so far the first radiocarbon data on gomphotheres in Chile.

Fig. 16.1 Southwest view from the elevated northern limit of the Pilauco site dominated by the Osorno and Calbuco volcanos (from left to right) corresponding to the Southern Volcanic Zone and the floodplain of the Damas and Rahue Rivers (Central Depresion) surrounded by the city of Osorno in the foreground

Monte Verde II (MV-II) is one of the few early sites with unambiguous archaeological evidence, which outlines a context marked by high technological diversity and residential complexity in southern Chile. For more than two decades this evidence was considered controversial. Nevertheless, currently it is widely accepted as an essential contribution to our knowledge regarding the first human settlements during the late Pleistocene in Patagonia and backed up by rigorous investigations over the last several years. These first settlers were apparently different from the "Paleo-Indians", a term that was probably first introduced by Roberts (1940). The recorded materials from MV-II hints towards the knowledge and living strategies of the early inhabitants of this forest-dominated zone and coastal region in southern Chile. The study of this site has identified an assemblage of lithics and technologies, so far unknown from this area, including two fragmented bifacial tips similar to the El Jobo type (Cruxent and Rouse 1956; Gnecco and Aceituno 2004). The MV-II site also revealed numerous unifacial lithics and unformatted artifacts from local rock types, as well as medium- to small-sized spheres shaped by means of cultural abrasion. The taphonomic processes in MV-II favored the extraordinary preservation of organic material such as bone and wood artifacts, knots from a rope made of twisted *Juncus*, anthropic selection of seeds, and medicinal plants, as well as some rudimentary domestic architecture (Dillehay 1997, 2009; Dillehay et al. 2008). The site Monte Verde II gave rise to a global paradigm in the discussion about the first settlements and associated lifestyle, characterized by an economy closely related to plant-based alimentary and medicinal resources. At the same time, it places the challenging task for comparison with the contemporaneous Pilauco site.

The last two decades are marked by several new discoveries originating from the re-evaluation of early South American settlements, which revealed a record of the great environmental diversity, to which the first hunters and collectors adapted when colonizing the continent. The early settlements in South American territory required a variety of strategies for the successful occupation of environments, such as tropical or cold-temperate rain forests (Gnecco and Aceituno 2004; López 2008; Dillehay et al. 2012a; Aceituno et al. 2013).

The debate about the validity of the archaeological contexts of early human settlements is well represented by Sistiaga et al. (2014), who stated: "although some claims have been made for pre-Clovis sites in North and South America, the putative earliest sites that are typically mentioned are still controversial". The skeptical attitude towards the "Clovis-first" hypothesis (Martin 1973) and resulting discussions are not focused on sites with scarce records of bifacial technology from the Holocene (younger than 11,700 cal yr BP; Fairbridge 1982; Cohen and Gibbard 2016),[1] and neither on those corresponding to the Younger Dryas chronozone (12,835–12,735 to 11,700 cal yr BP; Kennett et al. 2015), or the Clovis age (Waters and Stafford 2007, 13,125–12,925 cal yr BP). Instead, the debate arises around earlier sites on the American continent such as Monte Verde I, Monte Verde II, and some others dated between 19,000 and 14,500 cal yr BP (Dillehay et al. 2015 and references therein). The particular characteristics of the late Pleistocene climate have been considered

[1] https://dspace.library.uu.nl/handle/1874/354865.

responsible for the large dispersion of radiocarbon ages of different materials. Nevertheless, these ages converge after calibration, both in the Northern and Southern Hemispheres (Kelly 2003). In North America there are some emblematic and highly discusses pre-Clovis sites such as for instance, Topper, Debra L. Friedkin, Meadowcroft Rockshelter and Cactus Hill sites (Goodyear 2005; Waters and Stafford 2007; Waters et al. 2011; Jenkins et al. 2013; Collins 2014; Bourgeon et al. 2017).

The South American archaeological context includes several sites in Argentina (e.g., Miotti 2004; Paunero et al. 2007), Brazil (e.g., Parenti 2001; Aimola et al. 2014; Boëda et al. 2014; Neves and Hubbe 2005; Vialou et al. 2017), Chile (e.g., Núñez et al. 1994; Jackson et al. 2007; Pino et al. 2013; Latorre et al. 2013), Peru (e.g., Sandweiss et al. 1998; Dillehay et al. 2012a, b; Rademaker et al. 2014), and Uruguay (e.g., Fariña and Castillo 2007; Suárez 2018; Suárez et al. 2018). Most of these old sites contain a variety of lithic assemblages (Dillehay et al. 2015), whereas some them only present bones of megafauna with cutting marks that might be the result of human activity.

Expeditive lithics are generally present on early South American archaeological sites, but it is generally a challenging task to establish an unambiguous correlation with the extinct megafauna. Furthermore, due to their expeditive character, the cultural materials are usually scarce and lack diagnostic features. The debate around these materials and the arising controversy of their archaeological context originate from the insufficient evidence for their cultural assignment, particularly when a singular unifacial lithic technology is present and/or the corresponding radiocarbon ages are ambiguous (Fiedel 2000). The unifacial technology has been interpreted as serving for the preparation of tools designed for simple tasks of short duration, whereas other, more complex activities of the early hunters and collectors apparently required different techniques.

It is necessary to note something that has been very little discussed. In the site MV-II there is a wide range of artifacts made of wood, including spears with fire-hardened tips (Dillehay 1997). If the records, at any site, were only represented by expeditive artifacts and those made of wood were not preserved, this may hinder the possibility of an integral interpretation of the settlements. Furthermore, it would reduce the possibility of understanding human behaviors related to mobility patterns (foragers and collectors; Binford 1987) and the occupational strategies of these environments in the late Pleistocene (Aceituno and Loaiza 2010; Aceituno et al. 2013; Borrero et al. 2015).

The archaeological literature in America emphasizes the presence of bifacial technologies as one of the unambiguous evidences for human activity in the past. Nevertheless, there are several characteristics of both uni- and bifacial technologies that need to be considered in the archaeological context of their interpretation. The bifacial technology represented by fishtail projectile points is generally absent on sites older than 12,900 cal yr BP (Suárez 2018). The unifacial technology generally receives less attention, although it is frequently found and numerically dominant in early sites and demonstrates greater morphological variability. These unifacial techniques need to be fully considered and interpreted in detail in order to close existing gaps in our understanding of the first American societies and their behavior (Binford 1980, 1987; Aceituno and Loaiza 2010; Dillehay et al. 2012a, b).

Furthermore, it is important to draw the research attention towards a better understanding how the mobility strategies of the first settlers aimed to obtain raw materials for the production of lithic tools. Understanding the selection of rock types from local or distant origin, as well as the usage, discarding and/or maintenance of lithic artifacts would probably open the way for new hypotheses considering the recurrent and representative appearance of unifacial instead of bifacial technologies extending across the late Pleistocene—early Holocene transition (Dillehay 2000; López 2008). According to Dillehay et al. (2008), the uni- and bifacial technologies on the American continent represent different adaptations to contrasting eco-cultural conditions, and likely derive from an earlier, and so far, undefined, technology from East Asia.

The assemblage of expeditive and simple lithic remains is commonly referred to as "unifacial edge-trimmed tradition" (Dillehay 2000; Bryan and Gruhn 2003) and generally corresponds to unspecialized artifacts. The low-cost production of these tools likely implied reduced fabrication times of flakes and artifacts from raw materials with local origin that were discarded after usage directly at the working place (Aceituno and Loaiza 2010). Archaeological studies from Brazil also present a record of early stable occupations by groups of hunters and collectors who apparently visited the same places in a recurrent manner and had a lifestyle based on general subsistence strategies (Schmidt Dias 2004; Böeda et al. 2014; Vialou et al. 2017). An increased occurrence of expeditive unifacial technology has been documented in Colombia in artifacts knapped from local raw materials such as quartz and chert (López 2008; Dillehay 2000). Finally, the basal stratigraphic level at the Huaca Prieta site in Peru also reveals an assemblage of flakes and artifacts derived from the unifacial edge-trimmed tradition (Dillehay et al. 2012a, b).

The faunal record at Pilauco matches a not frequent late Pleistocene pattern of coexisting megafauna and extant micro-mammals (González et al. 2010, 2014, 2017, 2018; Recabarren et al. 2011, 2014, 2015; Pino et al. 2013; Labarca et al. 2013, 2014, 2016). The pollen record from layers PB-7 and PB-8 indicates a northern Patagonian landscape conformed by conifers and Myrtaceae forests along with wide grass pastures that provided the food source for major Pleistocene herbivores (Abarzúa et al. 2016).

Gomphotheres in Chile have been generally considered as particularly flexible mammals regarding their diet and are characterized as latitude-independent leaf-browsers, while these animals modeled the forest composition as landscape architects (González et al. 2018). In terms of their trophic discrimination (diet-tissue from $\delta^{15}N$ collagen), gomphothere bones from Pilauco do not match modern vegetation. The difference between the gomphotheres from Pilauco and those from other sites in the region between 38° and 42°S is apparently not related to climatic gradients, but rather to other factors such as wild fires, pasture intensity, coprophagy, or the fertilization of vegetation along regular migratory routes (González et al. 2018).

Some gomphothere bones exhibit teeth marks from a large carnivore (Labarca et al. 2014; see Chaps. 4 and 8, this volume). The analysis of fossil beetles from Pilauco indicates a rather dry wetland, at least on a seasonal basis (Tello et al. 2017; see Chap. 12, this volume), whereas sedimentological and taphonomic analyses from layers PB-7 and PB-8 indicate notable colluvial contribution. Only two gomphothere exemplars are preserved in situ, whereas the remaining gomphothere bones and remnants from other extinct and extant species are generally fragmented and isolated, which is concordant with transport by colluvial processes from the northern topographic high limiting the site (see Chaps. 3 and 8, this volume). Most of the fossil Pleistocene fauna and preserved cultural materials at Pilauco originate from layer PB-7. The overlying PB-8 layer is very similar in terms of sedimentological characteristics, but the amount of lithic materials and fossil fauna is significantly lower and accompanied by a total disappearance of gomphothere remains.

The different paleo-environmental analyses performed at Pilauco are generally coherent and identify a variety of living spaces where the first settlers took advantage of the present fauna during a period of challenging environmental changes. The present chapter focuses on the characterization and interpretation of cultural materials from Pilauco (with emphasis on knapped lithics) and aims to provide new insights and contribute to the present debate and controversies around early South American settlements.

16.2 The Archaeological Method

The Pilauco and Los Notros records derive from archaeological excavations of the two fossil-bearing stratigraphic layers PB-7 and PB-8. All excavation grids with archaeological remains have been documented by ground plotting of isolated specimens or assemblages of materials. This allows to map the position of each object within an overlaying grid and reconstruct the spatial distribution of cultural material and fossil fauna. All samples (both from organic and lithic origin) were listed in a single file following Renfrew and Bahn (1993).

The selection of knapped lithic artifacts from Pilauco and Los Notros sites implied the discarding of hundreds of cobbles with natural fracturing due sedimentary processes. The selection criteria followed Collins (1997) and considered the presence of continuous weathering surface covering the entire cobble or distribution of fractures along joint planes, among others. The final selection of material was performed in the laboratory and based on morphological and technological features that allowed distinguishing between those related to natural processes and manufactured specimens. Finally, the lithic artifacts were described and classified according to raw materials and morphological and technological criteria, such as their form (basic or derived from a core), dimensions (total length and width, maximum thickness of each specimen), size-reduction steps (primary, secondary, tertiary), degree of conservation (entire or broken), percentage of cortex, number of negatives on the dorsal side, shape of the active edges and angle of modified blades (natural or reshaped) (Inizan et al. 1999; Odell 2004; Andrefsky 1998, 2000).

The raw materials of all cores, artifacts, flakes, and debitage were analyzed to determine their local or exotic origin. To identify the raw materials suitable for knapping, a random sampling was performed in the layer PB-6 in the grids located in the southeastern part of the site. Sampling and classification of rock types were also executed in the natural surroundings from the Pacific coast in the west to the Andean foothills in the east, encompassing the catchment area of the Damas River and the beach region of Puyehue Lake (see Chap. 3, this volume). We applied a multidisciplinary approach integrating geological and geochemical analyses to study the provenance of materials used at the site. An ICP-MS (Inductively Coupled Plasma Mass Spectrometry) analysis was performed to characterize the composition of trace elements in artifacts, debitage, and rock samples from the surroundings.

16.3 The Radiocarbon Context of the Archaeology

The fifty-five stratigraphic radiocarbon ages obtained from the site indicate that the oldest layer bearing fossil fauna is PB-7. This stratigraphic horizon is temporally constrained between 13,650 $\pm$ 70 ^{14}C yr BP (16,681–16,155 cal yr BP, 2σ) at the base and 12,173 $\pm$ 42 ^{14}C yr BP (13,815–14,148 cal yr BP) at the top. The PB-8 layer shows decreased concentration of fossil megafauna and is constrained between 12,035 $\pm$ 50 ^{14}C yr BP (14,026–13,726 cal yr BP) at the base and 10,660 $\pm$ 30 ^{14}C yr BP (12,706–12,625 cal yr BP) at the top (see Chap. 3, this volume).

In Pilauco, there are two main areas with extinct and extant fossil fauna and cultural materials. The spatial and temporal distribution of most of the bone remains from two well-articulated gomphotheres and chipped stones are preserved mainly in two assemblages in the NE and SW areas. The latter are scarcer in the NE-area where they are represented by eleven artifacts between 400 and 430 cm of local elevation. The north-eastern area was excavated between 2007 and 2010, where a gomphothere bone, two horse teeth, and one tibia from cf. *Hemiauquenia paradoxa* provide age constraints of 12,540 $\pm$ 90, 12,035 $\pm$ 50, 11,457 $\pm$ 140, and 11,320 $\pm$ 90 ^{14}C yr BP, respectively (Figs. 16.2 and 16.3; grids 11H, 7G and 12G; see Chap. 1 and 3, this volume).

The SW-area bears two diachronic sectors—SW-T (top) and SW-B (base)—where most of the megafauna bones have been recovered and dated. Eight gomphothere bones provide ages constraints between 13,260 $\pm$ 70 and 12,725 $\pm$ 40 ^{14}C yr BP (grids 14AB, 14AC, 15AC, 15AD, 16AD, and 17AC; Figs. 16.2 and 16.3). The SW-B sector corresponds to the base of layer PB-7 between 330 and 385 cm of local elevation and displays most of the artifacts obtained from the site (for a taphonomic discussion see Chap. 8, this volume). The upper SW-T sector corresponds to layer PB-8 (435–445 cm of local elevation) and bears a group of lithic artifacts associated with bone remains of *Equus* (Fig. 16.4). Most of the lithics (95%) from both sectors show reduction by knapping along one of their lateral faces (unifacial) by initial and simple reduction, which suggests the dominance of primary and secondary debitage.

A series of independent dates on bulk sediment, wood and other plant materials confirm the age of the fossil bones (Fig. 16.3). Therefore, the age range of the lithic artifacts in the NE area and the SW-T sector varies between 14,600 and 13,500 cal yr BP, and between 16,400 and 15,500 cal yr BP., respectively. Furthermore, in terms of their topographic location and stratigraphic features, the NE area and SW-T sector clearly differ from the SW-B sector.

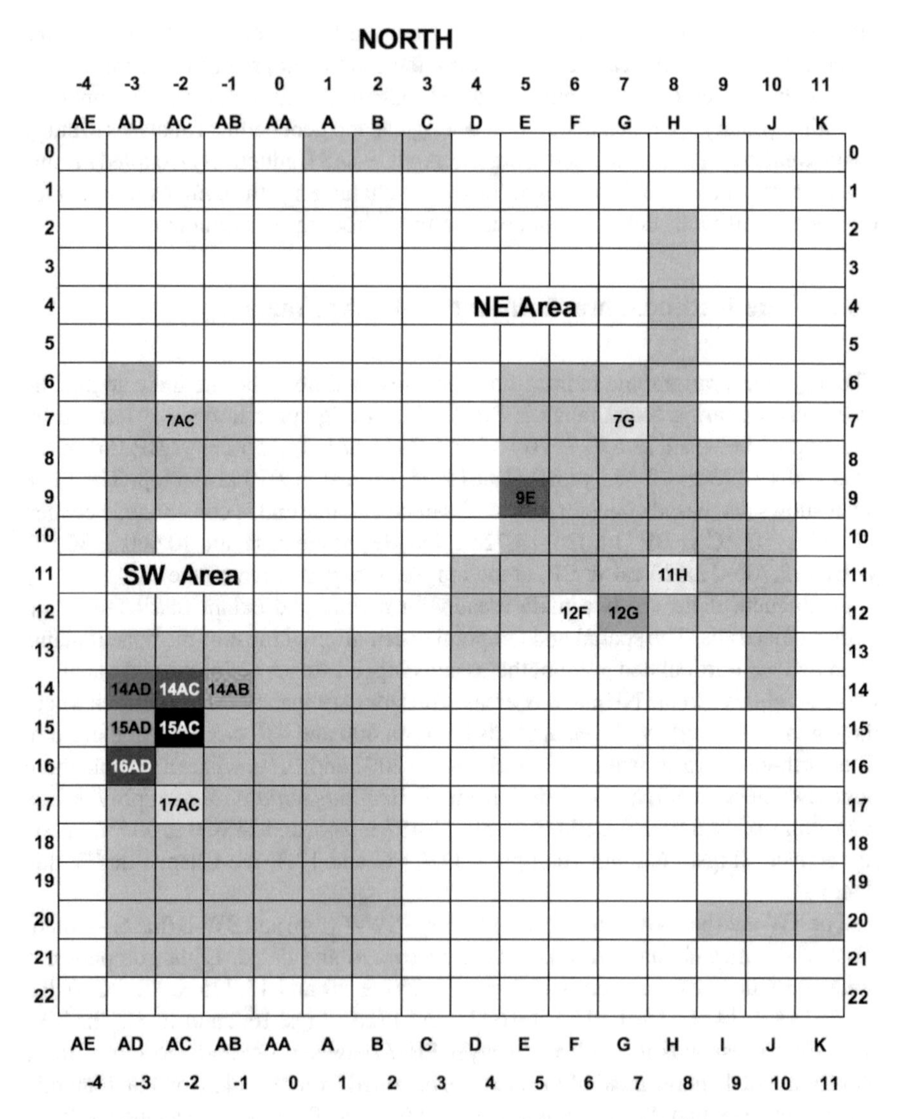

Fig. 16.2 Location of grids with radiocarbon ages from Pilauco site. The remaining grids are outlined in light brown. The color-coding of numbered grids correlates with data in Fig. 16.3

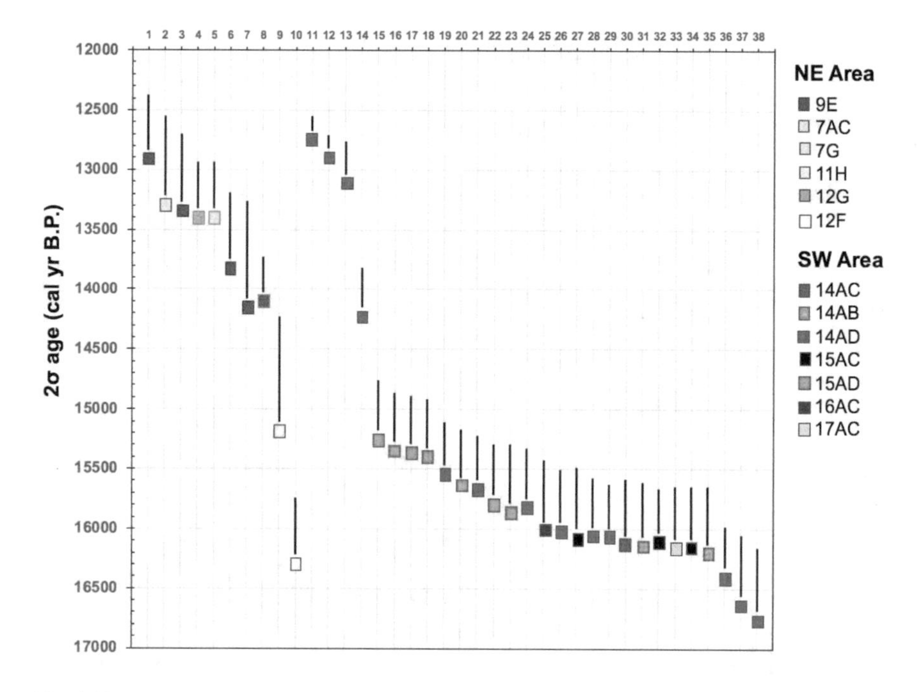

Fig. 16.3 Calibrated age range BP (2σ, 98.4%) in the NE and SW areas for stratigraphic layers PB-7 and PB-8. The red outlines of the squares correspond to ages from gomphothere bones. Data number nine corresponds to the first age dating in Pilauco from 2006 (note the large 2σ). For the spatial distribution of the data see Fig. 16.2

16.4 The Geochemistry of the Knapped Lithics

Similar to other early settlements, the lithics from this assemblage are expeditive. There are few exceptions that feature more advanced carving and processing along their margins and edges. The assemblage of lithics is formed by 140 knapped and 7 non-knapped stones shaped by battering, pecking, grinding or polishing (Collins 1997). The assemblage of non-naturally shaped lithics is formed by 53 specimens from the SW area of Pilauco that also represents the earliest level of occupation at the site.

The analysis of 44 archaeological samples from layers PB-7 and PB-8 identifies four different groups based on the proportions of trace elements (Stern 2018; Navarro-Harris et al. 2019). Group A shows the lowest content of Ti, Mn, V, and Sr and highest content of Rb, Y, Zr, Nb, Cs, Ba, Hf, Pb, Th, U, and rare earth elements (REE). Specimens from Group A correspond to the most acidic (high-SiO_2) rock samples, probably rhyodacite and rhyolite. In turn, Samples from Group D are characterized by the highest content of Ti, V, Mn, and Sr, and the lowest content of Rb, Y, Zr, Nb, Cs, Ba, Hf, Pb, Th, U, and REE, and correspond to mafic compositions between

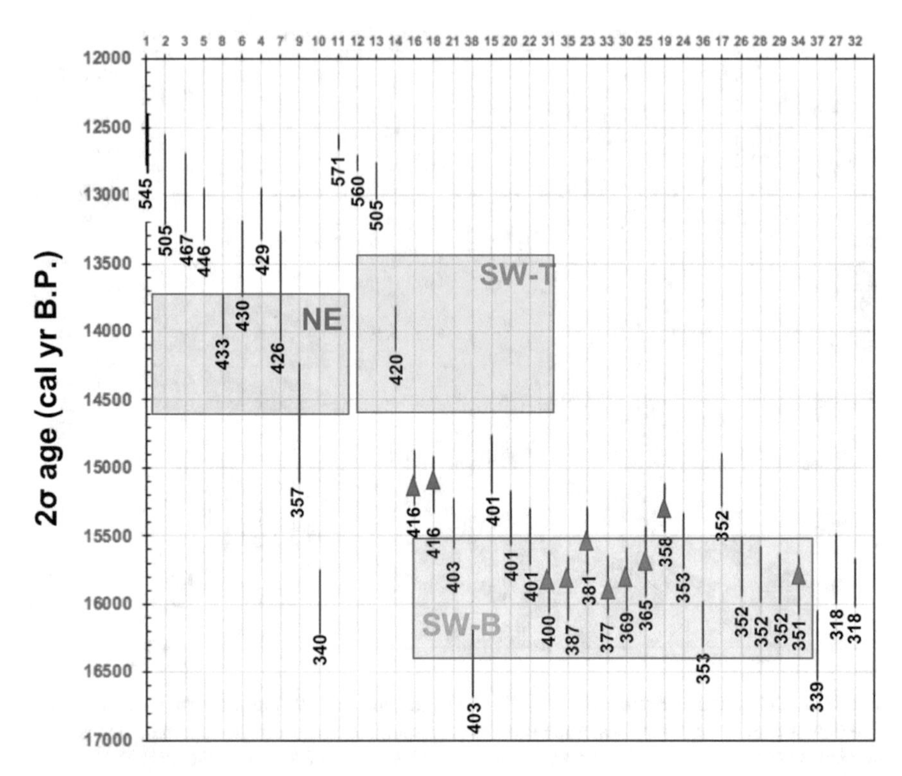

Fig. 16.4 Calibrated age range BP (2σ, 98.4%) according to area and subarea. The numbers indicate the local elevation of dated specimens. Blue triangles indicate gomphothere bones. The NE, SW-T, and SW-B rectangles outline indicate artifact assemblages located close to gomphotheres or horse bones

basalts and andesitic basalts. Specimens from groups B and C are characterized by intermediate trace element compositions corresponding to andesites or dacites.

However, in the Nb/Y versus Zr/Ti diagram the former chemical classification is only generally maintained, for example for the andesitic and andesitic-basaltic samples (group D, samples 36–44; Fig. 16.5). The group C matches the limit between dacite and andesite (29–35), whereas the group B corresponds to the field of the rhyodacite and dacite (25–28). In turn, group A is characterized by a gradual change in composition between rhyolite (samples 1 to 7, 14, 15, 17, 19, 23, and 24), rhyodacite (samples 8 to 13 15, 16, 18, 20 to 22, 24, and 27), and dacite (25, 26, 28, and 32–35) (Table 16.1; Figs. 16.5, 16.6, 16.7, 16.8 and 16.9). It terms of petrography, all rhyolitic, rhyodacitic, and dacitic samples correspond to volcanic glasses (obsidian and rhyodacitic and dacitic glasses, respectively). This group of lithics from Pilauco documents the first South American late Pleistocene artifacts and debitage from rhyodacitic and dacitic glasses and the oldest obsidians from the Americas.[2]

[2]The sample mentioned by Stern (2018) as the oldest obsidian used in Patagonia originates from the site Los Notros.

Table 16.1 Values of Zr/Ti and Nb/Y from 44 cores, artifacts, stone flakes, and debitage from Pilauco with correlative number used in Fig. 16.5. Type refers to the rhyolite to basalt composition proposed by Stern (2018)

Type	Lab code	Pilauco code	Number	Zr/Ti	Nb/Y	Type	Lab code	Pilauco code	Number	Zr/Ti	Nb/Y
A	CS5072	12I-P288-250308	1	0.116	0.236	A	MP0031	18AB-P001-310211	23	0.157	0.162
A	CS1411	15A-P006-230812	2	0.167	0.284	A	MP0021	6C-P586-221015	24	0.130	0.152
A	CS5005	12I-P026-180108	3	0.203	0.214	B	CS5003	14AD-P031-412010	25	0.052	0.176
A	CS5081	17AC-P059-131210	4	0.111	0.283	B	CS5074	6F-P010-190308	26	0.052	0.133
A	CS5085	14AA-P005-030812	5	0.116	0.375	B	CS1412	15AC-P047-161210	27	0.061	0.200
A	CS5073	12I-P289-250308	6	0.151	0.212	B	MP0014	16B-P032-110216	28	0.053	0.209
A	CS5079	15AC-P046-161210	7	0.098	0.254	C	CS5086a	15AC-P185-271011	29	0.029	0.152
A	CS5001	12I-P294-270308	8	0.093	0.346	C	CS5086b	15AC-P185-271011	30	0.029	0.148
A	CS5076	17AC-P057-131210	9	0.077	0.181	C	CS5078	17AC-P058-161210	31	0.034	0.113
A	CS5071	12I-P287-25Q308	10	0.065	0.291	C	CS5080	14AC-P045-091210	32	0.033	0.157
A	CS5002	12I-P030-270308	11	0.101	0.260	C	CS7002	14AD-P045-091210	33	0.032	0.250
A	CS5075	15AC-P186-271011	12	0,084	0.333	C	MP0023	14AB-P005-030215	34	0.029	0.286
A	CS5084	17AB-P020-010312	13	0.086	0.297	C	MP0035	14AB-P067-060212	35	0.036	0.176
A	CS7001	14AA-P005-030812	14	0,119	0,361	D	CS5077	15AD-P049-101210	36	0.016	0.090
A	CS7003	16AD-P019-091110	15	0.128	0.137	D	CS5082	14AD-P032-041210	37	0.014	0.100
A	CS7004	14AD-P048-091210	16	0.073	0.172	D	CS5006	14AD-P046-091210	38	0.010	0.133
A	MP003Q	18AB-P358-280312	17	0.201	0.176	D	CS5004	17AC-P038-231110	39	0.011	0.160
A	MP0032	14B-P075-180216	18	0.088	0.178	D	CS7003	16AD-P019-091110	40	0.016	0.200
A	MP0033	18AC-P077-230212	19	0.226	0.228	D	CS7004	14AD-P048-091210	41	0.016	0.240
A	MP0034	14AB-P143-290212	20	0.076	0.142	D	MP0022	17AC-P101 -220216	42	0.013	0.096
A	MP0036	14AB-P086-090212	21	0.110	0.130	D	MP0024	14AB-P173-070312	43	0.012	0.235
A	MP0037	14AB-P105-230212	22	0.060	0.179	D	MP0025	14AB-P066-060212	44	0.017	0.131

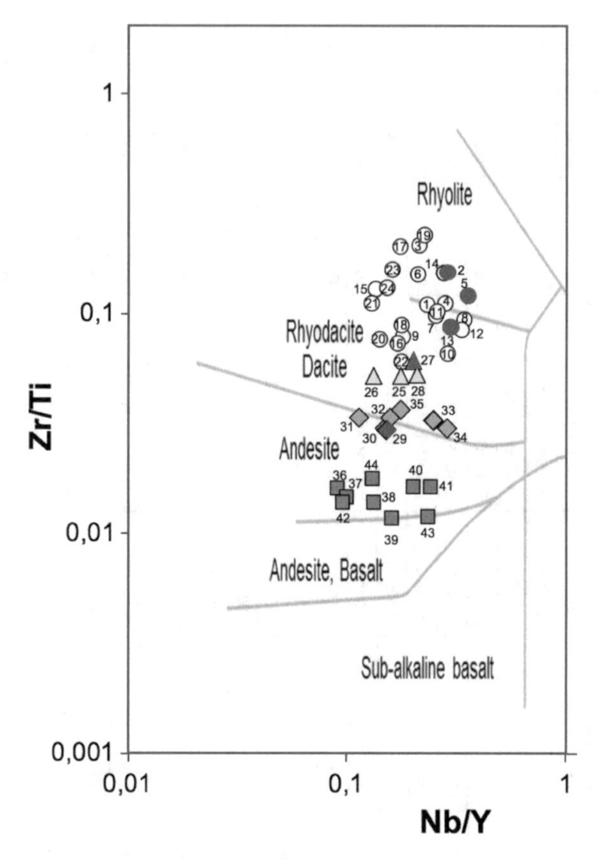

Fig. 16.5 Geochemical classification of lithics from the Pilauco site (44 cores, artifacts, stone flakes, and debitage). The circles, triangles, diamonds, and squares correspond to groups A, B, C, and D according to Stern (2018) and Navarro-Harris et al. (2019). The red symbols correspond to obsidian artifacts 15A-P006-230812 and 14AA-P005-030812 (samples 5 and 14, Figs. 16.6 and 16.7), dacitic glass 17AB-P020-010312 (sample 13, Fig. 16.8), rhyodacitic glass 15AC-P185-270111 (sample 27, Fig. 16.9), and dacitic-andesitic volcanic glass 14AD-P045-090112 (sample 29, Fig. 16.10). Modified from Navarro-Harris et al. (2019)

16.5 The Rock Composition of the Knapped Lithics Assemblage

There are two geological criteria that allow us to unambiguously discard naturally fractured cobbles and pebbles that could potentially be mistaken for cores, artifacts, and stone flakes. The first criterion considers that the proportions of different compositions among naturally broken clasts should be similar to those among generally available clasts in the surrounding area (as recorded in layers PB-1 to PB-6 and the fluvial equivalents to PB-8 located in the southern part of the site; see Chap. 3, this volume) (Fig. 16.10).

Fig. 16.6 Obsidian artifact 15A-P006-230812 (sample 2 in Fig. 16.5) from the stratigraphic layer PB-8, local elevation 444 cm. Primary flake with isolated retouching on the ventral distal side and restricted retouching on the left dorsal side. Cortex with very small and dense crescent markings. The dark mark on the cortex in the center of the dorsal surface corresponds to the sample extracted for geochemical analysis. The arrows indicate the location of one or more percussion points and lines indicate active edges. The striking platform on top

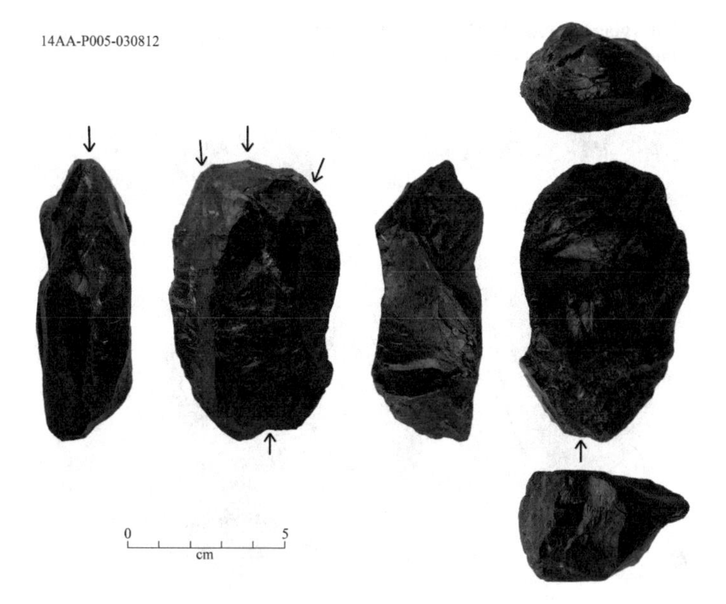

Fig. 16.7 Obsidian artifact 14AA-P005-030812 (sample 14 in Fig. 16.5) from the stratigraphic layer PB-8, local elevation 443 cm. Bifacial core from a small cobble. One face shows five extractions, while the other has only one negative. This second face exhibits a patina of unidentified origin, but probably correspond to the outer obsidian hydration rim. This piece was recorded in a vertical position

Fig. 16.8 Dacitic glass artifact (17AB-P020-010312) from stratigraphic layer PB-7 (local elevation 376 cm). Secondary flake of pebble with irregular edges

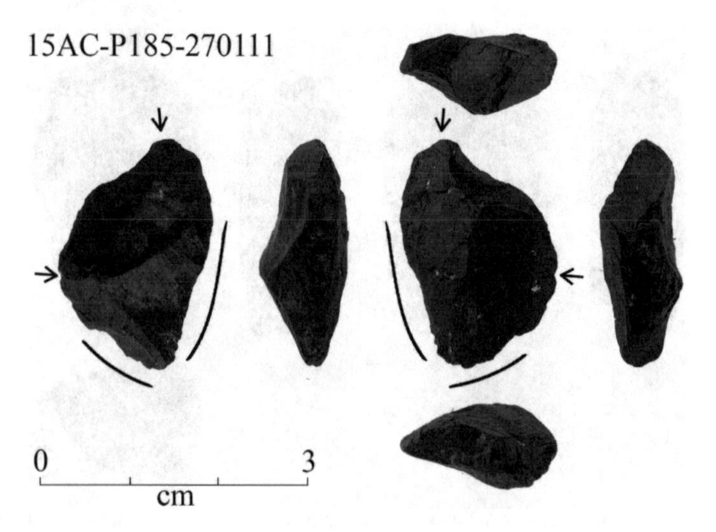

Fig. 16.9 Rhyodacitic glass artifact 15AC-P185-270111 from stratigraphic layer PB-7, local elevation 365 cm. Flake of pebble with knapping on both sides (bifacial), probably knapping debitage

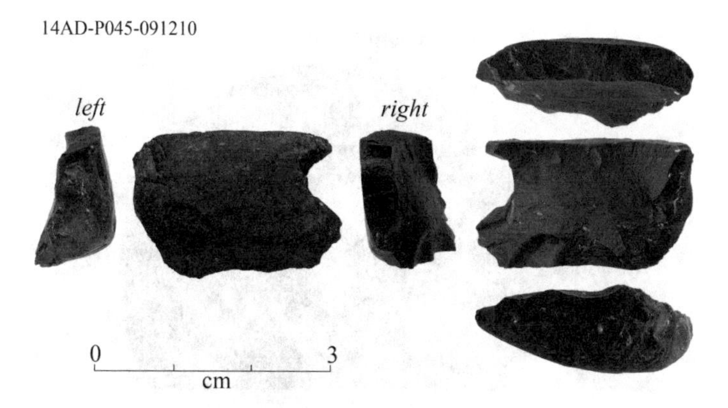

Fig. 16.10 Dacitic-andesitic glass specimen 14AD-P045-091210, classified as debitage. Fragment from a flake with transversal fracture. Medial and proximal portions, as well as the percussion point, cannot be recognized

Figure 16.11 shows the composition of 200 clasts randomly sampled along the Damas River riverbanks. This fluvial channel transports reworked sediments from glacial moraines along the eastern shores of Lake Puyehue, outwash deposits, and volcano-clastic sediments from the San Pablo unit. Eight types of igneous rocks have been recognized in the river deposits. They are represented by a dominant fraction of fine-grained basalts (36%) and then andesites (18%) and porphyritic andesites (18%). The coarse-grained rocks are represented by granite, K-feldspar granite, granodiorite, and diorite and constitute 25% of all igneous rocks. Interestingly, the Damas River sediments, as well as the outwash deposits, completely lack volcanic glass fragments (see Chap 3, this volume). Furthermore, among the recovered non-glassy artifacts the basalts are clearly prevailing above the andesites, although both lithologies are available in the natural environment. For example, in the SW-B sector the basaltic specimens dominate with 64% against the andesitic group (9%). The geochemical REE analysis indicates the Puyehue-Cordón Caulle Volcanic Complex as the most likely source for the volcanic artifacts at Pilauco (Navarro-Harris et al. 2019). The coarse-grained gravels on the site (granites or porphyry andesites with visible crystals) do not include fractured specimens that could mistakenly be classified as lithic artifacts.

The second criterion refers to the fact the archaeological artifacts on obsidian, dacite, and rhyodacite were knapped from disk-shaped pebbles, typical for shorelines affected by waves. The chemical composition of the glassy pebbles mostly corresponds to the Andean volcanoes and suggests a likely origin along the shorelines of the great Andean lakes. The wave action, in these settings, results in constant collisions between pebbles recording a dense distribution of crescent marks of their surface, which are responsible for the light-colored outer surface (Navarro-Harris et al. 2019) (Fig. 16.12).

The knappers of these lithics chose the local and exotic raw materials with fine-grained texture, which knowingly facilitates the formation of conchoidal fractures and naturally sharp edges. The usage of volcanic glass of rhyodacitic, dacitic, and andesitic-dacitic composition (27%, Fig. 16.13) is of particular interest and so far is the first South American record.

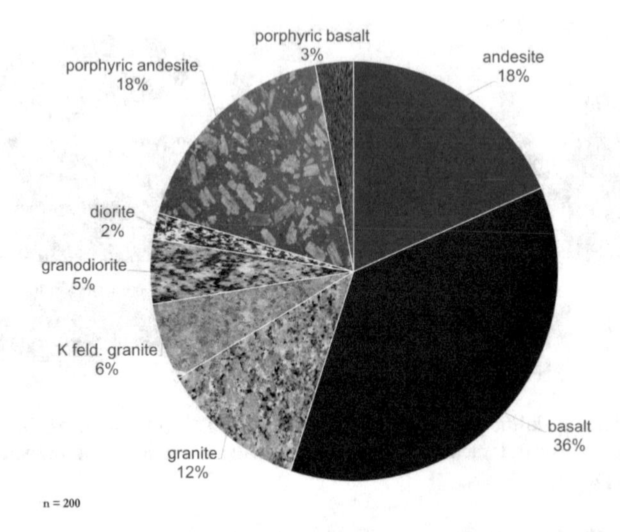

Fig. 16.11 Compositional distribution of cobbles and pebbles from recent Damas River deposits

Fig. 16.12 Left panel: Artifact 18AC-P065-311210. Right panel: Detail from cortex surface showing the density and size of crescent marks. The grooves (3 mm width) resulted from the extraction of sample for geochemical analysis and unveil the contrasting color of the rock

16.6 The Knapped Assemblage

The lithic assemblage of the sites Pilauco and Los Notros was mainly based on pebbles and cobbles as primary material. The set of 53 lithic specimens analyzed here documents different morphologies that share the same unifacial technology. These raw materials were subjected to some general intervention or more sophisticated reduction in size by selective knapping, particularly focusing on the lateral and distal margins of the rock fragments. The dominant group of artifacts encompasses those that experienced reduced formatting, as well as primary or secondary stone flakes with some minor removal of cortex. The artifacts exhibiting more advanced formatting are usually characterized by complete removal of the cortex and devel-

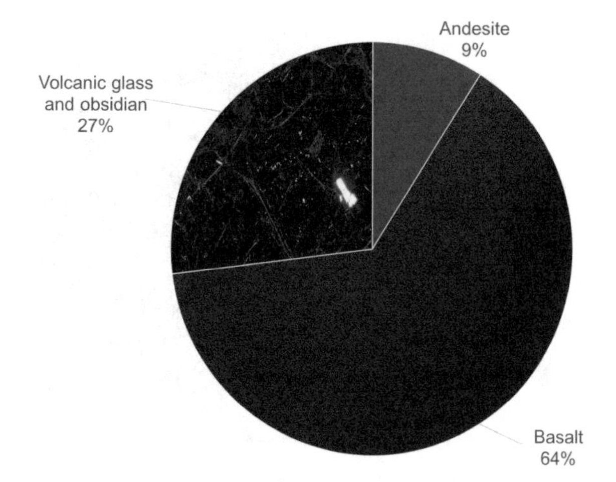

Fig. 16.13 Distribution of raw materials among the lithics from subarea SW-B

opment of retouched margins along the active sides. Some exceptional specimens reflect the profound knowledge of these ancient carvers of the unifacial technology, demonstrating their ability to produce complex instruments by fine and continuous retouching to increase the effectiveness of the stone blades.

The lithic assemblage from Pilauco has been generally classified as cores (group A, 6 specimens, 11%), knapped pebbles and other fragments (group B, 9 specimens, 17%) and unifacial artifacts (group C) including those derived from stone flakes and cores (14 specimens, 26%), active non-retouched stone flakes (15 specimens, 28%), and discarded stone flakes or debitage of 3.5 cm or more in size (10 specimens, 19%). At the site Los Notros 2 specimens from group A and 3 specimens from group C have been recorded.

This assemblage is associated with dozens of micro-flakes smaller than 2 mm in size. This accumulation is referred here to as microdebitage and might result from size-reduction of larger lithic specimens due to knapping or usage or trampling. Notably, most of the microdebitage corresponds to the typical black obsidian (Fig. 16.14) and is therefore associated with artifacts that are absent on the site.

The assemblage of lithics in Pilauco and Los Notros represents an important part of the operative technological chain, which allows us to hypothesize that most of the specimens have been manufactured directly at the site. The local raw materials are generally of good quality with clearly prevailing fine-grained or aphanitic textures (Table 16.2). The size of the gravel selected for the production of these tools appears to have ranged between pebble and cobble size.

Group A encompasses 6 mostly bifacial cores with different morphologies and unidirectional isolated extraction of flakes (Aschero 1983). The original material corresponds to pebbles and cobbles composed of glassy material or fine-grained, aphanitic basalts (Table 16.2). Five of these specimens were recovered from layer PB-7 (e.g. 13AA-P113-250217, 16AD-P019-091110, and 17AA-P092-300115; Table 16.2; Figs. 16.16 and 16.17) and one from layer PB-8. This latter specimen (14AA-P005-030812, Fig. 16.15) stands out from the remaining assembly of

Fig. 16.14 Example of debitage in obsidian, artifact 15B-P023-190116. The sample is 2 mm long

Table 16.2 Core specimens from Pilauco (layers PB-7 and PB-8) and Los Notros (layer LN-2). L = long, I = intermediate and S = short, after Folk (1980)

Cores	Morphology	Texture	Layer	L–I–S (mm)	Elevation (cm)
14AA-P005-030812	Bifacial sub-oval	Glassy	PB-8	84–52–29	443
14A-P027-250113	Unifacial irregular	Glassy	PB-7	34–22–11	385
14AC-P023-201110	Bifacial sub-pyramidal	Aphanitic	PB-7	44–31–28	378
17AA-P092-30015	Bifacial irregular	Aphanitic	PB-7	40–32–17	359
16AD-P019-091110	Unifacial sub-oval	Glassy	PB-7	43–36–22	366
13AA-P113-250217	Bifacial irregular	Glassy	PB-7	39–30–21	363
5BJ-N001-251116	Unifacial sub-oval	Glassy	LN-2	41–26–15	569
7BJ-N001-201016	Unifacial sub-oval	Glassy	LN-2	28–24–14	603
6BJ-N003-171116	Unifacial polyhedral	Aphanitic	LN-2	32–29–11	573

specimens due to its larger size, several extraction negatives ($n = 6$), and morphology (cf. cores 17AA-P092-300115 and 13AA-P113-250217, Fig. 16.16 and Fig. 16.17, respectively).

Only one core specimen seems to originate from a primary (not transported) rock source (16AD-P019-091110, Table 16.2). It is characterized by a cortex without weathering patina or crescent marks and was possibly extracted from a rock outcrop and transported to the site by a human individual. Except for pieces 17AA-P092-30015 and 16AD-P092-30015-17AA, all others are exhausted.

So far, only 3 active cores have been recovered from Los Notros. Two of them were knapped from pebbles—one from a good-quality dacite (5BG-N001-251116) and the other from obsidian (7BJ-N001-201016). The latter is characterized by some nega-

Fig. 16.15 Core specimen 14AA-P005-030812, see text and Table 16.2

Fig. 16.16 Core specimen 17AA-P092-300115. For details see text and Table 16.2. This lithic exhibit the typical surface patina interpreted as outer glass hydration rim

tives and flakes extractions (Table 16.2; Fig. 16.18). The third core consists of polyhedral aphanitic basalt and exhibits with a sub-rectangular form (6BJ-N003-171116).

Considering the artifacts knapped from pebbles (group B), the analyzed samples vary in terms of their morphology. The pebble- and cobble-sized particles used for the fabrication of artifacts were compact in terms of their original form (types C, Cp, Cb, and CE after Folk 1980). They were manufactured on andesites and basalts from aphanitic to slightly porphyritic textures. They characterize an industry with a simple reduction dedicated to producing edges for scraping or cutting and in lesser quantity for splitting or chopping (e.g., artifact 14AB-P019-060213, Fig. 16.19). In turn, all artifacts from this ensemble that were knapped from cobbles were encoun-

Fig. 16.17 Core specimen 13AA-P113-250217, for details see text and Table 16.2

Fig. 16.18 Core specimen 7BJ-N001-201016, typical secondary-source obsidian. The cortex exhibits crescent marks, whereas the fresh surface preserves the fluidal sub-parallel glassy texture

tered fractured. It was possibly used for striking on hard materials. Distal bifacial (14AB-P019-060213, Fig. 16.19) or laterally isolated fractures (14A-P030-120213) are documented on some of these specimens. Some artifacts from this group have continuous natural blades, whereas others do not allow the interpretation of specific functions.

Group C (unifacial artifacts derived from pebbles, stone flakes, and cores) is composed of 14 specimens recovered from Pilauco and 3 specimens from Los Notros, which are generally characterized by significant morphological and lithological variability (Table 16.3). The specimens exhibit more regular lateral and distal modifications. Some stone flakes are marked by primary and secondary reworking and limited retouching along the active margins, which can be continuous at least on one side. Other specimens have been significantly thinned out by carving in the process of extraction of the entire cortex. Overall, group C is characterized by the highest degree of formatting in the entire lithic collection at both sites. This group is described

Fig. 16.19 Artifact 14AB-P019-060213, porphyric basalt, splintered edges, possible used as a chopper

according to different factors such as lithology, shape of active edges, technology, functionality according to angles of use, and stratigraphic position (Table 16.3).

The C1 subgroup includes all artifacts with convex active edges. It consists of 3 specimens from Pilauco (15A-P006-230812, 18AC-P065-311210, and 15B-P052-010316) and 2 specimens from Los Notros site (5BI-N010-201016 and 6BG-N002-130916), all of them knapped from pebbles. These specimens were impacted by direct percussion and split into two halves producing natural blades. The specimen 15A-P006-230812 exhibits continuous retouching along the distal and lateral edges. The dorsal face is still covered by cortex and forms a platform formed by percussion. The active edges are convex, the distal and lateral angles vary between 45° and 70° and were likely used for cutting and scraping (Fig. 16.20). Specimen 18AC-P065-311210 (angle 45º–65º) and 15B-P052-010316 (angle 55º–110º) are also assigned to this group (Fig. 16.12 and Fig. 16.21, respectively).

The two artifacts from Los Notros have lengths between 32 and 56 mm (5BI-N010-201016 and 6BG-N002-130916, Figs. 16.22 and 16.23). Both were knapped in porphyric basalt and exhibit dorsal surfaces entirely covered by cortex except for the active areas along the edges with natural blades without retouching. The left distal side of specimen 5BI-N010-201016 has an active and continuous straight edge with convex shape. On the right side the edge is continuous with sinuous shape. The second specimen has a right-side edge with scalloped sinuosity, likely as a result of usage. The angles of specimen 5BI-N010-201016 (15° and 20°) suggest the edges were designed to cut, whereas artifact 6BG-N002-130916 (30° and 45°) was rather shaped for cutting and scraping tasks (Table 16.3).

Table 16.3 Group C artifacts from Pilauco and Los Notros sites, divided in subgroups (1–5) according to technology and functionality. Long (L), intermediate (I) and short (S) axis in mm, angles in degree

Artifact	Raw material	Shape	Technology	Functional interpretation	Conservation	Elevation	L–I– S	Distal angle	Right lateral angle	Left lateral angle
Pilauco Site										
*Subgroup C*1										
15A-P006-230812	Volcanic glass	Oval	Knapped in a pebble, with cortex, distal and lateral ultra marginal retouch	Cut, scrape, multif	Complete	444	27–22–7	45	35	70
18AC-P065-311210	Volcanic glass	Oval	Knapped in a pebble, with cortex, distal and lateral ultra marginal retouch	Cut, multif	Complete	400	24–22–7	65	45	50
15B-P052-010316	Volcanic glass	Oval	Knapped in a cobble, dorsal cortex, without retouch, distal and lateral continuous active edges	Multif	Complete	380	80–68–30	95	110	55
*Subgroup C*2										
17AA-P056-050213	Volcanic glass	Subrectangle	Knapped in a pebble, primary flake, distal and right lateral ultra marginal retouch	Multif	Complete	373	45–25–14	45	90	55
13A-PC06-050214	Volcanic glass	Suboval	Knapped ti a pebble, primary flake, distal ultra marginal retouching, distal and lateral notches	Multif	Complete	407	26–20–10	60	Does not apply	Does not apply

(continued)

Table 16.3 (continued)

Artifact	Raw material	Shape	Technology	Functional interpretation	Conservation	Elevation	L–I– S	Distal angle	Right lateral angle	Left lateral angle
Subgroup C3										
17AD-P035-191110	Basalt	Rectangular	Scraper knapped in a flake, without cortex	Scrape	Complete	373	38–28–8	75	Does not apply	50
17AC-P023-110111	Basalt	Subrectangle	Knapped ti a secondary flake, lateral edges possibly used, fractured distal (trampling?)	Multif	Fractured	353	39–33–9	Does not apply	50	35
17AA-P081-120115	Basalt	Polyhedron	Knapped in a core, without cortex, long flakes towards the edges	Scrape	Complete	365	33–31–12	65	65	70
Subgroup C4										
15AD-P126-250111	Basalt	Subrectangle	Knapped in a flake, without cortex, lateral and distal retouch	Cut	Complete	361	33–20–8	45	35	65
18AC-P079-010312	Basalt	Subrectangle	Knapped in a flake, without cortex, right lateral continuous retouch, fractured distal	Cut	Fractured	337	38–20–11	Does not apply	35	65
Subgroup C5										
17AB-P016-060212	Basalt	Subtriangle	Knapped in a flake, without cortex, retouch on both laterals. Distal active, point well developed	Multif	Complete	401	39–25–7	Does not apply	45	55

(continued)

Table 16.3 (continued)

Artifact	Raw material	Shape	Technology	Functional interpretation	Conservation	Elevation	L–I– S	Distal angle	Right lateral angle	Left lateral angle
16AC-P001-241010	Volcanic glass	Romboidal	Knapped in a flake, without cortex and retouch, both laterals and distal actives, point well developed, thinning central dorsal to apprehend	Multif	Complete	409	44–28–15	50	40	75
18AC-P085-050312	Basalt	Triangle	Knapped n a secondary flake, natural dorsal cortex, point well developed	Drill	Fractured	333	33–25–11	65	Does not apply	35
14A-P033-180213	Basalt	Subtriangle	Knapped n a secondary flake, natural dorsal. left lateral and distal modified, high lateral fractured	Multif	Fractured	385	58–47–11	40	Does not apply	35
Los Notros site										
*Subgroup C*1										
5BI-N010-201016	Basalt	Suboval	Knapped in a pebble, primary flake with active continuous edge, without retouch	Multif	Entera	422	32–20–8	15	Does not apply	20

(continued)

Table 16.3 (continued)

Artifact	Raw material	Shape	Technology	Functional interpretation	Conservation	Elevation	L–I– S	Distal angle	Right lateral angle	Left lateral angle
6BG-N002-130916	Basalt	Subrectangle	Knapped in a pebble, primary flake with active continuous edge	Multif	Entera	427	56–53–13	40	30	45
Subgroup C5										
IBH-N004-100516	Basalt	Subtriangle	Knapped in pebble, secondary flake, natural dorsal, lateral and distal modified, high lateral fractured, knife	Multif	fractured	200	52–25–15	40	40	45

Fig. 16.20 Artifact 15A-P006-230812 from subgroup C1. The grooves on the cortex resulted from the extraction of sample for geochemical analysis. For details see Tables 16.1 and 16.3; Fig. 16.5

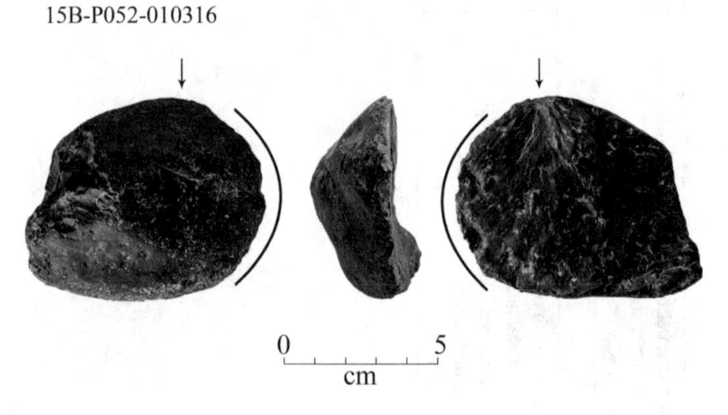

Fig. 16.21 Artifact 15B-P052-010316, dacitic volcanic glass, subgroup C1. The arrow indicates the point of percussion

The subgroup C2 includes two artifacts knapped in glassy rocks and characterized by limited active edges in a straight to convex shape with some notches (17AA-P056-050213 and 13A-P006-050214, Figs. 16.24 and 16.25). Similar to subgroup C1, these artifacts originate from small gravels broken in halves to produce primary flakes. The major length (L) varies between 26 and 45 mm (Table 16.3). In both specimens the active edges display some limited ultra-marginal retouching and two small notches. These active edges are rather short and located close to each other on the distal and right-lateral sides. Their possible function was to scrape or to cut precise incisions. The artifact 17AA-P056-050213 was beside to a gomphothere rib at 373 cm of local elevation (Fig. 16.25).

Fig. 16.22 Basalt artifact 5BI-N010-201016 from subgroup C1. For details see text and Table 16.3

Fig. 16.23 Basalt artifact 6BG-N002-130916 from subgroup C1. For details see text and Table 16.3

The C3 subgroup includes 3 artifacts with extended continuous active edges, without segments and yielding lengths of 33, 38, and 39 mm. The three specimens were knapped in basalts with different textures (artifacts 17AD-P035-191110, 17AC-P023-110111, and 17AA-P081-120115; Figs. 16.26, 16.27 and 16.28). According to the size, morphological and technological characteristics these specimens demonstrate the highest degree of formatting within the entire collection from Pilauco and Los Notros. They were knapped from flakes derived from pebbles and wedged out by direct percussion.

The artifact 17AD-P035-191110 (Fig. 16.26; Table 16.3) is a tool designed for scraping and was knapped from porphyry basalt. The dorsal side has been wedged out by successive carving and consequently lacks cortex. It exhibits continuous retouch-

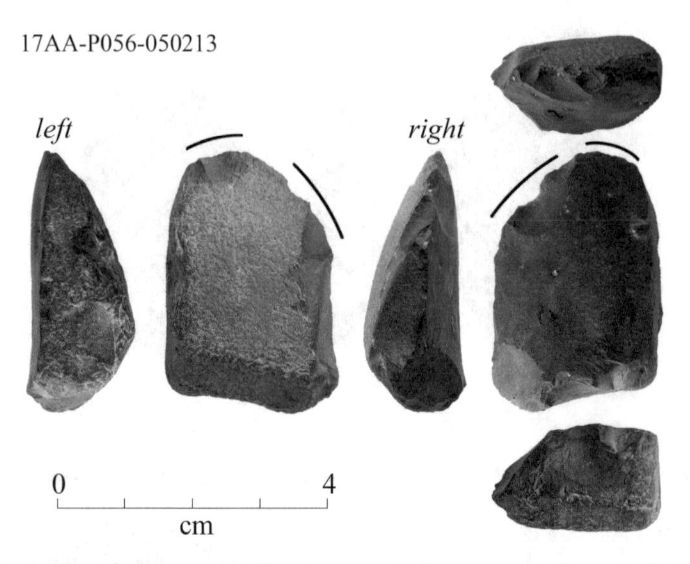

Fig. 16.24 Artifact 17AA-P056-050213, subgroup C2, Pilauco. For details see Table 16.3

Fig. 16.25 Artifact 13A-P006-050214, subgroup C2, Pilauco. For details see Table 16.3

ing on the left-lateral and distal portions. The dorsal face, as well as the proximal and central portions of this specimen were thinned out by carving, possibly to enable a better handgrip during usage. The artifact 17AC-P023-110111 is a secondary medium-sized stone flake (Fig. 16.27; Table 16.3). The proximal portion of its dorsal face exhibits a percussion mark by carving providing a better grip of the tool. This specimen has a distal irregular active edge with fracture in the central part. The

Fig. 16.26 Artifact 17AD-P035-191110 in porphyric basalt of poor-quality for knapping. For details see text and Table 16.3

Fig. 16.27 Artifact 17AC-P023-110111, secondary stone flake with knapping on its dorsal face. Secondary flake knapped in the dorsal face. For details see text and Table 16.3

sinuosity of the edge was probably the result from abrasion or scraping on some hard material.

The third specimen (17AA-P081-120115) consists of aphanitic high-quality basalt and corresponds to a brush-like tool derived from a polyhedral core. According to its origin and morphological features this artifact can be described as a coarse specimen with a sinuous and continuous active basal edge, produced by regular unidirectional extractions away from its center. The angular measurements are 65° and 70° and therefore coherent with the interpretation of its functionality as a brush-like tool (Fig. 16.28; Table 16.3).

Fig. 16.28 Artifact 17AA-P081-120115, fine-grained basalt. Corresponds to a brush-like tool derived from a polyhedral core

The group C4 includes two stone flake artifacts characterized by rough removal of the dorsal cortex and exhibiting one or more continuous active edges. The lateral edges are active (18AC-P079-010312, Fig. 16.29) and the distal portions of the specimen exhibit continuous marginal retouching (15AD-P126-250111, Fig. 16.30). The former specimen has a fracture in the distal portion. The angular measurements of these specimens are 35° and 65° identifying their active edges as designed for cutting.

The subgroup C5 is formed by artifacts from stone flakes with pronounces apex and without cortex (17AB-P016-060212, Fig. 16.31 and 16AC-P001-241010) or with natural dorsal face (18AC-P085-050312 and 14A-P033-180213, Fig. 16.32). This assemblage includes five specimens with variable size and lengths between 33 and 53 mm. All artifacts have basaltic composition, except for artifact 16AC-P001-241010, which was knapped in dacitic glass. The active edges are along the lateral and distal portions and the angular characteristics suggest a multifunctional usage. The distal portion is generally point-shaped and in some specimens exhibits fractures, probably as a result of usage (17AB-P016-060212 and 14A-P033-180213). All artifacts derive from rocks with excellent quality for carving. At Los Notros site the specimen 1BH-N004-100516 is also assigned to this subgroup and displays an active point-shaped distal portion (Fig. 16.33).

The assemblage of knapped lithics generally shares a uniform technology for size-reduction. It implies unifacial modification of natural active edges, as well as some ultra-marginal retouching or more advanced carving along standardized forms. The group of specimens is diverse in terms of their morphology and size but indicating preference for pebbles and cobbles as raw material. Fragments of good-quality basalt

Fig. 16.29 Artifact 18AC-P079-010312, basalt. Fractured distal portion. For details see text and Table 16.3

Fig. 16.30 Basalt artifact 15AD-P126-250111. Fractured distal portion. For details see text and Table 16.3

Fig. 16.31 Artifact 17AB-P016-060212, subgroup C5. For details see text and Table 16.3

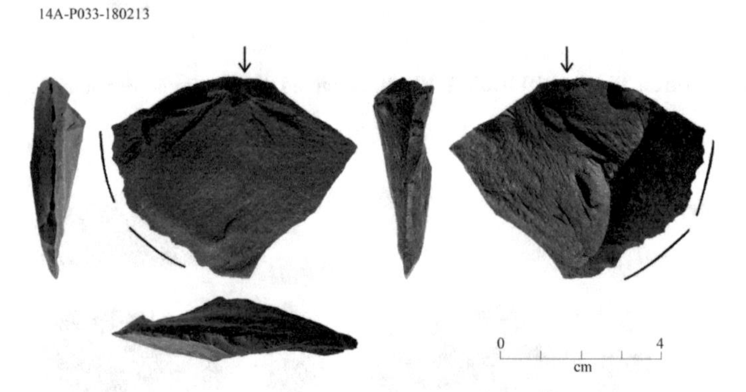

Fig. 16.32 Artifact 14A-P033-180213, very fine-grained basalt. For details see text and Table 16.3

and glassy rocks are present but are less frequent. With respect to the use of obsidian, it would seem that their selection is opportunistic, as the pebbles dominate over the cobbles.

The primary sources for obsidian and other rocks with glassy texture are still unknown (Navarro-Harris et al. 2019). There is no doubt that the Pilauco environment was particularly favorable for human presence given the abundance of local fauna and flora. The lithics knappers who arrived at Pilauco relied on alternative sources for good-quality raw material, such as the fine-grained basalts with local origin. Nevertheless, the acquisition of glassy raw materials with allochthonous origin for the production of tools and the transport to the site demonstrate strategies that also imply a large-scale familiarity with the landscape.

Fig. 16.33 1BH-N004-100516, basalt. For details see text and Table 16.3

Table 16.4 Sphericity values after Folk (1980); $(S^2/LI)^{1/3}$; L = long, I = intermediate and S = short. The first three characters of the nomenclature of each artifact indicates the grid

Artifact	L–I–S	Sphericity	Elevation	Lithology
07F-P020-200308	2.90–2.37–2.04	0.85	481	Vesicular aphanitic andesite
06G-P065-270308	3.10–2.55–2.34	0.88	463	Olivine basaltic-andesite
12E-P012-100308	3.57–3.33–2.57	0.82	440	Plagioclase—olivine basalt
15A-P034-190113	2.10–1.96–1.80	0.92	390	Plagioclase—olivine basalt

16.7 Polished Stones Assemblage

Pilauco and Los Notros sites also bear a record of lithics that were subjected to polishing or abrasion. Twelve artifacts with oval or circular morphology were recovered. Among them, four andesitic or basaltic specimens with porphyric or vesicular texture present rough surface apparently resulting from profound abrasion (Table 16.4; Figs. 16.34 and 16.35).

Collins (1997) analyzed and compared 475 cultural stones from the MV-II site against 1525 natural stones. He concluded that the cultural material generally includes a greater fraction of equidimensional and particularly flat shapes. For example, sphericity values greater than 0.85 are assigned to 5 to 7% of the natural gravels from Chinchihuapi Creek at Monte Verde and comprised 19–21% of the cultural specimens at this site (Collins 1997). In case of Pilauco, at least 4 polished stones present sphericity between 0.85 and 0.92 and evidences for abrasion due to picking that results in a characteristic macroscopically rough and opaque aspect. The micro-

Fig. 16.34 Polished lithics from Pilauco. 12E-P012-100308 (upper left); 07F-P020-200308 (upper right); 06G-P065-270308(lower left); 15A-P034-190113 (lower right). For details see Table 16.4

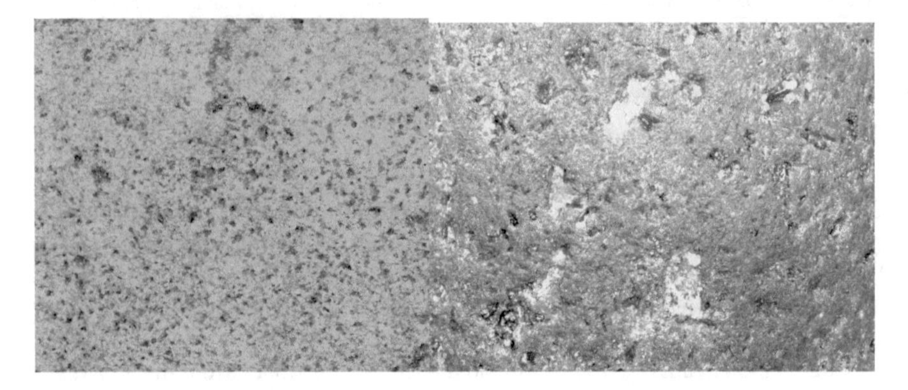

Fig. 16.35 Surface detail from artifacts 07F-P020-200308 (left, vesicular andesite, 0.5 mm vesicles) and 15A-P034-190113 (right, plagioclase–olivine basalt, 2 mm phenocrysts). Note the degree of polishing and the preferred abrasion of small phenocrysts compared to the surrounding matrix. Each image was rendered from 15 sub-images using the software Helicon Focus©

scopic examination of these samples reveals the differential abrasion and wearing of the crystals relative to the surrounding matrix (Table 16.4; Fig. 16.35).

Furthermore, two larger polished specimens were recovered and identified as pounding tools or hammers, next to 2 strikers and 3 specimens with unidentified function. Within this category, the artifact 5BH-LN05-110816 was recovered from a low-energy depositional environment corresponding to the wetland at Los Notros site, and can therefore be characterized as a manuport object (Fig. 16.36).

Finally, there is an assemblage of intentionally perforated relatively soft tuff samples from the San Pablo unit (Fig. 16.37). The first specimen (7AE-P027-050218) is a flat-shaped disk (VE from, Folk 1980) and measures 24 mm (L), 17 mm (I), and 4 mm (S), respectively. The second specimen (10C-P002-020218) is an irregular ellipsoid and measures 31 mm (L), 25 mm (I), and 11 mm (S). The artifact 7AE-P027-050218 exhibits sub-oval irregular picking and subsequent perforation with a sub-circular tool (2–3 mm diameter), whereas the artifact 10C-P002-020218 was perforated with a single sub-circular tool with diameter of 2 mm (Fig. 16.38). The perforation on artifact 10C-P002-0202 crosses the entire S-length, whereas the perforation of 7AE-P027-050218 traverses completely its longest portion (L-length). The perforations postdate the formation of the weathering surface on both tuff samples, which suggests they were selected for processing in their current shape.

Fig. 16.36 Dorsal view of artifact 5BH-LN05-110816, Los Notros site

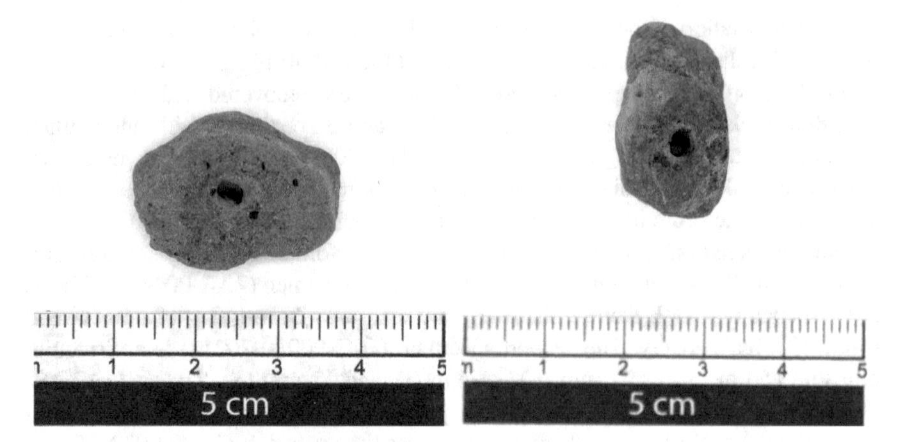

Fig. 16.37 Perforations on artifact 7AE-P027-050218 (left panel, perforation along S-length) and 10C-P002-020218 (right, perforation along L-length)

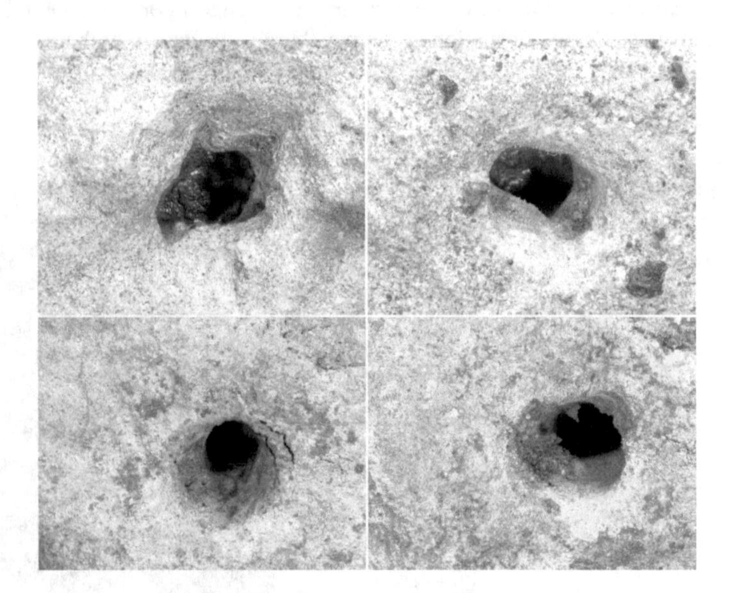

Fig. 16.38 Two-side view of the perforations of artifacts 7AE-P027-050218 (above) and 10C-P002-020218 (below). Note the post-weathering processing in on both tuff samples. The images were rendered from 15 photographs with differing focus using the software Helicon Focus©

16.8 Another Cultural Evidences

There are also some additional evidences that would confirm human presence at Pilauco. For instance, a seed of *Prumnopitys andina* (Poepp. ex Endl.) de Laub (specimen 8H-P024-271107) presents a bilateral perforation (Fig. 16.39). The perforation is almost symmetric (diameters of 2 and 3 mm), resembling a necklace bead, and it is not located along the scar uniting the seed with its placenta. This type of perforation is currently used by artisans of the Pehuenche tradition in the production of necklaces and bracelets.

On the other hand, the record of pollen and hundreds of seeds from the same conifer recovered at Pilauco mark the base of the layer PB-8 (see Chap. 10). Astorga and Ramos (2008) randomly collected 1300 seeds in a modern *P. andina* forests at the sites Los Lleuques in Ñuble (36°59′S/71°30′W) and Conguillío in the Araucanía Region (38°39′S/71°40′W) to characterize the origin of the symmetric perforation observed on the seed specimen described above. Thirty-five percent from the analyzed seeds show marks from rodents and about 5% evidence of insect activity. The seeds that were perforated by rodents were characterized by a single hole in a longitudinal direction, which is generally located in the apical portion of the seed along with the hilum. This part of the seed is relatively soft, which explains the preferred gnawing by rodents. In turn, perforations by insects are significantly smaller and randomly distributed on the seed shell. The perforations on seed sample 8H-P024-271107, as well as on two other seed specimens with rodent or insect perforations

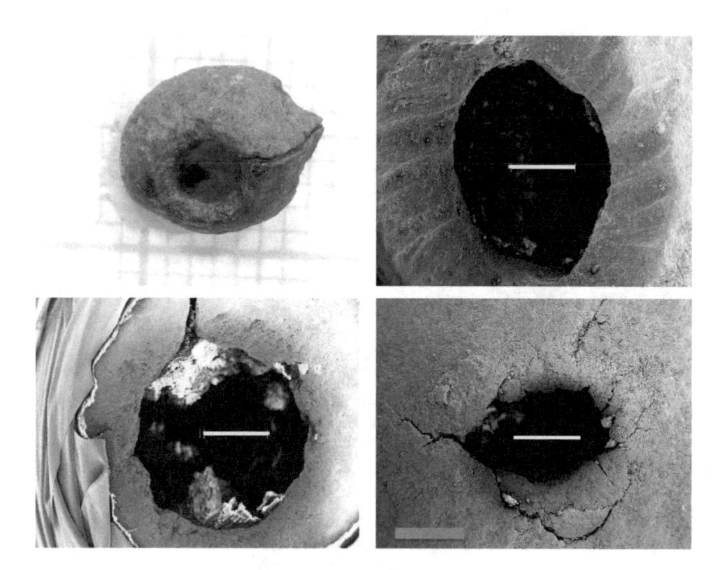

Fig. 16.39 Upper left panel: seed specimen 8H-P024-271107, photograph (each green squares have 1 mm) and SEM-image (lower left panel) of the perforation. Upper right panel: SEM-images of seed specimen perforated by a rodent and by an insect (lower right). Scale bar 1 mm

were carbon-coated and analyzed on SEM-images for comparison (Fig. 16.39). The marks from the rodent's teeth and the shape of the perforation by an insect are characteristic features and can be easily distinguished from the completely polished appearance of the bilateral perforation on the seed recovered at Pilauco (specimen 8H-P024-271107, Fig. 16.39) The marks from the tool used for perforation were likely obliterated because of polishing by the threaded cord of the necklace.

The bone artifact 11I-P019-281207 represents another finding indicating the human presence at Pilauco (paleontological collection MHMOP-PI/60, Chap. 4, this volume). The specimen is a diaphysis from an undetermined long bone from *Equus* sp. (Fig. 16.40). The assignment to this taxon is due to the small size of the specimen (Recabarren et al. 2011). The extreme distal portion of the bone is characterized by polishing and blunting of the beveled edge. This sharp artifact lacks natural fractures. Instead, the edge was cut into right angle and apparently smoothed on purpose. A group of rounded rock fragments as well as pieces of carbon were found stratigraphically below the tool.

The findings from Pilauco are not restricted to continental elements. During a practical course at the site, a master student from Universidad Austral de Chile (Ismael Ríncon Portero) recovered a small mytilid shell fragment from the invertebrate species *Choromytilus chorus* (Molina 1782). The species assignment was based

Fig. 16.40 Bone artifact 11I-P019-281207 in dorsal (upper panel) and ventral (lower panel) views

on comparison of shells from living specimens of *C. chorus* with those from the similar taxa *Mytilus chilensis* Hupe 1854 and *Aulacomya atra* (Molina 1782). *C. chorus* is a large invertebrate reaching 20 cm in length and its shells are thick, robust, and difficult to break. These invertebrates are widely distributed along the entire coast of Chile and southern Peru and represent a common part of the diet in those countries. Currently, the closest natural habitat of these animals is located at the Pacific coast 50 km west from Pilauco. The recovered the shell fragment preserves the nacre and the periostracum and exhibits the presence of epibionts, which could not be classified because of post-mortem abrasion. The shell samples from living species of *Choromytilus* exhibit epibionts including Bryozoa and Polychaete from the family Serpulidae (Eduardo Jaramillo, pers.com November 2018). fragment (8AD-P0008-021117) has a triangular shape with maximum side length of 1 cm (Fig. 16.41). The right-lateral side of the fragment exhibits a series of linear marks that cross the growth increments of the shell and are oriented perpendicularly to the wide edge of the specimen, as well as a brilliant patina on one edge possibly resulting from usage (Fig. 16.42). To our knowledge, this is the first evidence for a marine element from a late Pleistocene site associated with megafauna.

The MV-II site bears abundant evidences for woodwork (Dillehay 1989, 1997), whereas the record from Pilauco has just a few unambiguous evidences for modifications of wood pieces by humans. One of them is the specimen 15AC-P038-101210 (Fig. 16.43), which represents a piece of wood (20 cm × 6 cm) from an undefined latifoliate tree. It exhibits three parallel vertical incisions; the most distal incision is deepest and particularly straight. The basal portions of these incisions were likely smoothed with a tool similar to a gouge. Related woodwork is represented by a thin piece of wood from a latifoliate tree with an oval shape and sharp, spear-like termination on one side (artifact 9A-P001-010818; Figs. 16.44 and 16.45).

Fig. 16.41 Interior side of the triangular specimen 8AD-P0008-021117 *Choromytilus chorus* showing the state of conservation of the nacre and the distal sharpened edge. Small scale bars 1 mm

Fig. 16.42 Distal fragment of *Choromytilus chorus* (8AD-P0008-021117) showing the linear marks oriented with an angle of 80° to the edge and the patina zone with a darker color, likely resulting from usage. The maximum width of the edge measures 2 mm

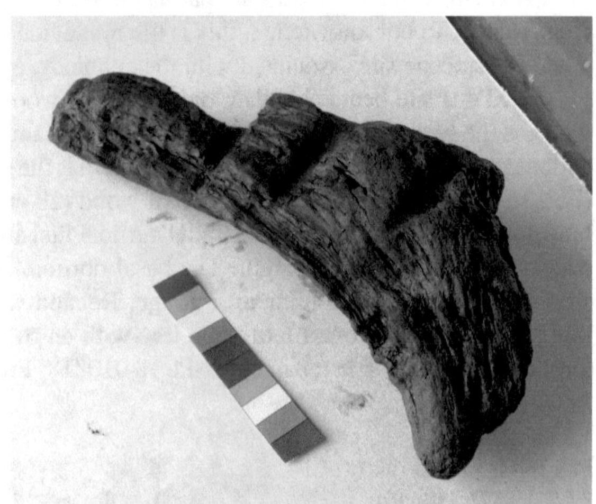

Fig. 16.43 Wood fragment from layer PB-7 showing three deep vertical incisions. (artifact 15AC-P038-101210), Each color bar equals 1 cm

Fig. 16.44 Wood fragment from layer PB-8 (artifact 9A-P001-010818). The longitudinal fracture developed during drying of the sample

Fig. 16.45 Right panel: macroscopic view of the sharp termination of artifact 9A-P001-010818. Left panel: microscopic view of the extreme distal of the sharp edge. Note that the wood (dark brown color) and the resin (silver color) have both been polished

16.9 Discussion and Conclusions

The Pilauco site is neither a residential site, nor a workshop for production of lithic tools. The presence and distribution of knapped lithics and their spatial association with numerous bones from extinct and extant fauna might indicate that the site was occasionally visited in a recurrent manner and over the course of at least 2000 years. The groups of humans visiting the site, likely relied mainly on direct supply of materials for the production of tools, part of which were abandoned on the site after their usage. The presence of some exceptionally knapped pieces found on the site, as well as the presence of glassy rocks and obsidian microdebitage, which show negative retouching or artifacts recycling, there must also have been curatorial strategies for certain instruments. This seems coherent with a possible transport of uni- or bifacially formatted tools back to the base camp or to the site of residence. This is supported by the scarce occurrence of tools in process from other raw materials, such as the only tool elaborated from a bone fragment. Unfortunately, most of the glassy lithic artifacts are covered with patina, which precludes the observation and characterization of possible micro-traces from their usage. The opaque character of this cover might be the result of devitrification of the raw material—a process that is well-known for obsidian but has not been described so far for less acid glasses.

The Pilauco site shares some general characteristics with other early settlements, such as the predominance of unifacial expeditive technologies on raw materials of local origin, the spatial association of these artifacts with bone remains of gomphotheres, or the presence of tools manufactured by abrasion or percussion. There are still many open questions to be answered and hypotheses to be tested at Pilauco and Los Notros sites. The Pilauco site is highlighted by a remarkable coherence between age data, stratigraphy and spatial distribution of cultural materials. The working process was apparently concentrated in certain areas of the site, whereas the concentration of lithic artifacts and debitage in some excavation units correlates with increased occurrence of fossils from extinct fauna (grids 14AA, 14AD, 15 A, 16 A, 17AB and 18AC).

Borrero (1989–1990) postulates different levels of appropriation of new places in a territory by early settlers. The first one is very light and opportunistic, using the place but leaving little evidence and bringing foreign lithic raw material. If they think the site is optimum, they come back, in which case is possible to observe higher levels of occupation. The sites are denser, there is use of local raw materials, until the most colonized stage which is already a residential site itself. According to this model, the Pilauco site is not representative for the first opportunistic stadium (Borrero 1989–1990; Borrero and Franco 1997). The recorded material suggests that Pilauco was a place visited occasionally over the course of many centuries. The site preserves evidences for an extended chain of production tasks of unspecialized tools, which suggests that most of the working process was executed and maintained at the same place. This chain includes sequences of tasks for the reduction of lithic raw materials and their usage. Due to the presence of debitage the artifacts would have been manufactured or their edges revived at the site, then used and later discarded at the same place (Briz 2010). Therefore, the site can be interpreted as repeatedly occupied by humans, carnivores, and herbivores animals resulting in overlapping archaeological and paleontological evidence (Pino et al. 2013).

In summary, the identification and characterization of human presence during the first South American settlements greatly benefits from a well-preserved stratigraphic record and precise age dating but is somehow challenged by the rather simple and unformatted lithics typical for this time period. There is a greater need to study and interpret these early contexts. They might appear simple at a first glance, because of their association with rather unspecialized tasks. However, it is necessary to search for new theoretical and methodological approaches in order to understand the extended application and maintenance of unifacial technologies and the uniform techno-morphological patterns in lithic artifacts dated between 15,000 cal yr BP and the Pleistocene-Holocene transition (Borrero and Franco 1997; Jackson and Prieto 2005).

References

Abarzúa AM, Lobos V, Martel-Cea A (2016) ¡Pequeño testigo de grandes cambios!: polen, semillas y carbones escondidos en Pilauco In: Pino M (ed) El Sitio Pilauco. Patagonia Noroccidental de Chile, Universidad Austral de Chile. TAQUACH, Osorno, pp 112–127

Aceituno J, Loaiza N (2010) Estructura interna y movilidad en el valle del rio San Eugenio en la Cordillera Central de Colombia. Revista de Arqueología del Área Intermedia 8:84–120

Aceituno FJ, Loaiza N, Delgado-Burbano ME, Barrientos G (2013) The initial human settlement of northwest South America during the Pleistocene/Holocene transition: synthesis and perspectives. Quat Int 301:23–33

Aimola G, Andrade C, Mota L, Parenti F (2014) Final pleistocene and early holocene at Sitio do Meio, Piauí—Brazil: Stratigraphy and Comparison with Pedra Furada. J. Lithic Studies 1(2):5–24

Andrefsky W (1998) Lithics. Macroscopic approaches to analysis. Cambridge manuals in archaeology. Cambridge University Press, Cambridge
Andrefsky W Jr (2000) Lithics. Macroscopic approaches to analysis. Cambridge University Press, Cambridge
Aschero C (1983) Ensayo para una clasificación morfológica de artefactos líticos aplicada a estudios tipológicos comparativos. Apéndices AC. Revisión, Unpublished Revisión. Cátedra de Ergología y Tecnología (FFyL-UBA), Buenos Aires
Astorga G, Ramos P (2008) Semillas de Lleuque en el sitio Pilauco. In: Pino M (ed) Pilauco, Un Sitio Complejo del Pleistoceno Tardío. Osorno, Norpatagonia Chilena. Universidad Austral de Chile, Imprenta América, Valdivia, Chile, pp 105–112
Binford I (1980) Wilow smoke and dog's tails: hunter-gatherer settlement systems and archaeological site formation. Am Antiq 45:1–17
Binford I (1987) Researching ambiguity: frames of reference and site structure. In: Kent S (ed) Method and theory for activity and research. Columbia University Press, NewYork, pp 449–512
Boëda E, Clemente-Conte I, Fontugne M, Lahaye C, Pino M, Felice G, Guidon N, Hoeltz S, Lourdeau A, Pagli M, Pessis AM, Viana S, Da Costa A, Douville E (2014) A new late Pleistocene archaeological sequence in South America: the Vale da Pedra Furada (Piauí, Brazil). Antiquity 341:927–941
Borrero L (1989–1990) Evolución cultural divergente en la Patagonia Austral. An Inst Patagonia 19:133–139
Borrero L, Franco N (1997) Early patagonian hunter-gatherers: subsistence and technology. J Anthropol Res 53:219–239
Borrero L, Politis G, Sandweis DH, Boëda E, Bracco Boksar R (2015) Con lo mínimo: los debates sobre el poblamiento de América del Sur. Intersecciones Antropol 16:5–14
Bourgeon L, Burke A, Higham T (2017) Earliest human presence in North America dated to the Last Glacial Maximum: new radiocarbon dates from Bluefish Caves. Canada. PLoS ONE 12(1):e0169486
Briz I (2010) Dinámicas producción-consumo en conjuntos líticos: el análisis de los conjuntos líticos de la sociedad yámana. Magallania 38(2):189–211
Bryan A, Gruhn R (2003) Some difficulties in modeling the original peopling of the Americas. Quat Int 109–110:175–179
Collins MB (1997) The lithics from Monte Verde, a descriptive-morphological analysis. In: Dillehay TD (ed) Monte Verde, a late Pleistocene Site in Chile: the archaeological context, pp 383–504
Collins MB (2014) Initial peopling of the America: contexts, findings and issues. In: Renfrew C, Bahn P (eds) The Cambridge world prehistory, vol 2. Cambridge University Press, New York, pp 903–922
Cruxent JM, Rouse I (1956) A lithic industry of Paleo-Indian type in Venezuela. Am Antiq 22:172–179
Dillehay TD (1989) (ed) Monte Verde: a late Pleistocene settlement in Chile: paleoenvironmental and site context, vol 1. Smithsonian Institution Press, Washington, DC
Dillehay TD (1997) (ed) Monte Verde: a late Pleistocene settlement in Chile, the archaeological context and interpretation, vol 2. Smithsonian Institution Press, Washington, DC
Dillehay TD (2000) The settlement of the Americas: a new prehistory. Basic Books, New York
Dillehay TD (2008) Profiles in pleistocene history. In: Silvermann H, Isbell WH, Handbook of South American archaeology. Springer, New York, pp 29–43
Dillehay TD (2009) Probing deeper into first American studies. Proc Natl Acad Sci USA 106(4):971–978
Dillehay TD, Ramírez C, Pino M, Collins M, Rossen J, Pino-Navarro JD (2008) Monte Verde: seaweed, food, medicine, and the peopling of South America. Science 320(5877):784–786
Dillehay TD, Bonavia D, Goodbred S, Pino M, Vásquez V, Rosales T (2012a) Chronology, mound-building and environment at Huaca Prieta, coastal Peru, from 13 700 to 4000 years ago. Antiquity 46:48–70

Dillehay TD, Bonavia D, Goodbred S, Pino M, Vásquez V, Rosales T (2012b) A late Pleistocene human presence at Huaca Prieta, Peru, and early Pacific Coastal adaptations. Quat Res (Orlando) 77:418–423

Dillehay TD, Ocampo C, Saavedra J, Oliveira Sawakuchi A, Vega RM, Pino M, Collins MB, Cummings LS, Arregui I, Villagran XS, Hartmann GA, Mella M, González A, Dix G (2015) New archaeological evidence for an early human presence at Monte Verde, Chile. PLoS ONE 10(11):e0145471, 1–27

Fairbridge RW (1982) The pleistocene-holocene boundary. Quat Sci Rev 1(3):215–244

Fariña RA, Castilla R (2007) Earliest evidence for human-megafauna interaction in the Americas. Bar Int Ser 1627:31

Fiedel SJ (2000) The peopling of the new world: present evidence, new theories, and future directions. J Archaeol Res 8(1):39–103

Folk R (1980) Petrology of sedimentary rocks. Hemphill Publishing Company, Austin, TX

Gnecco C, Aceituno J (2004) Poblamiento temprano y espacios antropogénicos en el norte de Suramérica. Complutum 15:151–164

González E, Prevosti FJ, Pino M (2010) Primer registro de Mephitidae (Carnivora: Mammalia) para el Pleistoceno de Chile. Magallania 38(2):239–248

González E, Labarca R, Chávez M, Pino M (2014) First fossil record of the smallest deer cf. Pudu Molina, 1792 (Artiodactyla, Cervidae) in the late Pleistocene of South America. J Vert Paleontol 34(2):483–488

González E, Domingo L, Tornero C, Pino M, Fernández MH, Sevilla P, Villavicencio N, Agustí J (2017) Late Pleistocene ecological, environmental and climatic reconstruction based on megafauna stable isotopes from northwestern Chilean Patagonia. Quat Sci Rev 170:188–202

González-Guarda E, Petermann-Pichincura A, Tornero C, Domingo L, Agustí J, Pino M, Abarzúa AM, Capriles JM, Villavicencio NA, Labarca R, Tolorza V, Sevilla P, Rivals F (2018) Multiproxy evidence for leaf-browsing and closed habitats in extinct proboscideans (Mammalia, Proboscidea) from Central Chile. Proc Natl Acad Sci USA 115(37):9258–9263

Goodyear AC (2005) Evidence of pre-Clovis sites in the eastern United States. In: Bonnichsen R, Bradley B, Stanford D, Waters M (eds) Paleoamerican origins: Beyond Clovis, Center for the Study of the First Americans, TX, A&M University, pp 103–112

Heusser C (1966) Late-Pleistocene pollen diagrams from the Province of Llanquihue, Southern Chile. Proc Am Philos Soc 110(4):269–305

Inizan ML, Reduron-Ballinger M, Roche H, Tixier J (1999) Technology and terminology of knapped stone. Nanterré: CREP. Translated by J Féblot-Augustins. Préhistoire de la Pierre Taillé, Tome 5, France

Jackson D, Prieto A (2005) Estrategias tecnológicas y conjunto lítico del contexto paleoindio de cueva Lago Sofía 1, Última Esperanza, Magallanes. Magallania 33(1):115–120

Jackson D, Méndez C, Seguel R, Maldonado A, Vargas G (2007) Initial occupation of the Pacific coast of Chile during late Pleistocene times. Curr Anthropol 48(5):725–731

Jenkins DL, Davis LG, Stafford TW, Campos PF, Connolly TJ, Cummings LS, Hofreiter M, Hockett B, McDonough K, Luthe I, O'Grady PW, Swisher ME, White F, Yates B, Yohe RM, Yost Ch, Willerslev E (2013) Geochronology, archaeological context, and DNA at the Paisley Caves. In: Graf KE, Ketron CV, Waters MR (eds) Paleoamerican odyssey. Texas A&M University Press, College Station, pp 485–510

Kelly RL (2003) Maybe we do know when people first came to North America; and what does it mean if we do? Quat Int 109:133–145

Kennett JP, Kennett DJ, Culleton BJ, Tortosa JEA, Bischoff JL, Bunch T, Daniel R Jr, Erlandson JM, Ferraroh D, Firestone RB, Goodyear AC, Israde-Alcántara I, Johnson JR, Jordá Pardo JF, Kimbel DR, LeCompte MA Goodyear AC (2015). Bayesian chronological analyses consistent with synchronous age of 12,835–12,735 Cal BP for Younger Dryas boundary on four continents. Proc Natl Acad Sci USA 112(32):E4344–E4353

Labarca R, Pino M, Recabarren O (2013) Nuevos antecedentes sobre los Lamini fósiles (Artiodactyla: Camelidae) del Pleistoceno final del extremo sur de Chile (38°–42°S). Estud Geol (Madr) 69(2):255–269
Labarca R, Recabarren O, Canales-Brellenthin P, Pino M (2014) The gomphotheres (proboscidea: Gomphotheriidae) from Pilauco site: scavenging evidence in the late Pleistocene of the Chilean Patagonia. Quat Int 352:75–84
Labarca R, Alberdi MT, Prado JL, Mansilla P, Mourgues FA (2016) Nuevas evidencias acerca de la presencia de Stegomastodon platensis Ameghino, 1888, Proboscidea: Gomphotheriidae, en el Pleistoceno tardío de Chile central. Estud Geol (Madr) 72(1):e46
Latorre C, Santoro C, Ugalde P, Gayo E, Osorio D, Salas-Egaña C, DePol- HR, Joly D, Rech J (2013) Late Pleistocene human occupation of the hyperarid core in the atacama desert, northern Chile. Quat Sci Rev 77:19–30
López CE (2008) Landscape development and the evidence for early human occupation in the inter-and tropical lowland of the Magdalena River, Colombia. Syllaba Press, Miami, FL
Martin PS (1973) The discovery of America. Science 179:969–974
Miotti L (2004) Quandary: the Clovis phenomenon, the first Americans, and the view from Patagonia. In: Lepper B, Bonnichsen R (eds) New perspectives on the first Americans. A Peopling of Americans Publication. A&M Texas University Press, USA, pp 35–40
Navarro-Harris X, Pino M, Guzmán-Marín P, Lira MP, Labarca R, Corgne A (2019) The procurement and use of knappable glassy volcanic raw material from the late Pleistocene Pilauco site, Chilean Northwestern Patagonia. Geoarchaeology 1:21. https://doi.org/10.1002/gea.21736
Neves WA Hubbe M (2005) Cranial morphology of early Americans from Lagoa Santa. Brazil: implications for the settlement of the new world. Proc Natl Acad Sci USA 102(51):18309–18314
Núñez L, Varela J, Casamiquela R, Villagrán C (1994) Reconstrucción multidisciplinaria de la ocupación prehistórica de Quereo. Centro de Chile. Latin Am Antiq 5(2):99–118
Odell GH (2004) Lithic analysis. Kluwer Academic/Plenum Publishers, New York
Parenti F (2001) Le gisement quaternaire de Pedra Furada (Piaui, Brésil): stratigraphie, chronologie, évolution culturelle. Editions Recherche sur les civilisations
Paunero RS, Frank AD, Skarbun F, Rosales G, Cueto M, Zapata G, Paunero M, Lunazzi N, Del Giorgio M (2007) Investigaciones arqueologicas en el sitio Casa del Minero 1, Estancia La María, Meseta Central de Santa Cruz. In: Morello F, Martinic M, Prieto A, Bahamonde G (eds) Arqueología de Fuego-Patagonia. Levantando Piedras, Desenterrando huesos y Develando Arcanos. Ediciones CEQUA, Punta Arenas, Chile, pp 577–588
Pino M, Chávez Hoffmeister M, Navarro-Harris X, Labarca R (2013) The late Pleistocene Pilauco site, south central Chile. Quat Int 299:3–12
Rademaker K, Hodgins G, Moore K, Zarrillo S, Miller C, Bromley GR, Leach P, Reid DA, Yépez Álvarez W, Sandweiss DH (2014) Paleoindian settlement of the high-altitude Peruvian Andes. Science 346(6208):466–469
Recabarren O, Pino M, Cid I (2011) A new record of Equus (Mammalia: Equidae) from the late Pleistocene of central-south Chile. Rev Chil Hist Nat 84:535–542
Recabarren O, Pino M, Alberdi MT (2014) La Familia Gomphotheriidae en América del Sur: evidencia de molares al norte de la Patagonia chilena. Estud Geol (Madr) 70(1):e001. https://doi.org/10.3989/egeol.41346.273
Recabarren O, Buckley M, García N Pino M (2015) Determinación de la secuencia de aminoácidos del colágeno (I) en restos de gonfoterio (Proboscidea, Gomphotheriidae), del Pleistoceno tardío del yacimiento Pilauco (sur de Chile). In: Current trends in paleontology and evolution. XIII EJIP conference proceedings, pp 212–214
Renfrew C, Bahn P (1993) Arqueología. Teoría, métodos y práctica. Akal, Barcelona
Roberts FHH Jr (1940) Developments in the problem of the North American Paleo-Indian. In: Essays in historical anthropology of North America, Published in honor of John R. Swanton, Smithsonian miscellaneous collections, vol 100, pp 51–116

Sandweiss DH, McInnis H, Burger RL, Cano A, Ojeda B, Paredes R, Sandweiss M, Glascock MD (1998) Quebrada Jaguay: early South American maritime adaptations. Science 281(5384):1830–1832

Schmidt Dias A (2004) Diversify to people. The archaeologic brazilian context during the Pleistocene-Holocene transition. Cumplutum 15:249–263

Seguel Z, Campana O (1975) Presencia de megafauna en la provincia de Osorno (Chile) y sus posibles relaciones con cazadores superiores. In: Abstracts of Actas y Trabajos del Primer Congreso de Arqueología Argentina, Buenos Aire, pp 237–242

Sistiaga A, Berna F, Laursen R, Goldberg P (2014) Steroidal biomarker analysis of a 14,000 years old putative human coprolite from Paisley Cave, Oregon. J Archaeol Sci 41:813–817

Stern CR (2018) Obsidian sources and distribution in Patagonia, southernmost South America. Quat Int 468:190–205

Suárez R (2018) The peopling of Southeastern South America: cultural diversity, paleoenvironmental conditions, and lithic technological organization during the late Pleistocene and early holocene. In: Robinson E, Sellet F (eds) Lithic technological organization and paleoenvironmental change. Studies in human ecology and adaptation, vol 9. Springer, Cham pp 281–300

Suárez R, Piñeiro G, Barceló F (2018) Living on the river edge: The Tigre site (K-87) new data and implications for the initial colonization of the Uruguay River basin. Quat Int 473:242–260

Tello F, Elgueta M, Abarzúa AM, Torres F, Pino M (2017) Fossil beetles from Pilauco, south-central Chile: an upper pleistocene paleoenvironmental reconstruction. Quat Int 449:58–66

Vialou D, Benabdelhadi M, Feathers J, Fontugne M Vialou AV (2017). Peopling South America's centre: the late Pleistocene site of Santa Elina. Antiquity 91(358):865–884

Waters MR, Stafford TW (2007) Redefining the age of Clovis: implications for the peopling of the Americas. Science 315(5815):1122–1126

Waters MR, Forman SL, Jennings TA, Nordt LC, Driese SG, Feinberg JM, Keene JL, Halligan J, Lindquist A, Pierson J, Hallmark C, Collins MB, Wiederhold JE (2011) The Buttermilk creek complex and the origins of Clovis at the Debra L. Friedkin site, Texas. Science, 331(6024):1599–1603

Weischet W (1958) Studien über den glazial bedingten Formenschatz der südchilenischen Längssenke im West-Ost-Profil beiderseits Osorno. Pet Geogr Mitt 161–172

Weischet W (1959) Geographische Beobachtungen Auf Einer Forschungsreise in Chile. Erkunde Band XII 6–22

Chapter 17
Planning and Managing a Palaeontological Tourism Destination: The Case of Pilauco-Osorno, Chile

Silvia Constabel and Katerina Veloso

Abstract Archaeological and palaeontological findings in the territory of the Los Lagos Region, Chile, such as those emerging from Monte Verde II and Pilauco sites require a science knowledge translation into a simple language. Palaeontological and archaeological scientific tourism, the processes of outreach trips, the exhibition of scientific results and the training of stakeholders or actors with links to the territory, together with the media, are key to scientific dissemination. This chapter deals with planning and recommendations for the successful management of scientific tourism, especially around the destination of Osorno. The initiatives are based on workshops, training courses and seminars carried out at the destination with stakeholders and local actors in the palaeotourism value chain.

Keywords Tourism · Stakeholders · Scientific tourism

17.1 Background

Science has produced innumerable advances in human knowledge throughout history. The majority of people are unaware of many of these discoveries, which are not directly related to scientific researches of the respective disciplines. Findings of this type such as those associated with Pleistocene archaeological sites require a transfer and diffusion process to make the science understandable by a diverse group of people. New forms of tourism such as palaeontological tourism and scientific tourism are key to scientific dissemination.

The findings of bones and other cultural vestiges that account for an important presence of Pleistocene mega- and microfauna at the Pilauco site, together with

S. Constabel (✉) · K. Veloso
Instituto de Turismo, Universidad Austral de Chile, Valdivia, Chile
e-mail: sconstabel@gmail.com

S. Constabel
TAQUACH, Transdisciplinary Center for Quaternary Research in the South of Chile, Universidad Austral de Chile, Valdivia, Chile

M. Pino and G. A. Astorga (eds.), *Pilauco: A Late Pleistocene Archaeo-paleontological Site*, The Latin American Studies Book Series,
https://doi.org/10.1007/978-3-030-23918-3_17

antecedents coming from the Monte Verde II site, allow us to infer the habits and feeding patterns of the first humans that populated this area. These populations of hunter-gatherers produce a cultural substrate that could afford the destination of Osorno with a distinguishing feature in Chile, allowing for example the incorporation of gastronomic concepts such as the palaeo-diet or eating as our ancestors did, which have already been exploited in other latitudes with recognized success.

Such a feature could through the devising of a gastronomic experience focused in palaeo-diet themes, for example by providing simple food identity methods around the site of Pilauco, suggested culinary preparations, locating decorative items, "merchandising", or simply sharing accurate and timely information about the wealth and importance of a given resource. These are just some antecedents that could invite us to innovate new forms of tourism in emerging destinations and encourage the diversification of available products and services, to increasingly demanding and experienced tourists.

17.2 Palaeontological Tourism

Special interest tourism describes a set of tourism types such as adventure tourism, ecotourism, agritourism, and includes scientific, educational, rural, whale-watching, birdwatching, archaeological and palaeontological tourism. Given their potentials for biodiversity, geo-territorial wealth and variety of human manifestations, they constitute an interesting opportunity for areas or regions with a promising tourist potential (Szmulewicz and Veloso 2014). They allow for the generation of economic resources through the sustainable use of a region's natural resources, cultural and natural heritage. They can also encourage conservation by promoting the formation of sustainable awareness involving local communities in this type of activity.

Palaeontological tourism or palaeo-tourism corresponds to one of the new forms of special interest tourism. Thus, the main objective is that visitors interact with the study of palaeontology, acquire scientific knowledge, understand the materials and methodologies used in the scientific study and, of course, visit palaeontological sites (Perini and Calvo 2008). For Vejsbjerg and Encabo (2005) a site of palaeontological tourism can be staged either in an urban center, where the most significant tourist offer is a visit to a museum with palaeontological collections, or in a rural area with palaeontological, geological and landscape viewpoints. In the latter case, the supply of services, equipment and infrastructure would be provided by a nearby urban support center.

If a palaeontological tourist destination would be acknowledged by visitors, it would be essential to have a concept or brand that differentiated it from competing palaeontological sites, while ideally complementing them at a regional level (Bigné et al. 2000).

According to Encabo and Vejsbjerg (2002), using palaeontology as a tourism resource enables the offering of a diverse range of tourism products, as well as generating other sources of income for destinations which establish museums, parks or palaeontological sites. Palaeontological environments constitute a natural heritage that can be used for tourism, either in museums, through the observation of fossils in situ, or contextually evidencing their existence within the ecosystem in which they lived. For this to be achieved, there is a requirement for the involvement of motivated stakeholders, the availability of the technical conditions necessary for the promotion of the tourism and they need to have endurability and sustainability in terms of time (Perini and Calvo 2008).

There are a diverse range international tourism experiences based on palaeontological scientific research, with various levels of development. The Royal Tyrrell Museum, Canada, located 6 km northwest of Drumheller, Alberta, in Midland Provincial Park, has a series of galleries that chronologically exhibit the wonders of 3.9 billion years of life on Earth. Among them are hundreds of dinosaur fossils, most of them found in Alberta and many in the nearby Dinosaur Provincial Park. Dinosaur skeletons from the Upper Cretaceous of North America are exhibited in a huge room. Also featured are a living Cretaceous garden with more than 600 species of plants and a window to the preparation laboratory, which allows visitors to observe technicians carefully preparing fossils for research and exhibition. The services and complementary activities offered by this museum include guided and self-guided visits to the Badlands, a science practice room, simulated fossil excavation and fossils identification activities, school programs and summer camps for children and families. There is a coffee shop with a capacity for 150 people and a gift shop offering a variety of souvenirs, toys, clothing and special books based on the contents of the museum. The funds raised by the store support the research and education programs at the museum. An audio guide service is available in English, French, Spanish, Japanese, German and Dutch, with tours lasting approximately 65 min.

Dinopolis is located northeast of Madrid, Teruel Province, Spain. The site is characterized by having a large variety of palaeontological sites with deposits consisting of extremely diverse ages and morphologies. This abundance has motivated the development of different initiatives developed for the dissemination of the location's palaeontological heritage. The Joint Palaeontological Foundation Teruel-Dinópolis is an institution of the Government of Aragon, which focusses on provincial development through the social use of localized palaeontological resources. Included in its activities are research in palaeontology and the conservation of palaeontological heritage and its dissemination, mainly through Dinópolis. This initiative aspires to be a working example of the compatibility of palaeontological knowledge with a diffusion based on entertainment and recreation. The palaeontological consortium operates as a society with a mixed public and private participation (Fig. 17.1). The Dinópolis theme park is a business initiative aimed at developing the tourism sector of Teruel from the existing palaeontological heritage. The theme park approaches the history of life on earth using various resources such as animation, thematic displays based on different geological periods, light and sound shows, video projections, a 3D cinema, games and animatronics, all of which try to capture the attention of the visitor. Its 10,000 m^2 Palaeontological Museum, which was created in 2004, is the largest existing complex in Spain as far as the history of life on Earth is concerned.

Fig. 17.1 Dinopolis territory (*Source* www.dinopolis.com)

Another example is the Egidio Feruglio Palaeontological Museum (MEF) and Bryn Gwyn Geopark, Argentina, located in the city of Trelew, Chubut Province. The MEF is one of the most important scientific museum institutions in Argentina and is internationally recognized. The museum is organized on two floors and has three dinosaurs commanding visitor attention: an Epachtosaurus, the most complete known Tyrannosaur, and the Argentinosaurus—the largest terrestrial animal known to date, which measured up to 40 m in length and weighed around 100 tons. Regarding visitor services, the museum has an auditorium, a coffee shop, a museum shop with a large number of merchandises including souvenirs, replicas of fossils and highly-realistic 3D models made with the same techniques as those used in the exhibition pieces. The museum also manages Bryn Gwyn Palaeontological Geopark, which gives visitors a glimpse into the past 40 million years. Located on the southern wall of the lower valley of the Chubut River, the Geopark has been working on discovering and exhibiting traces of the geological formation of Patagonia using several official walking trails for over 14 years. This palaeontological set offers some interesting programs to its visitors. MEF Volunteers is a more scientific example and attracts researchers from around the world with different levels of training. These volunteers are able to collaborate over various assignments at the museum: assisting in the preparation and preservation of fossils, supporting the work of museum guides, participating in producing casts and replicas for exhibition, contributing to the preparation and execution of outreach activities, performing support tasks during exhibitions, fairs and other events and assisting in general technical and/or administrative tasks. Other outreach activities include "Explorers in Pajamas", which was designed for children who are interested in spending a night at the museum and participating in fun activities, such as sleeping among the dinosaurs, learning about the prehistoric life of Patagonia, exploring the Museum by lantern-light and making their own fossil replicas. The Scientific Cofee program consists of an informal conversation between a scientist and the public, with a coffee in between.In conjunction with the Bryn Gwyn Geopark, there is also a Palaeontological Expedition on offer adventure tourism and scientific tourism combine with a low-risk excursion into the field of palaeontology through half-day, full-day or two-day trips. Accompanied by experienced guides, participants explore areas of the park that are normally restricted to the general public in order to search for and identify fossils, learn about the geological history of the region and learn about the fauna and flora of the area.

Chile has sites that are of national and international interest. They originate a huge potential to develop palaeontological and archaeological tourism and therefore diversify the tourist offer based on cultural heritage. However, a review of the current experiences reveals a lack of specialized tourist services and programs. This means that palaeontological resources are not articulated with tourist services such as accommodation, catering, or crafts, and do not have individual hallmarks differentiating them from one another. It is only possible to visit the museums and palaeontological sites, which have limited transport services and only a handful of artisans linked to the resources. This underlines a number of missed opportunities on the part of entrepreneurs to develop specialized ventures in this new form of tourism, and a lack of knowledge of and empowerment by the tourism industry in regard to Chile's palaeontological heritage.

The Palaeontological Museum of Caldera, which serves the nearby Cerro Ballena Palaeontological Site, is the first palaeontological museum of Chile, founded in 2006 at the edge of the of Caldera Municipality in Caldera city, Atacama Region. This is the only museum in Chile with a directly-focused palaeontological theme and promotes the mission of generating greater local and national protection, conservation and awareness in relation to the country's existing palaeontological heritage. The fossils of Caldera have been known about for approximately 30 years. However, scientific investigations of the local palaeontology have only been developed in the last 12 years. Caldera village is located on a marine sedimentary deposit called the Bahía Inglesa Formation. In this area the sea formerly extended inland for several kilometers and in fact the very reason why palaeontology is part of the region's contemporary cultural identity. The museum has not only taken on responsibility for the palaeontological heritage, but also the materials from a large number of archaeological sites near Caldera. Brochures and documentaries have publicized work on Caldera palaeontology from not only a scientific perspective but rather a social-anthropological or cultural one. They also promote the periodical visits of students to the museum. The museum also supports the training of artisans and artists in the commercial production of replica fossils made of resin and silicone in order to avoid the illegal trafficking of fossils. Cerro Ballena lies 12 km south of Caldera and is one of the most important in South America paleontological site with an age of 16–17 million years (Constabel and Veloso 2018). The Palaeontological Site is currently closed to the public because it does not have the necessary protection measures, or the facilities required to receive tourists. It is worth mentioning that although both tourist sites are included within the tourism offering of the Caldera–Bahía Inglesa coastal destination, there are no specialized tourism products for palaeo-tourism.

The Palaeontological Site of San Vicente de Tagua—Tagua has a design proposal for a Museum and Palaeontological Park, however there is still no start date for this work.

17.3 Benefits of Palaeontological Tourism

Cultural tourism generally shares content with other modalities such as ecotourism, adventure tourism and especially agri-tourism (Szmulewicz and Sahady 2011). The benefits of palaeontological tourism are very similar to those of cultural tourism.

There are potential rewards for palaeontological sites that find a way to disseminate their discoveries, both in the local community and further afield, granting a social character to palaeontological scientific research through the processes of receptive tourism. Another benefit can be the generation of new sources of income and resources to help finance and support palaeontological sites via various actions, such as fees for admittance, guided tours, and volunteer programs to collaborate with research.

A revaluation of the identity of a particular territory and its inhabitants could be beneficial. The empowerment of local and business communities over palaeontological cultural heritages would contribute to their conservation and dissemination. New opportunities can also be generated by the creation of projects and ventures linked to palaeontological sites and scientific excavations, which would promote the diversification of tourism for any territory that has these cultural resources.

Among the possibilities offered by palaeontological tourism for the enjoyment of visitors are site museums, interpretation and dissemination centers and theme parks. In general, tourists become part of an experience in which they can relate directly, through their senses, to a reality created and characterized artificially. Audiovisual effects bring species of the past into the present for the visitor.

The activities of tourist guides who are specialized in using palaeontological heritage resources, which also integrate the history and culture of a territory, are essential when developing this new form of tourism. It must be ensured that the scientific information of the palaeontological findings is delivered to different target markets using suitable language and interpretation. Specialized tourists, that means tourists with an interest in culture, children, adults and seniors have entirely different needs.

Educational activities, both for visitors and for the local community, should focus mainly on students of different levels who are interested in knowing in greater depth the findings and the work carried out by researchers in palaeontological sites. Most of the sites that develop palaeontological tourism incorporate this type of activity. Other activities include scientific volunteer programs, which are primarily aimed at undergraduates, graduates and palaeontology enthusiasts who are given the opportunity to conduct research placements at the sites, museums and palaeontological excavations. Participants aid on research, fossil excavations and other activities depending on their level of expertise and knowledge. Generally, these programs include lodging, food and complementary activities and last from a minimum of one week up to several months.

One of the main tourist services associated with palaeontological tourism experiences is the sale of handicrafts that relate to the patrimonial resource, which offers the possibility for artisans to develop new and unique product lines.

The palaeo diet phenomenon has emerged internationally in recent years and invites us to feed ourselves in the same way our prehistoric ancestors did, giving rise to different lines of specialized food products. This has motivated restaurants to incorporate preparations inspired by this diet in their menu and has even promoted the creation of thematic tourist routes based on palaeodieting, which incorporate palaeontological sites, stores, supermarkets, restaurants and other services that market these products.

17.4 Stakeholders in Palaeo-Tourism

The tourism industry has a complex production network comprised of various stakeholders from the public, private sector, non-governmental sector and other agents from the local community who intervene in the actions proposed for the development of tourism in a given territory. The interactions of these actors range from suppliers through to agents of the service process and ending with visitors is what is known as the value chain of tourism (Szmulewicz and Veloso 2013). When planning the development of new forms of tourism such as palaeo-tourism it is first necessary to identify and then analyse the degree of linkage between the stakeholders. This articulation depends in part on the feasibility of achieving the strategic objectives proposed by the destination.

In the case of palaeontological tourism, the main stakeholders are the institutions which manage the palaeontological resources, such as archaeo-palaeontological sites, museums, site museums, and interpretive parks; the principal attractions in terms of the creation of tourism products. In addition, there are accommodation businesses that would ideally have a thematic focus on the territory discoveries, gastronomic companies that incorporate palaeo diet foods into their menu, artisans who create associated craft pieces that utilize raw material or ancient techniques, tourist guides specialized in palaeontological and archaeological topics who are capable of interpreting for different target markets, transport and trade companies, and more. Together, these actors integrate the link of producers who constitute the supply chain made up of both formal and informal companies, and the tourism providers who deliver the final result in the form of tourist services and products receiving key support from the suppliers (Fig. 17.2).

Although the offer of palaeo-tourism can be positioned directly in the national and international markets, most of the sales are made by intermediaries; an area made up of international tour operators, national and local companies and wholesale and retail travel agencies who design and market excursions and palaeontological tourism products. Entities that play a strategic role in the development of emerging forms of tourism include public services, research and technical assistance institutions, education and training institutions at their various levels, finance institutions and trade, business and professional associations.

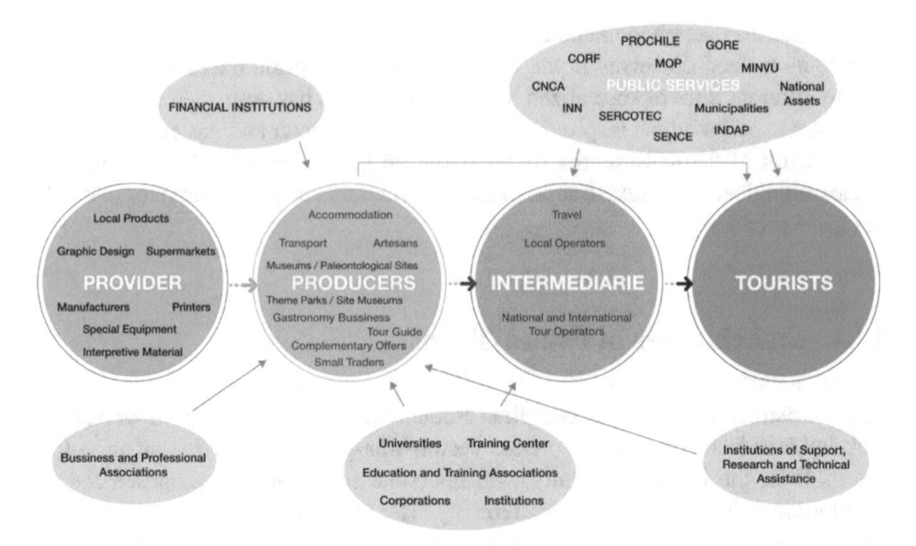

Fig. 17.2 Palaeo-tourism value chain

17.5 Development of a Palaeo-Archaeological Tourism Destination in Chile: The Osorno Case Study

The Province of Osorno has a huge variety of natural world resources and tourist attractions, mainly located in the Cordillera de los Andes and the Pacific coast sectors. In this sense, volcanoes, lakes, thermal baths and national parks are located on the former region, and beaches, coves and coastal landscapes on the latter. All these stand out for their millenary nature and their scenic beauty which were formed at the end of the last glaciation, an age similar to those of the Pilauco site. As a result, we could observe there a clear relationship between the natural resources of the region and the archaeo-palaeontological findings (Constabel and Veloso 2018).

In the coastal zone stands for the presence of indigenous Williche families who still maintain their traditions, festivities, and customs. The area is also home to one of the best-preserved indigenous cemeteries in the South of Chile. In this regard, the coastal zone is also linked to palaeo-tourism in the context of the historical and cultural origins of humans in the area.

In recent years, it has been possible to identify four micro-areas or tourism axes. They are developed with dissimilar levels of attractions, equipment, and infrastructure in the service of the tourist industry. The international route of the Cardenal Samoré Pass, which focuses on the Andes region (Puyehue), is the most important tourist axis because it is part of the binational corridor that connects Osorno with two major trans-Andean tourist centers: Villa La Angostura and San Carlos de Bariloche, which are both on the edge of Argentina's Nahuel Huapi National Park. A second axis comprises the route between the city of Osorno and the Pacific coast (San Juan de la Costa) that includes the coastal resorts of Bahía Mansa, Maicolpué and Pucatrihue. It in recent years have witnessed the emergence of ecotourism joint-ventures falling under the ethnographic private park 'Williche Mapulahual Network'. A third axis corresponds to the route joining Osorno city with the towns of Puerto Octay, Puerto Fonck and Las Cascadas on the northern slope of Llanquihue Lake, which are characterized by a marked presence of rural tourism ventures, agri-tourism and special interest tourism. Finally, the city of Osorno as the fourth axis fulfills the role of a center of services and a hub for the distribution of tourist flows. It is characterized by urban and cultural conditions that make it suitable for the development of business tourism and shopping, the latter being especially driven by the increasing arrival of trans-Andean tourists.

The archaeo-palaeontological site of Pilauco has an internationally recognized scientific value among researchers, and also among paleontologists and archaeologists. Its location makes it a special cultural tourism resource as one of the few archaeo-palaeontological sites in the world located in the middle of a city (Aconcagua River No. 159, Villa Los Notros, Osorno); for this reason it has continually excellent accessibility. The site is located in a municipal property, is open all year and is a point of continuous scientific activity thanks to the research from various disciplines of Universidad Austral de Chile, Universidad Católica de Temuco and other outstanding national and international associated universities. The investigations are financed mainly through competitive public funds and direct grants of the Municipalidad de Osorno.

The Pilauco site has a basic Visitor Center in the form of a pilot experience, with two spaces situated in containers. In the first, the visitor is welcomed, and the history of the site is explained along with the links between the various projects, before an exhibition of artisanal products based on findings at the site. Visitors next experience the conservation laboratory, where replicas of fossils from the site are displayed alongside palaeontological information. Finally, the Visitor Center offers an interpretative path that ends with a viewpoint overlooking the excavation. Visitors can sometimes also interact with the researchers working at Pilauco.

As a cultural tourism resource, Pilauco has significant attributes that carry a potential for the development of palaeontological tourism. However, there are a number of gaps that must be addressed in order to realize this value, namely: (a) to improve the scientific, historical and cultural interpretation program for different market segments; (b) to provide a greater quantity and better distribution of signposts to reach the site and navigate inside; (c) to ensure the availability of qualified guides; (d) to include other services such as a coffee shop, souvenir shop and basic facilities such as toilets; (e) to regularize the offer of activities, such as the program of

'Little Excavators', which is carried out in summer, among others. Some of these gaps could be addressed via a better relationship with the Chuyaca Pleistocene Park (see Chap. 1, this volume), another attraction near the Pilauco Museum.

17.6 Strategic Analysis of the Destination's Potential for Archeo-Palaeontological Tourism

There has been a gradual and sustained increase in demand for special interest tourism modalities, including for cultural, palaeontological and archaeological tourism across the national and international markets. As part of this increase, spikes have been observed in overnight stays, length of stays and levels of spending by tourists visiting the relevant destinations of the Los Lagos Region (Constabel and Veloso 2018). There is a fluid and expeditious connectivity with neighboring tourist destinations such as Puerto Varas, Valdivia, the 'Seven Lakes' circuit, as well as the trans-Andean zone of Bariloche and its surroundings. A substantial improvement of the Osorno—San Carlos de Bariloche binational corridor, with expansion that increased the safety and connectivity levels of the territory. It would further consolidate the most important binational tourist corridor in southern Chile. There is also an important proximity to other tourist distribution centers such as Puerto Montt city, which each year receives around 100,000 tourists by cruise ship. Another opportunity is posed by scientific positioning of the region in palaeontology, archaeology, geology, and related disciplines, thanks to the continuous generation of scientific knowledge e.g. on Geoparks that has been developing there for several years.

On the other hand, the proximity between the Pilauco archaeological-palaeontological site (Osorno) and the Monte Verde archaeological site (near Puerto Montt) would allow for the conceptualizing of an integrated tourist circuit connecting both sites and contributing to the diversification of the tourism offer, with differentiating attributes over competing tourist destinations. In order to take advantage of the identified opportunities, it would be necessary to work on overcoming some of the gaps or limitations that have impacted on the ability to capture and bring greater value to these destinations. Among them is weak appropriation and empowerment on the part of the community and its agents, which makes the protection, conservation and use of this space for education, tourism or recreation a slow process. In addition, the coordination between the actors of the destinations is limited and there is little convergence around addressing and agreeing on strategies. Furthermore, there has been lack of investment in infrastructure around the development of the destinations and for the enhancement of palaeo-archeological heritage, along with a lack of the resources needed to promote private investments aimed at capturing this value and incorporating it into the tourist offer. In the territory of the Chilean Republic there are a large number of state instruments for productive development, which do not necessarily conform to the needs of the sector in this initial stage of palaeontological tourism development.

A weak social capital is also evident in tourism around the destinations. The importance of this lies in the fact that social capital—measured by the strength of organizations, their relationships, their trust and interaction, the participation and associativity of their formal and informal institutions—constitutes the fundamental substrate for the empowerment and appropriation of the concept of palaeo-tourism and its potential enhancement. These aspects make it possible to chart the high dispersion of local actors, the low representation of their organizations, the low associative capacity among them and the lack of interest in participating in collective activities (Pearce et al. 2017).

In order to underline the necessity of proposed future actions, the limitations of the tourism offer in the territory are reflected in a problem tree, which synthesizes the main deficiencies and obstacles hindering an effective insertion into the tourist market (Fig. 17.3). The main causes of the absence of adequate conditions for the reception and attention of visitors to the destination can be seen in red, which in turn lead to a series of unwanted effects (in green), such as a weak tourist image of the destination and the resulting moderate contribution of the tourist industry toward the development of this territory.

As a medium-term objective, the territory is set to become a recognized palaeontological and archaeological tourism destination, with marketable products ensuring a quality and satisfactory experience of the visitor. The target image is defined as:

> Osorno: Land of the Gomphotheres, invites you to experience palaeontological and archaeological tourism in a setting of exuberant nature and cultural heritage. An unforgettable journey beckons, beginning in the middle of the city with the Pilauco site and Chuyaca Interpretive Park, and continuing from the Andes to the Pacific with a gastronomic enjoyment and the Williche culture, allowing the visitor will to learn, in situ, through the integration of prehistory, history, and science.

Strategic Objectives:

- To implement the necessary conditions in the Chuyaca Pleistocene Interpretative Park and the Museum of Pilauco that would allow for scientific discoveries to be brought closer to the local community and tourists visiting the destination.
- To design and market palaeontological and archaeological tourism products that articulate the various actors of the identified value chain.
- To strengthen the technical capacities for human capital and deliver the necessary tools to ensure the satisfactory experience of the visitor.
- To establish and support the consolidation of the Private Corporation for the Development of Palaeontology and Palaeotourism of the Province of Osorno.

17.7 Palaeontological—Archaeological Tourism Products at the Destination

In regard to the current scientific tourism offer, this destination only provides services such as trained tourist guides who accompany visitors along the path of the Pilauco site and through a circuit that incorporates the Chuyaca Pleistocene Park.

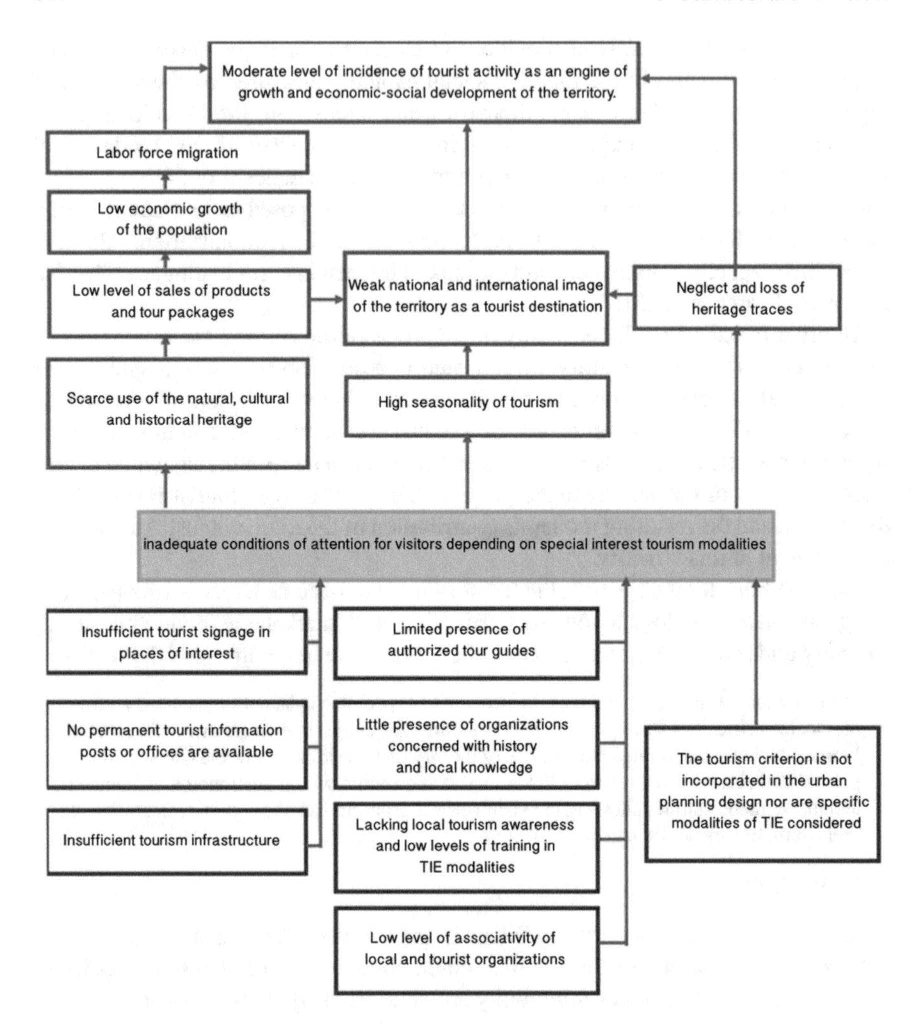

Fig. 17.3 Problem tree of the tourist offer of the territory (modified from Constabel and Veloso 2018)

Additionally, some nearby restaurants offer dishes inspired by the palaeo diet and a handful of artisans have developed basketry products and replicas of Pleistocene animals in felt.In the design process there are several products that were ideated during training workshops (Fig. 17.4), some of which have already been product tested, such as: the small excavator program carried out during the summer holiday seasons and as part of the Cordillera-to-Sea Program of Osorno. In order to develop the proposal for the palaeontological tourism management model in the Osorno destination (see Fig. 17.5), an adaptation of the CANVAS methodology was used (Osterwalder and Pigneur 2010), which proposes a logical order for the creation of business models.

- Half-day program aimed at children between 6 and 13 years old. They accompained with an adult will live the experience of knowing about paleontology and archeology, participate in a simulated excavation and become Little scientists.

Little Excavators

- Tourist program of 3 days in wich the tourist travels the province of Osorno from the Andes mountain range to the Pacific coast, learning about its geography, rain forest temperate biodiversity, Mapuche culture, Osorno's historyands Pilauco site.

Osorno from mountain to coast

- One-week cientific volunteer program for researchers and enthusiast whit interest in paleontology and archeology traveling alon or in groups, who wish to collaborate with scientific work carried out at the Pilauco site.

- Half day tour that seeks to incorporate the paleontological and archeological knowledge of the Pilauco site in the training of elementary an middle school students. Articulated with the history and culture of Osorno. Includes visit to Chuyaca Park, Historical Museum of Osorno and the Pilauco site.

Fig. 17.4 Palaeo-archaeological tourism products for the Osorno destination

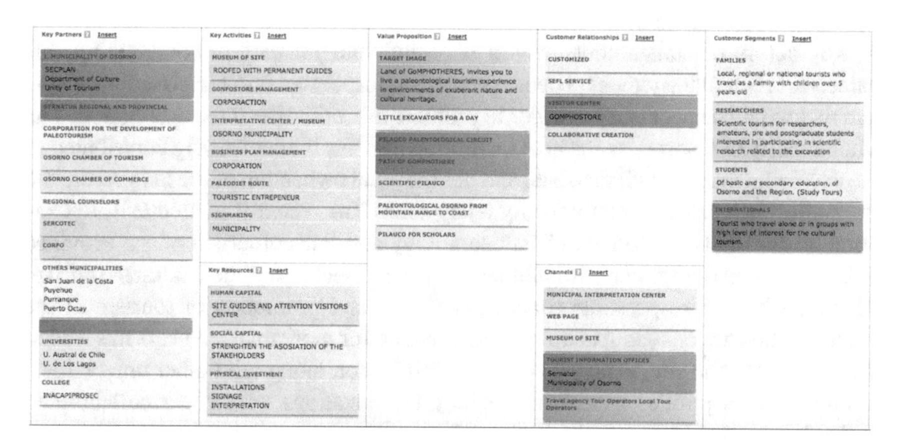

Fig. 17.5 Palaeontological and archaeological tourism management model for the Osorno destination

First, segments for whom value is going to be created needed to be defined. Based on the market study and the experience of the experts, four segments were prioritized: national tourists traveling with their families, researchers, students and international tourists. Subsequently, a value proposal was made for each of the segments, which correspond to the palaeontological tourism products described in the previous chapter. These products are connected to the defined segments through direct marketing channels such as the Interpretation Center, Museum, Instagram, tourist information offices, and indirect channels such as travel agencies and tour operators. It is expected that the customer loyalty would be achieved via a balanced combination of personalized attention and self-service, since there is already a self-guide manual for the main circuit. It will also important to maintain a direct relationship of feedback and even promote the co-creation of new products with tourists, which would allow for continual improvement with a new touristic form that does not have major counterparts at the national level, such as palaeontological tourism.

In order to achieve the proposed value of the conditions offered to visitors, it is necessary to have key resources such as specialized human capital trained in tourism and especially in palaeontological—archaeological tourism as well as an empowered social capital. Of course, the public and private economic resources needed to create the conditions and an economy for the operation of the proposed tourism products, are assumed.

The key activities for the management of palaeo-tourism in the destination of Osorno include the implementation of the Interpretive Center, the creation of a roofed site museum at the Pilauco site, the provision of specialized tour guides throughout the year, the establishment of hours and the management of a craft and souvenir shop where artisans can market products inspired by the findings at Pilauco. Information and signposting toward the excavation and the Chuyaca Pleistocene Park, and at the main tourist access points (Osorno and Puerto Montt airports, bus terminal, OITS) are also needed. The implementation of the business plan with the proposed tourism products is fundamental.

Key actors or stakeholders would play and integral part in concretizing this palaeo-tourism management proposal and are incorporated into the governance system that would be necessary in order to promote this initiative at the destination. The Municipality of Osorno, as the host institution that has fundamentally contributed to the development of the Pilauco Site and has the legal ownership of the land where the excavation is located, is undoubtedly a key actor. The recently created Private Corporation for the Development of Palaeontology and Palaeotourism of the Province of Osorno is also important. It includes the private sector and the National Tourism Service, which is responsible for the regulation and promotion of tourism in the country. The Universidad Austral de Chile, which for more than ten years has leaded the research at the site, is also a strategic contributor, along with other universities, training centers, public services and business organizations, which have collaborated in the development and positioning of palaeontological tourism (Fig. 17.5).

These actors must be linked to a level that allows a better regional and interregional coverage, similar to the meso-regional tourism program. The development of a new type of emerging tourism such as palaeo-tourism requires having the greatest

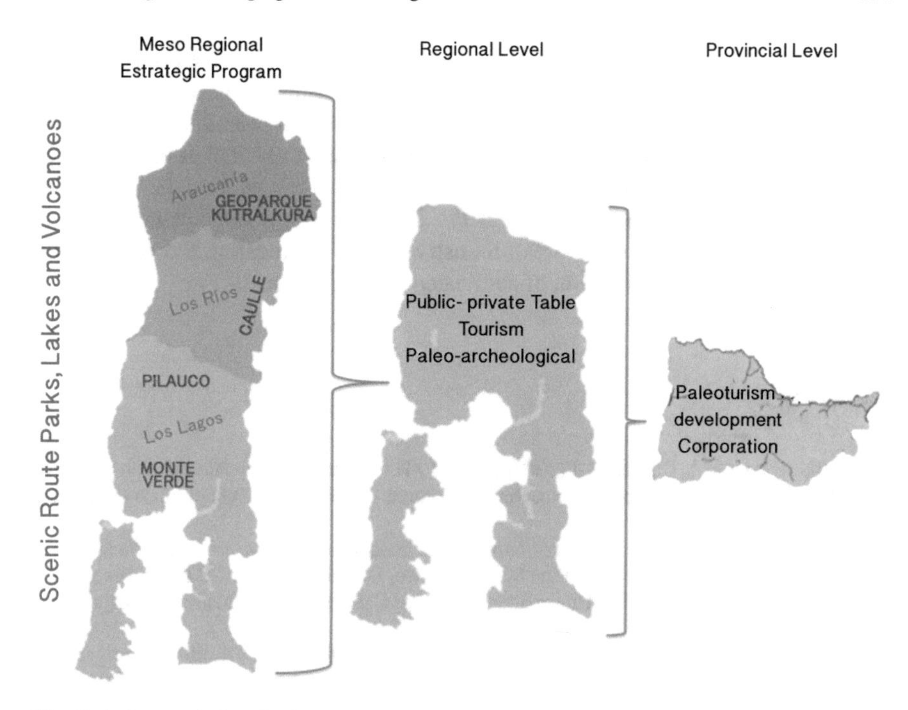

Fig. 17.6 Proposal of territorial articulation

number of strategic alliances and the determination of the stakeholders, so that the accomplishment of outlined medium-term goals can be observed. Thus, the creation of a regional level public-private table led by the Regional Government is suggested, which would articulate the sites of Pilauco and Monte Verde with regional public and private sector actors. This would strategically align with the meso-regional 'Scenic Route Parks, Lakes and Volcanoes' (Fig. 17.6) tourism initiative.

17.8 Conclusions

Successful experiences with palaeontological and archaeological tourism, nearby to this territory, reveal the possibility of diversifying the traditional tourism offer by taking advantage of patrimonial resources. Undoubtedly, Osorno as a destination has the potential to develop this new form of tourism and a simultaneous opportunity to integrate a wider circuit linked with the Monte Verde Site near Puerto Montt.

The Pilauco archeo-palaeontological site has an undeniable tourist potential, which is why it is essential to work on overcoming the shortcomings identified in order to optimize its tourist value. The generation of information along with the dissemination and transfer of these findings to stakeholders and local actors in the palaeo-tourism value chain should continue to be developed and supported.

Planned and articulated work between the various actors involved in palaeontological tourism management is key to ensuring the sustainable development of the destination in the medium and long terms. Having willing scientific, political and business entities will lead to the achievement of common and individual objectives regarding the conservation and empowerment of Chile's cultural heritage. The creation of enabling conditions within the destination is a fundamental element in regards to motivating the tourism industry toward aggregating the value of the tourism offer and the transaction of associated tourism products.

References

Bigné JE, Moliner MA, Callarisa LJ (2000) El valor y la fidelización de clientes: una propuesta de modelo dinámico de comportamiento. Revista Europea deDirección y Economía de la Empresa 9(3):65–78

Constabel S, Veloso K (2018) Planificación y gestión de un destino de turismo paleontológico: el caso de Pilauco-Osorno, Chile In: Pino M (ed) Pilauco, Un sítio complejo del Pleistoceno tardío. Osorno, Norpatagonia Chilena. Universidad Austral de Chile, Imprenta América, Valdivia, Chile, pp 200–228

Encabo M Vejsbjerg L (2002) El Paleoturismo. Anuario de Estudios en Turismo – Investigación y Extensión. Facultad de Turismo, Universidad Nacional del Comahue; Neuquén, Argentina

Norman D, Ngaire D (2001) Special interest tourism. Wiley, Australia

Osterwalder A, Pigneur Y (2010) Business model generation: a handbook for visionaries, game changers, and challengers. Wiley, New Jersey

Pearce D, Guala C, Veloso K, Llano S, Negrete J, Rovira A, … Reis A (2017) Destination management in Chile: objectives, actions and actors. Int J Tour Res 19(1):50–67

Perini MM, Calvo JO (2008) Paleontological tourism: an alternative income to vertebrate paleontology. Arq Mus Nac Rio J 66(1):285–289

Pino M (ed) (2008) Pilauco, Un Sitio Complejo Del Pleistoceno Tardío. Osorno, Norpatagonia chilena. Universidad Austral de Chile, Imprenta América, Valdivia, Chile

Salemme M (2001) El Patrimonio Arqueológico y Paleontológico. Curso de Ecoturismo Facultad de Turismo. Universidad Nacional del Comahue – Fundación Biósfera

Szmulewicz P, Sahady A (2011) Ruta Turística Patrimonial del Valle del Cachapoal. Universidad Austral de Chile, UACh-Corfo, Santiago, Chile

Szmulewicz P, Veloso K (2013) Diseño de rutas turísticas en áreas rurales y naturales: Orientaciones metodológicas. In: González M, León C, de León J, Moreno S (eds), El Turismo Rural y en Áreas Protegidas: una panorámica. Síntesis, pp 99–117

Szmulewicz P, Veloso K (2014) Oportunidades y desafíos de la investigación en turismo en Patagonía. An Inst Patagonia 41(2):27–42

Vejsbjerg L, Encabo M (2005) Gestión Ambiental en Sitios Paleoturísticos. Facultad de Turismo, Universidad Nacional del Comahue; Neuquén, Argentina

Chapter 18
The Pilauco and Los Notros Sites: A Final Discussion

Mario Pino

Abstract The findings published on this book span eleven years of active field research investigations at the Pilauco site and the first results of the neighboring site Los Notros. However, most of the archaeological and paleontological findings presented in this volume have been obtained from Pilauco, because the site Los Notros is yet at an early stage of stratigraphic definition. In spite of the different stages of research development, Los Notros is a promising paleontological research site and has already provided the oldest and youngest fossil remains of the Pilauco–Los Notros Complex. This chapter describes and discusses the main results obtained by each of the specialists who have worked at the Pilauco and Los Notros sites. It also discusses some aspects that escape in part the scientific knowledge of the sites.

Keywords Discussion · Monte verde · Clovis

The Pilauco site, on the other hand, has provided a wide range of biological evidence through several years of excavation that have permitted to assert new discoveries in important areas of late Pleistocene paleontological and archeological research in southernmost South America. These areas include general characterization and context of the site, taxonomic, taphonomic, and paleoenvironmental reconstructions (Recabarren et al. 2011, 2014; Pino et al. 2013; González et al. 2014; González-Guarda et al. 2017; Labarca et al. 2014; Tello et al. 2017), lithics assemblages and other cultural materials from Pilauco (Navarro-Harris et al. 2019; Moreno et al. 2019), causes of environmental and climate disruption during the Younger Dryas Chronozone (Pino et al. 2019), autoecology of gomphotheres (González-Guarda et al. 2018).

The geology of the site presented in Chap. 3 (this volume) has provided insights to understand the stratigraphy and sedimentological context of Pilauco together with the geomorphological context of the Osorno area. This investigation has provided not only a general overview of the landscape evolution of the area since the last interglacial, but also the processes of site formation during different stages of the last glaciation. Most importantly, this research indicates that the sedimentary fabric

M. Pino (✉)
Instituto de Ciencias de la Tierra, Universidad Austral de Chile, Valdivia, Chile
e-mail: mariopinoquivira@gmail.com

M. Pino and G. A. Astorga (eds.), *Pilauco: A Late Pleistocene Archaeo-paleontological Site*, The Latin American Studies Book Series,
https://doi.org/10.1007/978-3-030-23918-3_18

of layer PB-7 and PB-8 points toward colluvial transport from the northern hills with no fluvial structures observed in the layers. This observation is important for taphonomic and archaeological interpretations, because it rules out river redeposition, the production of geofacts and the long distant transportation of the fossil materials.

The general frame to understand the conformation of the South American faunas, and therefore, the paleontological diversity of Pilauco has been presented in Chap. 2. Additionally, this book includes several chapters (Chaps. 4–7) devoted to the description and interpretation of extinct and extant animals remains including not only the iconic giant gomphotheres, but also camelids, equids, and micromammals similar to extant rodents, and other animals such as the smallest deer cf. *Pudu* sp. and the skunk *Conepatus* sp. (González et al. 2010, 2014).

In Chile, the most abundant megafauna fossil types from the late Pleistocene—early Holocene limit are gomphotheres. This was the only group of proboscideans that reached South America during the Great American Biotic Interchange (GABI) counting today with one of the southernmost fossil records that extends to Chiloé Island. However, the classification of the South American gomphotheres has been under strong taxonomic discussion.

At Pilauco most of the gomphothere fossil bones lack diagnostic characteristics to assign them to specific species, but nevertheless the tusk of the site Los Notros allows us to assign this remain to *Notiomastodon platensis* (Chaps. 4 and 8, this volume). Based on this information is certainly possible that the whole assemblage of gomphothere remains from Pilauco belong to the same species. Ongoing studies of the proteins of these animals will provide new insights for further discussion. Additionally, recent studies using stable isotopes, dental microwear and dental calculus microfossils of *Notiomastodon platensis* molars indicated that gomphotheres at Pilauco were leaf browsers leaving in relatively closed forest environments. Such investigations have not only verified the consumption of arboreal and scrub vegetation by these animals, but also that the food pattern of South American gomphotheres appears to be more constrained by resource availability than by the potential dietary range of individual taxa (González-Guarda et al. 2017, 2018). The authors noted that in terms of the collagen $\delta^{15}N$ diet-to-tissue trophic differentiation, the Pilauco values are significantly larger than those found in modern vegetation and other paleontological sites north of Pilauco. After González-Guarda et al. (2018) factors such as fires, pasture, and coprophagy intensity or fertilization of flora located on regular migratory routes could explain the latitudinal differences.

Additionally, to the fossil record of gomphotheres the taxonomy of fossil remains assigned to Artiodactyla: Camelidae and Perissodactyla: Equidae discussed in Chap. 5 (this volume) provides the first record in Chile of the camelid cf. *Hemiauchenia paradoxa,* a species previously described only for the Pampean region of Argentina. In addition, equid remains provide evidence of the presence of a relatively small-sized animal assigned to *Equus andium.* Other dental fragmented pieces allow proposing the presence at Pilauco of a second large-sized equid form possibly different to *Equus*, although is not yet clear whether the two morphotypes may represent different developmental stages of the same species. Confirmatory evidence of the presence of herbivore megafauna at Pilauco has also been provided by studies of the dung fungi *Sporormiella* sp. (Chap. 6, this volume). Such studies have observed that *Sporormiella* sp. decreases to zero exactly in the contact between layers PB-8 and PB-9.

The analysis and identification of small animal remains at Pilauco discussed in Chap. 7, provides the interesting finding of occurrence of microfauna remains associated to extant species, whereas previous chapters presented evidence of extinct megafauna remains. If the pressure from human hunters was the main driver of animal demise, then highly palatable and easily hunted animals such as *Myocastor coypus* (up to 10 kg) and *Pudu puda* (up to 14 kg) might have also become extinct with the megafauna. This is important because it provides independent support toward environmental and climate disruption as one of the main contributing factors to the extinction of megafauna.

The series of chapters dealing with the paleontology of mega- and microfauna culminates with a taphonomic study of the megafaunal remains encountered at Pilauco (Chap. 8, this volume). This study concludes that most of the bone marks were produced by trampling and discards the existence of marks made by cultural artifacts. As pointed out by Navarro-Harris et al. (2019), it is well known from experimental and observational work on African elephants that experienced butchers prevent damaging the leaves of their tools. Even when knives are made of steel, cut marks are seldom seen on elephant carcasses (Crader 1983). In addition, cut marks from lithic artifacts on bones containing meat, such as the femur, are unusual (Frison and Todd 1986; Haynes and Klimowicz 2015). Another important conclusion of this investigation is the indication of in situ death of a gomphothere specimen to which later, other bone remains of different taxa were added by colluvial redeposition and vertical trampling. In addition, to all the animal records presented through the series of paleontological chapters, another megafauna representative is indirectly present at Pilauco. This has been inferred based on *postmortem* tooth marks of a large carnivore in several gomphothere bones non-corresponding to marks attributed to saber tooth tigers. The marks most likely correspond to the Pleistocene southern Patagonian jaguar [*Panthera onca mesembrina*, described by Martin (2013)]. The chapter also concluded that carnivores might have been responsible for the movement and possibly subtraction of the lacking gomphothere bones.

Several biological proxies (i.e., pollen, seeds, charcoal, diatoms, phytoliths, beetles, and coprolite parasites; Chaps. 9–13, this volume) have been used at Pilauco to gain an understanding of the environmental conditions that prevail between ca. 16,000 and 12,800 years ago during the time of deposition of the fossil-bearing sediments of the layers PB-7 and PB-8. In particular, pollen, seeds, cuticles, and charcoal investigations have provided a solid and generally concordant high-resolution picture of the vegetation and fire disturbance for the period. Thus, pollen and seeds assemblages show similar vegetation patterns before the transition from layer PB-8 to PB-9 indicating mainly the predominance of non-arboreal vegetation (i.e., Poaceae) along with the presence of a wide variety of North Patagonian forest taxa (e.g., *Nothofagus dombeyi*-type, Myrtaceae, and the conifers *Pilgerodendron*/*Fitzroya* and *Podocarpus nubigena*) growing under cooler conditions and high precipitation/humidity. Additionally, in layer PB-8 the combined presence of pollen and seeds of *Prumnopitys andina* at the end of the last glaciation provide the southernmost record for this conifer. Near the base of layer PB-8, is also recorded the presence of palustrine and aquatic ferns (Cyperaceae and Isoetes) indicating shallow water level and oligotrophic conditions some 16,000 years ago. This scenario is also concordant with interpretations derived from diatom and fossil coleopterans.

At the transition from PB-8 to PB-9 (ca. 12,800 years ago) a sharp decrease in abundance and diversity of plant species is recorded, while the concentration of charcoal particles reaches a maximum during this interval. The very low concentration of pollen suggests scarce vegetation and low productivity of the area marked by intense biomass burning during the interval. After ca. 12,800 years vegetation records indicated high abundance of pioneer/colonizing species in response to more disturbed open environments, and affinity of the vegetation to the Valdivian rainforest under seasonally drier and warmer conditions. The pollen assemblage from Pilauco is very different from other records in the region and may be related to high disturbance regime induced by the activity megafauna, most likely gomphotheres. This is the first time that in a paleoclimatic vegetation based reconstruction, the role of disturbances other than forest fires are open to discussion and will likely influence other pollen investigations in the region.

The diatom record from Pilauco presented in Chap. 10 (this volume) provides solid indication of relative variations in temperature and humidity pointing toward the occurrence of two distinctive periods in the diatom succession. Between ca. 16,000 and 14,800 cal. yr BP the dominant presence of benthic diatom species with affinities to cold water can be related to the end of the last glaciation and characterized by shallow waters and reduced humidity. The second period (between 14,800 and 14,200 cal. yr BP) is interpreted as having warmer conditions, and prompts toward the likely proliferation of macrophytes that may have served as living media for the epiphytic diatoms present in the record. Lastly, *Aulacoseira granulata* has been identified as an indicator of repeated flooding over a wetland. The layer PB-7 has an average deposition rate of 13 yr cm^{-1} (Pino et al. 2019). Based on the temporal scale used by the study of fossil diatoms (13 samples in 1800 yr), the occurrence of prolonged dry periods, longer than a couple of years, can be discarded.

The existence of at least seasonal dry periods was one of the most interesting conclusions of the study of Coleoptera (Chap 12, this volume). Thirteen families and 21 species of Coleoptera were described. They indicate the presence of a wide variety of forest habitats, grasslands, stagnant waters, and streams. Especially important is the presence of dung beetles, mainly because they have a strong association with large mammals, as they depend on fecal matter, exposed to the atmosphere and not submerged under water, to feed and nest. Therefore, it is possible to infer that the old Pilauco wetland was at least seasonally free of water. This situation allows at the same time to infer that the megafauna that died at the old Pilauco wetland, at least one gomphothere, was not permanently under water providing an opportunity to humans and other animals to scavenge meat.

More than a dozen of coprolites were recovered from layer PB-8. Four of them were analyzed to identify parasites eggs and phytoliths (Chaps. 11 and 13, this volume). The confirmatory presence of parasite eggs is reliable evidence that these are true coprolites and not simple accumulation of plant material. The parasite eggs identify two different host animals, a horse and a ruminant, possibly a deer or a camelid as suggested by other animal records in layer PB-8. In at least two of the coprolites the dominant phytoliths morphology is associated with C_3 grasses, pointing toward the possibility of a horse as the animal that produced it. In the third coprolite the difference in C_3 and C_4 grasses contents was very small. So it is possible to infer that the animal could have been an opportunistic feeder, also as *Equus andium*.

Already in 2011 in Pilauco a set of observations and isotopic analyses indicated that (1) there was a geological unconformity between the layers PB-8 and PB-9; (2) the lower part of layer PB-8 recorded hundreds of mega and microfauna fossils and cultural remains, whereas in the upper layer PB-9 fossil or cultural remains were absent; (3) layer PB-9 is characterized by presenting large cobbles and pebbles of volcanoclastic consolidated sediments originated from an intense process of erosion and gravitational transport from the northern hill; (4) layer PB-9 present high charcoal content and multiple broken fragments of wood and (5) the dates performed above and below the unconformity indicated an age of ca. 12,800 yr, equivalent to the base of the Younger Dryas chronozone. Clearly the charcoal and the colluvial erosion could be interpreted as the consequences of a forest fire, but how big? To help in resolving this question pollen and charcoal records from two columns containing the unconformity between PB-8 and PB-9 was analyzed (Chap. 9, this volume). This dataset was the basis for starting an intense collaboration with a group of asteroid impact experts, led by Dr. Allen West, which allowed the production of Chap. 15 (see also Pino et al. 2019). In this study, a lamina ca. 12,800 years old is reported at Pilauco exhibiting maximum concentrations of high-temperature iron, platinum, gold, chromium, and native iron particles uncommonly founded in continental deposits. These findings are consistent with evidence of impact at the base of the Younger Dryas found at multiple sites on four continents. Moreover, the disappearance of megafauna remains and dung fungi (Chap. 6, this volume) seen in Pilauco at the impact layer correlates very well with megafauna extinctions in the Americas (Pino et al. 2019). However, there are several detractors of the impact hypothesis, and especially many of them among North American archaeologists. This is mainly because in North America the lamina produced by the impact not only marks the end of the existence of the megafauna, but also is coincident with the end of the Clovis culture. The "Clovis police" (Deloria 2004) or the Clovis mafia[1] have never accepted that the geoarchaeological evidence of the site Monte Verde II in southern Chile demolished the hypothesis "Clovis First," and now they have troubles to accept that Clovis may have disappeared by an impact related event.

Chapter 16 (this volume) has discussed key observations and interpretations complementing the recently published lithic and other cultural material from Pilauco by Navarro-Harris et al. (2019). This investigation integrates all the archaeological and geo-archaeological evidence of human presence from 16,000 to 12,800 years ago, posing a new challenge to the Clovis supporters and other skeptical archaeologists.

---

[1] https://www.earthmeasure.com/on_suppression.html.

The presence and distribution of knapped lithics and their spatial association with bone remains of extinct fauna provides support to the hypothesis that humans occasionally, but recurrently visited the site. The Pilauco site also shares some general characteristics with other early settlements, such as the predominance of unifacial expeditive technologies on raw materials of local and extra local origin. Finally, the investigation of artifacts at Pilauco also provides the first South American late Pleistocene artifacts from rhyodacitic and dacitic glasses and the oldest obsidians from the Americas.

From the first publications of the site Monte Verde II (Dillehay and Collins 1988; Pino and Dillehay 1988) and later those that analyze the site Monte Verde I (Dillehay et al. 2015), were exposed to intense criticisms usually associated to wrong arguments such as the lack of experience of professionals making the discoveries, the no recognition of the stone artifacts as true, wrong radiocarbon dates or chronologies dating non-cultural events, etc. If the sites we are investigating were to have 5000 years instead of 15,000, there would be no criticism at all. A recent example came from Latin American colleagues who used the arguments of Clovis b and d (Politis and Prates 2018), without having visited the site or seen the collections. Chapter 16 includes some archaeologic evidence at Pilauco that go beyond the classic knapped artifacts. In Pilauco there are polished spherical stones like the ones described in Monte Verde, but also soft volcanic stones and a seed of *Prumnopitys andina* with bidirectional perforations. There are also two pieces of wood with clear human interventions and finally a small fragment of a marine mytilid shell (*Choromytilus chorus chorus* Molina 1782).

Finally, throughout our research investigations at Pilauco we have been involved with the natural and social disciplines described in this book, but also we have been interested in to promote the development of tourism initiatives using the scientific findings at Pilauco. The Universidad Austral de Chile has a long tradition of inserting tourism as one of the research topics developed in Economic Sciences. Chapter 17 (this volume) summarizes the results of a project granted by the Chilean State to compare international tourism experiences and to propose a route for paleontological and archaeological tourism, now also astrophysical in Osorno.

Short- and middle term plans for Pilauco and Los Notros sites include continuing excavations, employing molecular techniques to distinguish proteins at the sharp edges of archaeological tools, and try to implement molecular taxonomy using proteins and DNA from extinct and extinct animals. Also, we will continue to communicate the local and regional population about our findings and interpretations to enhance heritage related processes.

References

Crader DC (1983) Recent single-carcass bone scatters and the problem of "butchery" sites in the archaeological record. In: Clutton-Brock J, Grigson C (eds) Animals and archaeology, hunters and their prey, vol 1. BAR International Series, Cambridge, UK, pp 107–141

Deloria V Jr (2004) Marginal and submarginal. In: Mihesuah DA, Cavender A (eds) Indigenizing the academy. Wilson University of Nebraska Press, Lincoln and London, pp 16–30

Dillehay TD, Collins MB (1988) Early cultural evidence from Monte Verde in Chile. Nature 332(6160):150

Dillehay TD, Ocampo C, Saavedra J, Oliveira Sawakuchi A, Vega RM, Pino M, Collins MB, Cummings LS, Arregui I, Villagran XS, Hartmann GA, Mella M, González A, Dix G (2015) New archaeological evidence for an early human presence at Monte Verde, Chile. PLoS ONE 10(11):e0145471, 1–27

Frison GC, Todd LC (1986) The colby mammoth site. University of New Mexico Press, Albuquerque

González E, Prevosti FJ, Pino M (2010) Primer registro de Mephitidae (Carnivora: Mammalia) para el Pleistoceno de Chile. Magallania (Chile) 38(2):239–248

González E, Labarca R, Chávez M, Pino M (2014) First fossil record of the smallest deer cf. Pudu Molina, 1792 (Artiodactyla, Cervidae) in the Late Pleistocene of South America. J Vert Paleontol 34(2):483–488

González-Guarda E, Domingo L, Tornero C, Pino M, Hernández Fernández M, Sevilla P, Villavicencio N, Agustí J (2017). Late Pleistocene ecological, environmental and climatic reconstruction based on megafauna stable isotopes from northwestern Chilean Patagonia. Quat Sci Rev 170:188–202

González-Guarda E, Petermann-Pichincura A, Tornero C, Domingo L, Agustí J, Pino M, Abarzúa AM, Capriles JM, Villavicencio NA, Labarca R, Tolorza V, Sevilla P, Rivals F (2018) Multiproxy evidence for leaf-browsing and closed habitats in extinct proboscideans (Mammalia, Proboscidea) from Central Chile. Proc Natl Acad Sci USA 115(37):9258–9263

Haynes G, Klimowicz J (2015) Recent elephant-carcass utilization as a basis for interpreting mammoth exploitation. Quat Int 359:19–37

Labarca R, Recabarren O, Canales-Brellenthin P, Pino M (2014) The Gomphotheres (Proboscidea: Gomphotheriidae) from Pilauco site: scavenging evidence in the late Pleistocene of the chilean Patagonia. Quat Int 352:75–84

Martin FM (2013) Tafonomía y paleocología de la transición Pleistoceno-Holoceno en Fuego Patagonia. Interacción entre humanos y carnívorosy su importancia como agentes en la formación del registro fósil. Ediciones de la Universidad de Magallanes, Punta Arenas

Moreno K, Bostelmann JE, Macías C, Navarro-Harris X, De Pol-Holz R, Pino M (2019). A late Pleistocene human footprint from the Pilauco archaeological site, Northern Patagonia, Chile. PLOS ONE PONE-D-18-03927R2 (in press)

Navarro-Harris X, Pino M, Guzmán-Marín P, Lira MP, Labarca R, Corgne A (2019). The procurement and use of knappable glassy volcanic raw material from the late Pleistocene Pilauco site, Chilean Northwestern Patagonia. Geoarchaeology 1–21. https://doi.org/10.1002/gea.21736

Pino M, Dillehay T (1988) Monte Verde, south-central Chile: stratigraphy, climate change and human settlement. Geoarch 3(3):177–191

Pino M, Chávez-Hoffmeister M, Navarro-Harris X, Labarca R (2013) The late Pleistocene Pilauco site, Osorno, south-central Chile. Quat Int 299:3–12

Pino M, Abarzúa A, Astorga G, Martel-Cea A, Cossio N, Navarro RX, Lira MP, Labarca R, LeCompte M, Adedeji AV, Moore C, Bunch T, Mooney C, Wolbach WS, West A, Kennett JP (2019) Sedimentary record from Patagonia, southern Chile supports cosmic-impact triggering of biomass burning, climate change, and megafaunal extinctions at 12.8 Ka. Nature Scientific Reports Paper #SREP-18-09617C

Politis GG, Prates L (2018) Clocking the arrival of Homo sapiens in the Southern cone of South America. In: Harvati K, Jäger G, Reyes-Centeno H (eds) New perspectives on the peopling of the Americas. Words, bones, genes, tools: DFG center for advanced studies series. Kerns Verlag, Tübingen. ISBN 978-3-935751-28-5

Recabarren OP, Pino M, Cid I (2011) A new record of *Equus* (Mammalia: Equidae) from the Late Pleistocene of central-south Chile. Rev Chil Hist Nat 84:535–542

Recabarren O, Pino M, Alberdi MT (2014) La Familia Gomphotheriidae en América del Sur: evidencia de molares al norte de la Patagonia chilena. Est Geol (Madrid) 70(1):1–12
Tello F, Elgueta M, Abarzúa A, Torres F, Pino M (2017) Fossil beetles from Pilauco, south-central Chile: an upper pleistocene paleoenvironmental reconstruction. Quat Int 449:58–66

Glossary

Allochthonous accumulation The accumulated elements that experienced lateral displacement from the original site.

Before present (BP) For radiocarbon ages "present" is defined as year 1950.

Biological proxy Any paleoenvironmental record used in the inference of past events and whose indicator value is derived from a methodical tool based on the analysis of currently observable relationships.

Biotope Geographic space or part of the biosphere with uniform environmental conditions favouring the development of a particular community of living species.

Brachycephalic Skull shape that is shorter (in forward-backward direction), typical for gomphotheres.

Brachydont A low-crowned tooth with thick enamel covering, neck just below the gingival line, and at least one root without enamel.

Bunodont Teeth with low and rounded cuspids and wide occlusal surface, fulfilling the function of squashing or grinding the food.

Clade In phylogeny a clade is a group of organisms that consists of a common ancestor and all its lineal descendants.

"Clovis mafia" When professional archeology resistance to reliable evidence of humans in America before the Clovis Culture became so intransigent, John Adovasio (Meadowcroft Rockshelter dig director) coined a name for the resistance: the Clovis Mafia. SOURCE: Newsweek article The First Americans by Andrew Murr, April 25, 1999. https://www.newsweek.com/first-americans-164950.

Devitrification Is a type of weathering (hydration) over obsidian and others natural glasses, transforming the amorphous glass in a tiny mineral crystals, like feldspars.

M. Pino and G. A. Astorga (eds.), *Pilauco: A Late Pleistocene Archaeo-paleontological Site*, The Latin American Studies Book Series,
https://doi.org/10.1007/978-3-030-23918-3

Diastema A gap between teeth, because of evolutionary tooth loss or prolongation of the jawbone.

Epipleura External margin of the elytra.

Epiphytic Epipelic, episamon, referring to the living style of diatoms as attached to plants, muddy or sandy substrate, respectively.

Eutherian Placental mammals that keeps the embryo inside the uterus during the entire period of gestation.

Holotype A single individual exemplar or illustration of a species, known to have been used by the author to describe and assign a type nomenclature.

Intraclasts Sediment fragments or soft rocks originating from a local source or the same basin.

Lophodont Teeth with striated cuspids.
Ma years = million years.

Monophyletic In terms of evolution, a group of organisms descending from the same common ancestor, and all descendants of this ancestor share characteristics that differentiate them from other group of organisms.

Niche A niche is the sum of all environmental factors that act on an organism and, therefore, condition the position and persistence of it.

Pellets Mass of partly digested bone remains, horns, hairs and feathers orally expulsed by some predator birds.

Potamon areas Subdivisions of riverbeds establish a fundamental distinction between the high slope and torrential (ritron) and the low plain of slow current (potamon). Potamon environments include wide and flat channels, flanked by floodable lands, with muddy bottoms and abundant vegetation with roots and floating plants. In the potamon there can be environments of running water (lotic) or motionless (lentic).

Preserved element Any remain or body part that is a sign for the past existence of an organism.

Pronation Rotation of the forearm or foot so that in the standard anatomical position the palm or sole is facing.

Pronotum Plural pronota, is a structure that cover all or part of the thorax.

Sporopollenin Is a type of polyterpene that has the potential for conservation in a sedimentary deposit over several millions of years.

Subhypsodont Not entirely hypsodont. Teeth that do not grow during the entire life-time.

Symphysis Articulation with a fibrocartilaginous fusion between the bones.

Synonymous In taxonomy, the synonymous usage occurs when there is more than one name for the same taxon. When this occurs, the International Code for Zoological Nomenclature takes into account the law of priority, which means that the oldest name is considered valid and the remaining names are considered synonyms.

Taphfonomic resedimentation A process of taphfonomic disturbance consisting of the displacement, before being buried, and the deterioration, of remains and signs of previously deposited biological entities.

Zygodont Teeth with lophos that are united in a simple-transversal manner, continuing and without conules.

Printed in the United States
By Bookmasters